T0189786

MECHANICS AND PHYSICS OF POROUS MATERIALS

*Novel Processing Technologies
and Emerging Applications*

MECHANICS AND PHYSICS OF POROUS MATERIALS

Novel Processing Technologies and Emerging Applications

Edited by
Chin Hua Chia, PhD
Tamara Tatrishvili, PhD
Ann Rose Abraham, PhD
A. K. Haghi, PhD

First edition published 2024

Apple Academic Press Inc.
1265 Goldenrod Circle, NE,
Palm Bay, FL 32905 USA

760 Laurentian Drive, Unit 19,
Burlington, ON L7N 0A4, CANADA

CRC Press
2385 NW Executive Center Drive,
Suite 320, Boca Raton FL 33431

4 Park Square, Milton Park,
Abingdon, Oxon, OX14 4RN UK

Library and Archives Canada Cataloguing in Publication

Title: Mechanics and physics of porous materials : novel processing technologies and emerging applications / edited by Chin Hua Chia, PhD, Tamara Tatrishvili, PhD, Ann Rose Abraham, PhD, A.K. Haghi, PhD.
Names: Chia, Chin Hua (PhD), editor. | Tatrishvili, Tamara, editor. | Abraham, Ann Rose, editor. | Haghi, A. K., editor.
Description: First edition. | Includes bibliographical references and index.
Identifiers: Canadiana (print) 20230587755 | Canadiana (ebook) 2023058778X | ISBN 9781774914656 (hardcover) | ISBN 9781774914649 (paperback) | ISBN 9781003414469 (ebook)
Subjects: LCSH: Porous materials.
Classification: LCC TA418.9.P6 M43 2024 | DDC 620.1/16—dc23

Library of Congress Cataloging-in-Publication Data

Names: Chia, Chin Hua (PhD), editor.
Title: Mechanics and physics of porous materials : novel processing technologies and emerging applications / edited by Chin Hua Chia, PhD [and three others].
Description: First edition. | Palm Bay, FL, USA : AAP, Apple Academic Press, [2024] | Includes bibliographical references and index. | Summary: "Porous media exist in different modern materials. It presents great surface areas with small pore size distribution. These types of materials with controllable and adjustable pore diameters are given considerable attention due to their suitable properties and applications in several fields. Porous materials have many applications in our daily life. We use different types of porous materials to clean our drinking water, for instance. This new research-oriented volume focuses on exploring the wide range of porous materials. In this new volume, original contributions from international authors along with case studies on the synthesis, design, characterization, and applications of different types of porous materials and solids are presented in detail. The book covers different types of porous materials in the broad sense by considering experimental and theoretical aspects of materials science related to porous materials and solids. The book aims to help approach characterizing a particular types of materials for more in-depth analysis. This book is divided into three parts to determine the best techniques for solving particular porous materials problems, and in each part, the fabrication and characterization of porous materials are explored with applications, describing new methodologies to gain the required information along with limitations of various methods. To make this new title a practical reference book for research students and for engineers and scientists of different disciplines working with porous materials and solids, the editors have selected a very comprehensive range of case studies as well, designed to cover the basic concepts of porosity. These case studies also describe different types of pores and surfaces for readers"-- Provided by publisher.
Identifiers: LCCN 2023055242 (print) | LCCN 2023055243 (ebook) | ISBN 9781774914656 (hbk) | ISBN 9781774914649 (pbk) | ISBN 9781003414469 (ebk)
Subjects: LCSH: Porous materials.
Classification: LCC TA418.9.P6 M44 2024 (print) | LCC TA418.9.P6 (ebook) | DDC 620.1/16--dc23/eng/20231205
LC record available at https://lccn.loc.gov/2023055242
LC ebook record available at https://lccn.loc.gov/2023055243

ISBN: 978-1-77491-465-6 (hbk)
ISBN: 978-1-77491-464-9 (pbk)
ISBN: 978-1-00341-446-9 (ebk)

About the Editors

Chin Hua Chia, PhD
Professor, Materials Science Programme, School of Applied Physics, Universiti Kebangsaan Malaysia, Selangor, Malaysia

Chin Hua Chia, PhD, is a Full Professor at the Department of Applied Physics, Faculty of Science and Technology at Universiti Kebangsaan Malaysia (National University of Malaysia), Malaysia. He is a recipient of the Young Scientist Award from the National University of Malaysia and the Malaysian Solid State Science and Technology (MASS) in 2012 and 2014, respectively. He also received the Distinguished Lectureship Award from the Chemical Society of Japan (CSJ) in 2017. He is a member of several professional organizations and has published several book chapters and more than 200 articles in professional journals. He has presented at many professional meetings as well. His core research focuses on metallic nanomaterials synthesis, with a particular emphasis on structure control to tailor the properties of the materials in determining their performances for environmental remediation applications, such as wastewater treatment, supercapacitor, and hydrogen production.

Tamara Tatrishvili, PhD
Main Specialist, Office of the Academic Process Management,
Ivane Javakhishvili Tbilisi State University (TSU);
Director of the Institute of Macromolecular Chemistry and Polymeric
Materials at TSU, Tbilisi, Georgia

Tamara Tatrishvili, PhD, is the Main Specialist at the Office of Academic Process Management (Faculty of Exact and Natural Sciences) at Ivane Javakhishvili Tbilisi State University, as well as Director of the Institute of Macromolecular Chemistry and Polymeric Materials at TSU, Tbilisi, Georgia; DAAD alumni, and a member of the Georgian Chemical Society. Her research interests include polymer chemistry, polymeric materials, and chemistry of silicon-organic compounds. Dr. Tatrishvili is the author of more than 190 scientific publications, 12 books, and monographs.

Ann Rose Abraham, PhD

Assistant Professor, Department of Physics, Sacred Heart College
(Autonomous), Thevara, Kochi, Kerala, India

Ann Rose Abraham, PhD, is currently an Assistant Professor at the Department of Physics, Sacred Heart College (Autonomous), Thevara, Kochi, Kerala, India. Her PhD thesis was titled, Development of Hybrid Mutliferroic Materials for Tailored Applications. She has expertise in the field of condensed matter physics, nanomagnetism, multiferroics, and polymeric nanocomposites, etc. She has research experience at various reputed national institutes like Bose Institute, Kolkata, India, SAHA Institute of Nuclear Physics, Kolkata, India, UGC-DAE CSR Centre, Kolkata, India and collaborations with various international laboratories. She is the recipient of a Young Researcher Award in the area of physics and Best Paper Awards—2020, 2021, a prestigious forum for showcasing intellectual capability. She served as assistant professor and examiner at the Department of Basic Sciences, Amal Jyothi College of Engineering, under APJ Abdul Kalam Technological University, Kerala, India. Dr. Abraham is a frequent speaker at national and international conferences. She has a good number of publications to her credit in many peer-reviewed high impact journals of international repute. She has authored many book chapters and edited more than 10 books with Taylor and Francis, Elsevier, etc. Dr. Abraham received her MSc, MPhil, and PhD degrees in Physics from School of Pure and Applied Physics, Mahatma Gandhi University, Kerala, India.

A. K. Haghi, PhD

Research Associate, Department of Chemistry,
University of Coimbra, Portugal

A. K. Haghi, PhD, is a retired professor and has written, co-written, edited or co-edited more than 1000 publications, including books, book chapters, and papers in refereed journals with over 3800 citations and h-index of 32, according to the Google Scholar database. He is currently a research associate at the University of Coimbra, Portugal. Professor Haghi has received several grants, consulted for several major corporations, and is a frequent speaker to national and international audiences. He is Founder and former Editor-in-Chief of the *International Journal of Chemoinformatics and*

Chemical Engineering and *Polymers Research Journal*. Professor Haghi has acted as an editorial board member of many international journals. He has served as a member of the Canadian Research and Development Center of Sciences & Cultures. He has supervised several PhD and MSc theses at the University of Guilan (UG) and co-supervised international doctoral projects. Professor Haghi holds a BSc in urban and environmental engineering from the University of North Carolina (USA) and holds two MSc degrees, one in mechanical engineering from North Carolina State University (USA) and another one in applied mechanics, acoustics, and materials from the Université de Technologie de Compiègne (France). He was awarded a PhD in engineering sciences at Université de Franche-Comté (France). He is a regular reviewer of leading international journals.

Contents

Contributors

Devrim Balkose
Department of Chemical Engineering, İzmir Institute of Technology, Urla, İzmir, Turkey

Aruna Kumar Barick
Department of Chemistry, Veer Surendra Sai University of Technology, Siddhi Vihar, Burla, Sambalpur, Odisha, India

S. G. Bystrov
The Udmurt Federal Research Centre of the Ural Branch of the Russian Academy of Sciences, Izhevsk, Russia

Dipankar Chattopadhyay
Department of Polymer Science and Technology, University of Calcutta, Kolkata, West Bengal, India
Center for Research in Nanoscience and Nanotechnology, Acharya Prafulla Chandra Roy Sikhsha Prangan, University of Calcutta, Kolkata, West Bengal, India

Chin Hua Chia
Science Program, Department of Applied Physics, Faculty of Science and Technology, Bangi, Selangor, Malaysia

Siew Xian Chin
ASASIpintar Program, Pusat GENIUS@Pintar Negara, Universiti Kebangsaan Malaysia, Bangi, Selangor, Malaysia
Department of Physics, Faculty of Science, Kasetsart University, Bangkok, Thailand

V. Diamant
Vernadsky Institute of General & Inorganic Chemistry NAS, Kiev, Ukraine

Adrija Ghosh
Department of Polymer Science and Technology, University of Calcutta, Kolkata, West Bengal, India

Merin Joby
Department of Physics and Centre for Research, St. Teresa's College (Autonomous), Ernakulam, Kerala, India

Cintil Jose
Newman College, Thodupuzha, Kerala, India

Minu Jose
Department of Botany, St, Teresa's College (Autonomous), Ernakulam, Kerala, India

Sonia Khanna
Advanced Polymeric Materials Research Laboratory, Department of Chemistry and Biochemistry, Sharda University, Greater Noida, Uttar Pradesh, India

V. I. Kodolov
Kalashnikov Izhevsk State Technical University, Izhevsk, Russia

Sunsu Kurian
Department of Physics and Centre for Research, St. Teresa's College (Autonomous), Ernakulam, Kerala, India

Nataliya Kutsevol
Taras Shevchenko National University of Kyiv, Kyiv, Ukraine

Yuliia Kuziv
Taras Shevchenko National University of Kyiv, Kyiv, Ukraine

O. Kychkyruk
Ivan Franko Zhytomyr State University, Zhytomyr, Ukraine

Bindu M.
Department of Environmental Studies, Kannur University, Mangattuparamba Campus, Kannur, Kerala, India

R. A. Makhnev
Kalashnikov Izhevsk State Technical University, Izhevsk, Russia

T. M. Makhneva
The Udmurt Federal Research Centre of the Ural Branch of the Russian Academy of Sciences, Izhevsk, Russia

Maria Mathew
C-MET, Thrissur, Kerala, India

Pooja Mohapatra
Department of Chemistry, Veer Surendra Sai University of Technology, Siddhi Vihar, Burla, Sambalpur, Odisha, India

O. Nadtoka
Taras Shevchenko National University of Kyiv, Kyiv, Ukraine

Anju K. Nair
Department of Physics and Centre for Research, St. Teresa's College (Autonomous), Ernakulam, Kerala, India

Rony Rajan Paul
Department of Chemistry, CMS College, Kottayam, Kerala, India

Vadim Pavlenko
Taras Shevchenko National University of Kyiv, Kyiv, Ukraine

Pradeepan Periyat
Department of Environmental Studies, Kannur University, Mangattuparamba Campus, Kannur, Kerala, India

Biju Peter
Newman College, Thodupuzha, Kerala, India

M. Reshetnyk
National Nature—Historical Museum, NAS, Kiev, Ukraine

Nurul Hazwani Aminuddin Rosli
Materials Science Program, Department of Applied Physics, Faculty of Science and Technology, Bangi, Selangor, Malaysia
Physics Department, Centre for Defence Foundation Studies, Universiti Pertahanan Nasional Malaysia, Kuala Lumpur, Malaysia

I. Savchenko
Taras National Taras Shevchenko University of Kyiv, Kyiv, Ukraine

I. N. Shabanova
The Udmurt Federal Research Centre of the Ural Branch of the Russian Academy of Sciences, Izhevsk, Russia

Lipsa Shubhadarshinee
Department of Chemistry, Veer Surendra Sai University of Technology, Siddhi Vihar, Burla, Sambalpur, Odisha, India

Narender Singh
Advanced Polymeric Materials Research Laboratory, Department of Chemistry and Biochemistry, Sharda University, Greater Noida, Uttar Pradesh, India

Ajeesh Kumar Somakumar
Institute of Physics, Polish Academy of Sciences, Warsaw, Poland

D. Starokadomsky
Chuiko Institute of Surface Chemistry, National Academy of Sciences (NAS), Kiev, Ukraine; Institute of Geochemistry & Mineralogy, NAS, Kiev, Ukraine

D. Sternik
Maria Curie-Skłodowska University, Lublin, Poland

Soumya Suresh
International School of Photonics, Cochin University of Science and Technology, Cochin, Kerala, India

N. S. Terebova
The Udmurt Federal Research Centre of the Ural Branch of the Russian Academy of Sciences, Izhevsk, Russia

Sabu Thomas
School of Energy Materials, Mahatma Gandhi University, Kottayam, Kerala, India

Sheenu Thomas
International School of Photonics, Cochin University of Science and Technology, Cochin, Kerala, India

Pavlo Virych
Taras Shevchenko National University of Kyiv, Kyiv, Ukraine

Anupama Viswanathan
Sree Narayana College Nattika, Triprayar, Kerala, India

L. Vretik
Taras National Taras Shevchenko University of Kyiv, Kyiv, Ukraine

Chatchawal Wongchoosuk
Department of Physics, Faculty of Science, Kasetsart University, Bangkok, Thailand

E. Yanovska
Taras National Taras Shevchenko University of Kyiv, Kyiv, Ukraine

Senem Yetgin
Department of Food Engineering, Kastamonu University, Kastamonu, Turkey

Abbreviations

3D-GCNTs	3D graphene/carbon nanotubes
AFM	atomic force microscopy
ALH	aluminum hydroxide
APP	ammonium polyphosphate
ARROW	antiresonance reflecting optical waveguide
AuNPs	gold nanoparticles
Ce6	chlorine e6
ChGs	chalcogenide glasses
CL	Cerenkov luminescence
CNTs	carbon nanotubes
Cu–C MC	copper–carbon mesocomposite
CV	cyclic voltammetry
CVD	chemical vapor deposition
DCDMS	dimethyl dichlorosilane
DMR	diagnostic magnetic resonance imaging
D-PNIPAM	dextran-graft-poly-N-isopropylacrylamide
DRIFT	diffuse reflectance infrared Fourier transform
ECM	extracellular matrix
EIS	electrochemical impudence spectroscopy
EP	epoxy polymer
FP	Fabry–Perot
FRCs	fiber-reinforced composites
FRP	fiber-reinforced polymer
FTIR	Fourier transform infrared
GCD	galvanic charge and discharge
GO	graphene oxide
HF	hydrofluoric acid
KGN	kartogenin
LCST	lower critical solution temperature
LIBs	lithium ion batteries
MFL	melamine formaldehyde lacquer
MGr	metal-oxide-decorated graphene
MHT	magnetic hyperthermia

MIR	mid-infrared
MMT	montmorillonite
MOF	microstructured optical fibers
MOFs	metal–organic frameworks
MRI	magnetic resonance imaging
MTCS	methyl trichlorosilane
MTIR	total internal reflection
NC	negative curvature
NFMs	nanofibrous membranes
NIR	near-infrared
P(LLA-*co*-PCL)	poly(L-lactide)-*co*-poly(ε-caprolactone)
PA	polyamide
PAN	polyacrylonitrile
PBG	photonic bandgap
PC	photonic crystal
PCFs	photonic crystal fibers
PCL	polycaprolactone
PCL-HA	polycaprolactone-hydroxyapatite
PDT	photodynamic therapy
PEO	polyethylene oxide
PES	poly(ether sulfone)
PGS	polyglycerol sebacate
PLLA	poly(L-lactic acid)
PMA	pulsed magneto-acoustic
PMETAC	poly[2-(methacryloyloxy)-ethyl] trimethyl ammonium chloride
PNIPAM	poly(*N*-isopropylacrylamide)
PQASCA	perfluorooctylated quarternary ammonium silane coupling agent
PRP	platelet-rich plasma
pTSA-PANI	*p*-toluenesulfonic acid-doped polyaniline
PVA	polyvinyl alcohol
PVP	polyvinylpyrrolidone
RBs	rechargeable batteries
RI	refractive index
SCD	supercritical drying
SEM	scanning electron microscopy
SPIOs	superparamagnetic iron oxides

SPR	surface plasmon resonance
TC	tetracycline
TCS	tetrachlorosilane
TMCS	trimethyl chlorosilane
XPS	X-ray photoelectron spectroscopy

Preface

Porous media exist in different modern materials. They present great surface areas with small pore-size distribution. These types of materials with controllable and adjustable pore diameters have attracted a considerable amount of attention due to their desirable properties and applications in several fields. Porous materials have many applications in our daily lives. For instance, we use different types of porous materials to clean our drinking water.

This new research-oriented volume focuses on exploring a wide range of porous materials. In this volume, original contributions from international authors along with case studies on the synthesis, design, characterization, and applications of different types of porous materials and solids are presented in detail. The scope of this book is to cover different types of porous materials in a broad sense by considering experimental and theoretical aspects of materials science related to porous materials and solids. The book will help research students to approach characterizing a particular type of material for more in-depth analysis.

This book is divided into three parts to determine the best techniques for solving particular porous materials problems and in each part of this new book, fabrication and characterization of porous materials are explored with applications. This new reference book describes new methodologies to gain required information along with the limitations of various methods.

To make this new title, we have selected a very comprehensive range of case studies to create a practical reference book for research students, engineers, and scientists of different disciplines working with porous materials and solids. These case studies are designed to cover the basic concepts of porosity. They also describe different types of pores and surfaces to the readers.

This book will be useful in the design and synthesis of porous materials.

PART I
Porous Materials for Water Technology

CHAPTER 1

Porous Chitosan Adsorbents for Wastewater Treatment

SIEW XIAN CHIN[1,4], CHIN HUA CHIA[2],
NURUL HAZWANI AMINUDDIN ROSLI[2,3], and
CHATCHAWAL WONGCHOOSUK[4]

[1]ASASIpintar Program, Pusat GENIUS@Pintar Negara,
Universiti Kebangsaan Malaysia, Bangi, Selangor, Malaysia

[2]Materials Science Program, Department of Applied Physics,
Faculty of Science and Technology, Bangi, Selangor, Malaysia

[3]Physics Department, Centre for Defence Foundation Studies,
Universiti Pertahanan Nasional Malaysia, Kuala Lumpur, Malaysia

[4]Department of Physics, Faculty of Science, Kasetsart University,
Bangkok, Thailand

ABSTRACT

Water pollution has become a major problem in the world. This is due to the rapid development and industrialization which have resulted in a huge amount of wastewater being discharged into the water system. Wastewater treatment is crucial to protect our environment, health of mankind, and maintain a balanced ecosystem. The development of a cost-effective and sustainable wastewater treatment method is necessary. Among all, adsorption method is the most used and investigated by researchers. Hydrogel is a three-dimensional network structure material that could provide a good pollutant adsorption capacity, good swelling, and reversible swelling property. Chitosan is a biopolymer that produced via the de-acetylation of chitin. It

Mechanics and Physics of Porous Materials: Novel Processing Technologies and Emerging Applications.
Chin Hua Chia, Tamara Tatrishvili, Ann Rose Abraham, & A. K. Haghi (Eds.)
© 2024 Apple Academic Press, Inc. Co-published with CRC Press (Taylor & Francis)

is abundantly available, low cost, highly porous, biodegradable, and can be functionalized easily due to hydroxyl and amino functional groups. Chitosan has been widely used as the starting materials to produce chitosan-based adsorbents, such as chitosan-based composite hydrogels, chitosan-based hydrogels beads, copolymeric chitosan-based hydrogels, membranes, and clays. This chapter summarizes the current trend and development of chitosan-based adsorbents for wastewater treatment toward different types of pollutants (heavy metal ions, dye, organic pollutants, etc.). Various modified forms of chitosan-based adsorbents are discussed with particular emphasis on their use toward different types of pollutants in wastewater. This chapter also provides a brief perspective on the potential of chitosan-based adsorbents for wastewater treatment.

1.1 INTRODUCTION

Water is essential and vital for all life. Basically, it could be said that it is virtually impossible for biological life on Earth to live without aqueous media. However, wastewater contaminates with pollutants that are harmful to human health and biological life and ecology. The rapid globalization and industrialization are the main contributors. For instance, industrial wastewater (pharmaceutical industries, textiles, paper mills, metal plating and cleaning, battery manufacturing, wood and pulp, and food industries) comprised various hazardous heavy metal ions (such as mercury, copper, chromium, arsenic, lead, nickel, cobalt, etc.), pesticide residues, microplastics, dyes, and other organic or inorganic compounds.[55,58,106] Hence, it is crucial to remove pollutants from wastewater or at least minimize its amount before discharge into watercourse. Various wastewater treatment techniques are used for heavy metal ions and other pollutants' removal, such as ion exchange,[30,132] membrane filtration,[142] electrochemical methods,[32] membrane separation,[143] chemical precipitation,[63] chemical oxidation,[124] photocatalytic degradation,[114] and adsorption.[125] Each method has its own advantages and disadvantages such as in term of high production of sludge, cost-effective, low or high removal efficiency, operation methods/procedures, and cost of maintenance.[30] Recently, adsorption treatment has been studied and used widely due to its advantages, such as low cost and simple operation, good efficiency, and environment friendly. This treatment method could also offer reusability property via adsorption-desorption process in the wastewater treatment process.[11]

The commonly used adsorbent materials are activated carbon, clay, alumina, and agricultural by-products, such as oil palm empty fruit bunch fiber, rice husks, corn cob, and fruit shell.[21,52] In addition, the utilization of chitosan-based hydrogels or cellulose-based hydrogels as adsorbents for wastewater treatment has also received great attention from researchers. Natural-based polymers can also serve as alternative adsorbent due to their availability, renewability, and reusability.[39,57,126]

Chitosan is the second largest natural polymer with high molecular weight. It is a biodegradable polysaccharide with -(1-4)-linked glucosamine units which consist of $(1 \rightarrow 4)$-2-acetamido-2-deoxy-β-D-glucan (*N*-acetyl D-glucosamine) and $(1 \rightarrow 4)$-2-amino-2-deoxy-β-D-glucan (D-glucosamine) units.[4,29] Chitosan is produced by the deacetylation of chitin.[94] Chitin can be found mainly from invertebrates and is present in mushroom, green algae, yeast, and fungi. Marine products are the main contributors of chitin due to the increasing demand for seafood and causing the increase of seafood processing industry (especially from the exoskeletons of crab, shrimp).[92] Ocean ecosystem produces about 10^{12}–10^{14} tons of chitin annually.[31] Commonly, chitosan is obtained from the alkaline de-acetylation (alkaline hydrolysis) of chitin. In de-acetylation process, the acetyl groups of chitins are hydrolyzed and turned into free amine groups. The degree of de-acetylation is highly depending on the parameters for hydrolysis process such as the concentration of sodium hydroxide, temperature, and reaction time. Hence, different grades/molecular weights of chitosan that can be found ranged from 10,000 to 1,000,000 Da.[3]

Chitosan is a biodegradable polymer with polycationic nature where it can be biodegraded by lysosomes via decomposition similar to other types of natural polymer such as collagen, cellulose, hemicellulose, and lignin.[36] In the view of biodegradability, renewability, and safety, chitin/chitosan have attracted a lot of attention as the raw materials for hydrogels, micro/nano-composites, membranes, and sheets which could provide huge structural modification possibilities to tailor in broad scope of applications ranging from the field of material science to medicine.

As mentioned above, chitin is one of the polysaccharides that is available abundantly in nature, due to its poor solubility in solvent. Therefore, *N*-deacetylation of chitin to produce chitosan is necessary. Chitosan-based hydrogels (membrane, film, bead, imprinted, blended, composite, copolymeric, etc.) have excellent properties such as good water-holding, reversible swelling, biocompatibility, and biocompatibility with many amino and hydroxyl groups. The presence of hydroxyl and amino groups provides the capacity to introduce different functional groups to improve the absorption capacity toward different pollutants through ion exchange mechanism. However, to further enhance the disadvantages of poor water solubility and selectivity of pollutants, modifications of chitosan-based adsorbents are necessary to produce a suitable adsorbent for adsorption process in wastewater treatment.[136]

1.2   PREPARATION OF CHITOSAN/SYNTHESIS AS ABSORBENT MATERIALS

Chitosan has been widely used as a bio-based material for various applications, such as bio-based ceramics and bio-adsorbent. Numerous innovative studies have been extensively conducted to study and modify the surface and structure of chitosan-based biomaterials to tailor their performance to suit different applications. The overview of the modifications (structural and chemical) of chitosan is discussed as follows:

Chitosan has been widely used as a starting material for the synthesis of hydrogels for different applications, such as tissue engineering, drug delivery, and wastewater treatment.[28] Chitosan is rarely used singly due to some limitations. It is often being used as a starting material for the synthesis of hydrogel. The modification of chitosan-based hydrogels is important to improve its properties and improve its performance for specific applications. The chitosan-based hydrogels can be modified physically and chemically as shown in Figure 1.1. The structure of chitosan with abundant reactive amino and hydroxyl groups is thus making it suitable for both physical and chemical modifications. Hydrogel is a three-dimensional network structure with high porosity and specific surface area.[123] However, all materials have its pro and cons, same goes to chitosan hydrogels which usually possess low mechanical strength, low thermal stability, and low stability in acidic medium. Therefore, chemical and physical networking methods have been proposed by researchers to enhance the properties of chitosan hydrogels for various applications. Chemical and physical networking methods can be summarized in Table 1.1. In general, physical networking method (formation of ionic cross-link) has lower stability as compared to chemical networking (covalently cross-linking).[90] Types of cross-linkers, synthesis parameters, and degree of cross-linking of chitosan hydrogels are essential not only to improve the mechanical properties of chitosan hydrogels, without affecting the adsorption capacity for wastewater treatment. Researchers also need to take the balance between both mechanical properties and adsorption capacity of chitosan-based adsorbents because the degree of cross-linking could affect the amount of free amino and hydroxyl group in the chitosan.[59] The development of suitable functionalized hydrogels is essential to suit different types of pollutant absorption from the aqueous solution.

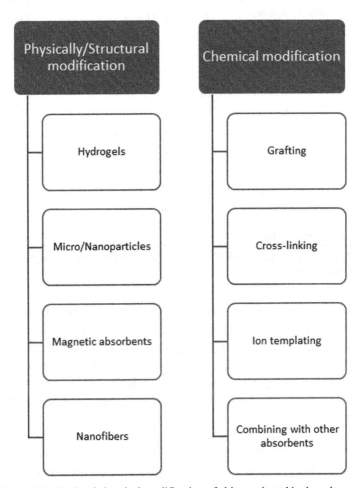

FIGURE 1.1   Physical and chemical modification of chitosan-based hydrogels.

The presence of amino groups on chitosan has been proven to be effective sorption for organic dyes and heavy metals ion. However, the cross-linking process to clarify hydrogels could lower down the amount of amino group in chitosan, thus lower down its absorption capacity.

TABLE 1.1   Chemical and Physical Networking Methods for Formation of Chitosan Hydrogels.

Networking	Mechanism	Example (chemical networking agents/physical networking agents)	References
Chemical networking	Formation of cross-linked polymers (irreversible covalent bonds)	N,N'-MBA	[13]
		Terephthalaldehyde	[37]
		EGDGE	[119]
		Formaldehyde	[9,73]
		GLA	[79,98]
		Genipin	[35]
		ECH	[76]
		Phenothiazine luminophore	[12]
Physical networking	Molecular entanglements, ionic interactions (ionic cross-linked chitosan hydrogels, anionic cross-linkers), hydrogen bonding	Sodium tripolyphosphate	[131]
		Sodium citrate	[49,54]
		Sulfosuccinic acid	[53]
		Oxalic acid	[97]

MBA: Methylenebisacrylamide; EGDGE: ethylene glycol diglycidyl ether; GLA: glutaraldehyde; ECH: epichlorohydrin.

On the other hand, several modifications of chitosan hydrogels had been introduced by researchers, such as decoration with polyethyleneimine,[71,79] polyvinyl alcohol,[5] acrylic acid,[51] and poly(N-methylaniline)[88] to produce a hybrid biodegradable adsorbent. These modifications have then increased the desired functional groups/resin's active sides to enhance the adsorption capacity of different types of pollutants via hydrogen bonding, electrostatic interaction, van der Waals force, and so on.

Meanwhile, impregnation of chitosan hydrogels with surfactants had been studied widely by many researchers. Mainly, there are three categories of surfactants, for example, anionic, cationic, and nonionic. The commonly used surfactants are hexadecyl trimethyl ammonium bromide,[22,40] cetyl-trimethyl ammonium bromide, sodium dodecyl sulfate,[33] dioctyl sulfosuccinate sodium salt,[55] hexadecylamine,[121] triton X-100,[22,23] 3-aminopropyl triethoxysilane,[74,120] and tea saponin.[82] The modification of chitosan-based hydrogels should be done carefully by studying the solubility of surfactants, the ionic (anionic or cationic) head group in the hydrogels, and the adsorption capacity. The pros and cons of each material need to be studied carefully to maximize the adsorption capacity of pollutants as well as their reusability.

1.3 CHITOSAN-BASED HYDROGELS FOR WASTEWATER TREATMENT

As mentioned, chitosan-based hydrogels with high porosity and specific surface area are making them suitable candidates for wastewater treatment. Therefore, the chitosan-based hydrogels for wastewater treatment, including chitosan-based composite hydrogels, chitosan-based hydrogels beads, and copolymeric chitosan-based hydrogels for the adsorption of several types of pollutants, are discussed to provide a summary on the current trend of chitosan-based hydrogel for wastewater treatment.

1.3.1 CHITOSAN-BASED HYDROGEL FOR WASTE-WATER TREATMENT VARIOUS POLLUTANTS

1.3.1.1 HEAVY METALS

One of the concerning environmental contaminants is heavy metals, which are present in most industrial wastewaters and water supplies. Natural soil erosion, atmospheric precipitation, volcanic eruptions, and wastewater disposal from a variety of industries, such as smelting, foundries, tanning, photography, plastic manufacturing, production, and consumption of metal-containing materials, are some of the processes by which heavy metals enter natural ecosystems.[20,65] The utilization of chitosan-based hydrogels for heavy metals ion has been studied extensively. The amount of heavy metals ion present in wastewater is usually in low concentration. However, in long periods, it can cause extended harm to human health and aquatic ecology. Chitosan-based hydrogels have several advantages to be used as adsorbent materials due to high swelling capability in aqueous solution, biodegradability, and abundant functional groups, which are recyclable.[19] Therefore, the removal of heavy metal ions from the wastewater has become a hot topic in scientific research area. Very often, the adsorption efficiency of chitosan-based hydrogels toward heavy metals ion is very much dependent on the pH of the solution.[117] Hence, fabrication and modification in preparing the chitosan-based hydrogels are crucial. Table 1.2 presents some recent experiment findings on chitosan-based adsorbents for the removal of various types of heavy metal ions. Mainly, the adsorption mechanism of chitosan-based adsorbents toward heavy metals ion is via electrostatic attraction due to the protonation of the amino groups under acidic condition.

TABLE 1.2 Partial Representative Chitosan-Based Adsorbents for Removal of Heavy Metal Ions.

Materials (chitosan based-adsorbents)	pH	Adsorbate	Maximum adsorption capacity	Reusability	Reference
Magnetic (Fe_3O_4) poly(ethylenimine)/chitosan (PEI/CS) hydrogel (with gallic acid)	2	Cr (VI)	476.2 mg/g	Up to 5 cycles	[79]
Chitosan, polyvinyl alcohol (PVA), zeolite nanofibrous composite membrane	–	Cr (VI), Fe (III), and Ni (II)	–Cr(VI), 100% (0.04 mmol/g) –Fe(III), 99% (0.019 mmol/g) – Ni(II), 100% (0.02 mmol/g)	Up to 5 cycles	[42]
Acrylic acid–chitosan blend hydrogel	5.5 5.7	Pb(II), Cu(II)	192 mg/g 171 mg/g	Up to 5 cycles	[85]
Silicate–titanate nanotubes chitosan hydrogel beads	6.5	Cadmium, Cd^{2+}	656 mg/g	7 cycles	[100]
Graphene oxide–chitosan–polyvinyl hydrogels	8	Cd^{2+} Ni^{2+}	171.11 mg/g 70.37 mg/g	Up to 6 cycles	[67]
Monolithic supermacroporous hydrogels (acrylic acid chains grafted onto the backbone of chitosan)	4–6	Cu^{2+} Pb^{2+}	302.01 mg/g 611.94 mg/g	Up to 5 cycles	[141]
Glucan/chitosan (GL/CS) hydrogels	7	Cu^{2+}, Co^{2+}, Ni^{2+}, Pb^{2+}, Cd^{2+}	342 mg/g, 232 mg/g, 184 mg/g, 395 mg/g, 269 mg/g	–	[51]
Hydrogel chitosan beads cross-link with glutaraldehyde	4–5	Cu(II) Phosphate	129 mg/g 250 mg/g	–	[7]
Carboxymethyl chitosan–kaolinite composite hydrogel	2.5–6	Cu^{2+}	206 mg/g	Up to 4 cycles	[45]
Acrylic glucose–chitosan hydrogel	7	Co^{2+}	202 mg/g	Up to 5 cycles	[77]
Chitosan, sodium alginate, calcium ion double-network hydrogel	3–7	Pb^{2+}, Cu^{2+}, Cd^{2+}	176.50 mg/g, 70.83 mg/g, and 81.25 mg/g	–	[116]

TABLE 1.2 *(Continued)*

Materials (chitosan based-adsorbents)	pH	Adsorbate	Maximum adsorption capacity	Reusability	Reference
Magnetic bentonite, carboxymethyl chitosan, sodium alginate hydrogel beads	5	Cu(II)	56.79 mg/g	Up to 4 cycles	[135]
Thiophene–chitosan (TCS) hydrogel	–	Hg(II)	20.53 mg/g	Up to 3 cycles	[83]
Fluorescent chitosan-based hydrogel (titanate and cellulose nanofibers modified with carbon dots)	7	Cr(IV)	228.2 mg/g	–	[81]
Gallic acid–modified carboxymethyl chitosan/iron ion (GA–CMCS/FeIII) complex hydrogels	7	Pb^{2+}, Cd^{2+}, Cu^{2+}	97.15 mg/g, 99.75 mg/g, 98.50 mg/g	Up to 10 cycles	[130]
Glucose–chitosan–acrylic hydrogel	7	Cu(II), Co(II)	286 mg/g, 185 mg/g	Up to 5 cycles	[138]

As can be seen from Table 1.2, extensive research works have been conducted. The development of chitosan-based adsorbents/hydrogels is still a hot topic with many attractive advantages, such as biocompatibility, fast removal rate, high absorption capacity, and reusability. It can be seen clearly that the performance of the adsorbents could be enhanced by different functionalizations (grafting of more functional groups) to suit different types of heavy metal ions adsorption.

1.3.1.2   DYE AND PIGMENT SORPTION

Organic dyes are one of the main types of contaminants emitted into wastewater from textile and other industrial activities.[44] The removal of dye or pigment from the effluents from the industrial wastewater (such as textile, pharmaceutical, printing, and battery manufacturing) is an issue that needs to be dealt. Leaving this issue untreated could led to severe problem to living organisms and heath deterioration on human health such as skin allergies (dermatitis or irritation reaction), jaundice, heart defects, and even cancer.[122] Numerous traditional materials, including metal oxides, organic substances, activated carbon, and zeolites, have so far been shown to be highly effective adsorbents in the removal of various dyes under specific circumstances.[47] Adsorption by using chitosan-based hydrogels has again gained great attention of researchers due to its ability to have interaction with dyes via electrostatic attraction, hydrogen bonding, or van der Waals force.[96] The performance of adsorption is highly depending on the pH, functional groups of the adsorbents, nature of the dyes (cationic and anionic), and so on. Modification of chitosan with the combination of other materials is necessary to increase the adsorption capacity. Thus, each combination of different materials needs to be investigated carefully to achieve the synergistic effect to maximize the adsorption capacity of the adsorbent.

From Table 1.3, chitosan-based adsorbents show good adsorption capability and capacity toward different types of dyes (anionic and cationic dyes). Modification of the chitosan-based adsorbents is necessary not only to improve their mechanical strength, adsorption capacity, and reusability.

1.3.1.3   OIL–WATER SEPARATION

Oily wastewater is one of the major problems in wastewater treatment. The main reason causing this pollution is due to illegal discharge of oil-containing

TABLE 1.3 Partial Representative Chitosan-Based Adsorbents for Removal of Dyes.

Chitosan-based hydrogels	pH	Adsorbate	Maximum adsorption capacity	Reusability	Reference
Carboxymethyl chitosan/phytic acid hydrogel	7	Methyl orange (MO) and Congo red (CR) dyes	13.62 mg/g 8.49 mg/g	Up to 5 cycles	[43]
Chitosan/hematite nanocomposite hydrogel capsules	2–7	Congo red (CR)	4705.6 mg/g	–	[95]
Chitosan/GO/Fe$_3$O$_4$ and acyl chloride hydrogel beads	3	Cationic ethylene blue (MB) and anionic eriochrome black T (EBT)	289 and 292 mg/g	Up to 6 cycles	[50]
HKUST-1/cellulose/chitosan aerogel	7	Methylene blue (MB)	526.3 mg/g	Up to 5 cycles	[76]
Graphene oxide (GO) cross-linked carboxymethyl cellulose and chitosan nanocomposites hydrogels	3–7	Methylene blue (MB) and methyl orange (MO)	655.98 mg/g 404.52 mg/g	Up to 12 cycles	[87]
Carboxymethyl cellulose and chitosan hydrogel	2–11	Acid orange (AO) and methylene blue (MB)	100 mg/g 120 mg/g	Up to 5 cycles	[127]
Carboxymethyl cellulose and chitosan–montmorillonite nanosheets composite hydrogel	2–10	Methylene blue (MB)	283.97 mg/g	Up to 5 cycles	[128]
Trimellitic anhydride isothiocyanate-cross-linked chitosan hydrogels	4	Congo red (CR)	63.05 mg/g	Up to 3 cycles	[89]
TiO$_2$ nanoparticles dispersed in chitosan-grafted polyacrylamide matrix hydrogels	2	Sirius yellow K-CF dye	1000 mg/g	Up to 6 cycles	[15]
Lamellar polysilicate magadiite encapsulated in chitosan hydrogel	4.5–6.5	Congo red (CR) and methylene blue (MB) dyes	135.77 and 45.25 mg/g	–	[91]
Iron(III) hydroxide doped chitosan via sodium dodecyl sulfate (SDS)	5	Alizarin red S (AR)	294 mg/g	–	[75]

TABLE 1.3 *(Continued)*

Chitosan-based hydrogels	pH	Adsorbate	Maximum adsorption capacity	Reusability	Reference
Magnetic iron oxide nanoparticles embedded in chitosan, graphene oxide hydrogels	4–11	Methylene blue (MB)	74.93 mg/g	Up to 4 cycles	[110]
Poly(vinyl imidazole) cross-linked chitosan hydrogel	4	Organic dye reactive black 5 (RB 5)	478.4 mg/g	Up to 4 cycles	[86]
Scaffold chitosan hydrogel with activated carbon	3	Food blue 2 (FB2) and food red 17 (FR17)	1428 and 1370 mg/g	Up to 5 cycles	[38]
Chitosan–montmorillonite hydrogel beads	6	Methyl green (MG)	303.21 mg/g	Up to 5 cycles	[66]

wastewater into the water system, well drilling for natural gas extraction and mining, oil spilling during extraction and processing, as well as transportation. This could cause severe harm to marine wildlife and thus destroy aquatic ecosystem.[105,108] Hydrogels have been widely used in oil–water separation because of its excellent properties, such as good water holding capacity and superoleophobic.[27,102] Liu and coworkers reported the preparation of supramolecular hydrogel network (acrylamide, acrylic acid copolymers and chitosan, cross-linked by amino-functionalized silica nanoparticles). The hydrogel demonstrated that very good mechanical properties, low swelling rate, and the underwater superoleophobicity can be retained even at corrosive condition, such as acidic, alkaline, and salt solution with good oil–water selectivity above 98.7%.[78] The same research group also reported a rubber-like, interpenetrating network poly(acrylamideco-acrylic acid) (P(AM-*o*-AA)/chitosan/methacryloxy propyl trimethoxyl silane (MPS)-modified SiO_2 nanocomposite hydrogel for oil–water separation. MPS-SiO_2 cross-linked P(AM-*co*-AA) covalent network could improve the mechanical properties of the hydrogel and thus improve its swelling properties and able to retain its superoleophobicity, and can be used after 50 cycles of separation with good separation efficiency (>99%).[112]

On the other hand, Lee and coworkers developed a chitosan–alginate hydrogel-coated mesh with excellent oil–water separation efficiency (>99%) under high flux. The produced hydrogel could offer some advantages, such as low cost, simple materials, and fabrication method while able to retain a good water superoleophobicity and good mechanical properties (stable and durable).[72] Another strategy for oily wastewater treatment was proposed by Feng and coworkers through the formation of hydrogel layer by layer on poly-acrylic acid grafted PVDF (PAA-g-PVDF) filtration membrane surface self-assembly of sodium alginate, chitosan, and Fe^{3+}. The composite membrane exhibited good oil–water separation efficiency (>99%) at separation flux of 974 L/m^2 h with good recyclability.[34] Zhang and coworkers proposed another approach fabricating the membrane-based hydrogel on stainless steel mesh as the matrix, namely, PAAm/CS@SSM (polyacrylamide/chitosan@stainless steel mesh). The chitosan acts as the hydrophilic component, polyacrylamide as the superoleophobic membrane on the stainless steel mesh. The oil–water separation efficiency is more than 99% with good selectivity.[139] A lot of existing studies about modification of chitosan-based hydrogels for oil–water separation for water treatment can be found, ranging from grafting with monomer/polymer, formation of hydrogels on membrane, cross-linking with different types of cross-linkers, and so on.[61,69,104] It can be concluded that chitosan-based hydrogel is a good candidate for oil–water separation with good durability and recyclability.

1.3.1.4   SEAWATER DESALINATION

Water scarcity, or the lack of access of fresh and clean water, has become a concerning issue due to rapid industrialization. About 75% of the earth is covered with water. However, only a fraction of water can be used directly as a result of various types of pollution and the content of salt is high in the seawater.[26,46] Hence, seeking of suitable solution is necessary to tackle this issue. Desalination is one of the suitable solutions for this. However, the process of desalination could be very complex and require expensive equipment with high energy consumption.[6,70] Hydrogel has been used for the reverse and forward osmosis. For instance, forward osmosis, hydrogel, is modified with different draw agents to produce desalinated water. Hydrogel could offer some advantages, such as good water holding capacity, high swelling and reversible swelling behavior, low cost, and biodegradable. Sorour and coworkers fabricated chitosan alginate hydrogel grafted with acrylamide and used as softeners for saline solution and reverse osmosis desalination brine. They found that the produced hydrogel shows a promising adsorption capacity of calcium (54 mg/g) and magnesium (316 mg/g) as compared to commercial resins.[111] Li and coworkers investigated the use of hydrogel particles (carbon filler mixed with hydrogel) as the draw agents for forward osmosis. The addition of carbon filler in the hydrogel is able to improve the swelling pressures and directly boost the water flux for FO desalination.[68]

Distillation is one of the key approaches for desalination process. However, distillation process involves a high amount of energy for heating and vaporization process which could result in high operating expenses with low return margin.[107] Solar evaporator has attracted great attention from researchers due to its low cost and simple operation procedure and its sustainability.[41,101] Yu and coworkers fabricated a molybdenum carbide/carbon-based chitosan hydrogel for solar absorber and water evaporation accelerator. The excellent salt-resistant property can be achieved due to the unique pore structure of the hydrogel, resulting in high daily evaporation yield (13.68 kg/m^2). The hydrogel provides a good solar–thermal conversion efficiency (96.15%) with a water evaporation rate up to 2.19 kg/m^2 h.[133] Another approach is proposed by Ren and coworkers by fabricating an aerogel-based solar evaporator (chitosan-based aerogel, with gold deposited through sputtering). This aerogel could be used for desalination as well as heavy metal ions removal. This aerogel evaporator provides an excellent evaporation rate up to 3.12 kg/m^2 h with good stability (maintain over 88.2%) and high salt resistance (maintain 80% at 200 g/kg) even after 10 days of water evaporation process. This is a great alternative for desalination and water purification process.[101] On the other hand, Xu and coworkers reported a chitosan-based hydrogel containing acrylamide, *N,N*'-methylene bisacrylamide, and ammonium persulfate with pyrrole monomer, namely, chitosan/PAAm/PPy hydrogel evaporator. This hydrogel evaporator provides a high water evaporation rate of 2.41 kg/m^2 h without a significant decay in term of its performance even after 14 days of continuous evaporation experiments in simulated marine water.[129]

Obviously, there are still a lot of opportunities and applications for chitosan-based hydrogels in water applications. However, the process could be challenging and requires more investigations for a better enhancement of the materials.

1.4 OTHER TYPES OF CHITOSAN-BASED ADSORBENTS

1.4.1 CHITOSAN-BASED MAGNETIC ABSORBENTS

Magnetic absorbents have attained high attention due to their magnetic nature, which makes it easier to be separated and removed the adsorbent via an external magnetic field.[115,140] Magnetic adsorbents have been developed to remove heavy metal, oil, dyes, hazardous organic compounds, and pharmaceuticals.[60] There are three categories of magnetic materials: ferromagnetic, paramagnetic, and diamagnetic materials. The magnetic properties of these materials are mostly influenced by their particle size.[115] Iron oxides, such as magnetite (Fe_3O_4) and maghemite (Fe_2O_3), have been investigated extensively among magnetic materials because of their robust superparamagnetic activity, ease of synthesis, high thermal stability, and low toxicity. Due to their sensitivity to acidic environments, these magnetic particles' capacity to be separated magnetically may be decreased.[118]

Furthermore, chitosan-based magnetic absorbance is used for the removal of diverse types of pollutants, such as heavy metals, dyes, and organic contaminants from water. With the application of a magnetic field, magnetic adsorbents can be separated from an aqueous solution, giving them a substantial advantage over other adsorbents.

There are many types of pollutants that can be found in the water system. The rapid development and industrialization have caused the urge for researchers to study various types of pollutants in wastewater and explore the harm of these pollutants to living organisms. Antibiotics pollution in water has become the concerning issue as the high content of antibiotic could cause increased emergence of resistant pathogenic bacteria; this will directly affect the sustainability of the environment.[62] The process to separate or remove antibiotics from the contaminated water could be very challenging. The development of suitable treatment methods can raise the water quality. Adsorption is again the most common approach investigated by many researchers. Lu and coworkers demonstrated the grafting of AMPS onto the CTS using the Pickering-HIPEs test plate. A porous chitosan-g-poly(2-acrylamide-2-methylpropanesulfonic acid) (CTS-g-AMPS) hydrogel. The hydrogel shows a good adsorption capacity of tetracycline (TC) and chlorotetracycline (CTC) (806.60 and 876.60 mg/g, respectively) in a pH range ranging from 3.0 to 11.0. The adsorption equilibrium can be attained within 90 min for TC and 50 min for CTC. After five successive adsorption cycles, the magnetic porous adsorbent still managed to maintain its adsorption capacities without reducing significantly (TC: 759.82, CTC: 842.99 mg/g).[80]

Ahamad et al.[2] fabricated a magnetic chitosan composite hydrogel from chitosan, diphenylurea, and formaldehyde with magnetic nanoparticles containing manganese iron oxide ($MnFe_2O_4$) to create a magnetic composite (CDF@MF) through condensation polymerization in an acidic medium. The CDF@MF provides a rapid removal of TC from aqueous solutions with the maximum adsorption capacity of 168.24 mg/g. From the obtained result, the adsorption kinetics and process followed a pseudo-first-order model. After seven cycles of regeneration, the CDF@MF exhibits an adsorption capacity of 157.97 mg/g at room temperature. The porous magnetic nanocomposite, CDF@MF, is a cost-effective adsorbent and could be a potential candidate for the removal of TC due to its stability and reusability. On the other hand, Zhang et al.[137] demonstrated the grafting of copolymerization onto the surface of chitosan/Fe_3O_4 composite particles and produced magnetic composite adsorbents with the core–brush structure (CS–MCP). The produced magnetic composite hydrogel was then used to remove diclofenac sodium and TC hydrochloride from water. A good adsorption ability can be observed, and it was attributable to the anionic contaminant species and the positively charged brushes' electrostatic affinity to one another.

Besides antibiotics, there are many other types of organic pollutants (microplastics, pesticides, phenol, organic acid, etc.) in the freshwater resources. A lot of existing and ongoing studies with the aim to develop a renewable and reusable absorbent are using chitosan as the raw material.

1.4.2 CHITOSAN AS MODIFIER FOR BIOCHARS

Biochar is a solid material with a high carbon content produced when biomass is thermochemically converted at temperatures below 700°C in the absence oxygen.[99] Typically, leaves, crop stalks, animal carcasses, manure, and wood chips are used as the raw materials for making biochar. Potentially hazardous elements (PTE) in water can be removed using biochar as an adsorbent by physical adsorption as well as chemical reactions, such as ion exchange, electrostatic attraction, complexation, precipitation, and redox.[134] Though biochar has a limited adsorption capacity, the density of the functional groups on its surface depends on the pyrolytic temperature, with a general loss of these functional groups as the pyrolytic temperature rises (da Silva Alves et al., 2021). To produce more effective adsorbents, substantial effort has been given to modify the biochar synthesis process and alter the surface by oxidation and/or functionalization. Chitosan has been used as a dispersing and stabilizing agent to create chitosan-modified biochar composites (CMBCs) by coating the biochar surfaces. The biochar particles are hard to be removed from aqueous solutions, but when the biochar is mixed with the chitosan, the solution phase can be easily separated (da Silva Alves et al., 2021).

Chitosan-modified biochar was created by Shi et al.[109] and utilized to remove dissolved organic matter (DOM) from biotreated coking wastewater (BTCW) that was obtained from a full-scale coking wastewater treatment facility. The maximum removal effectiveness of DOM in BTCW by biochar was enhanced after chitosan modification and increased from 12% to 52%. The biochar demonstrated a better affinity for all DOM components than unmodified.

Afzal et al.[1] developed chitosan/biochar hydrogel beads (CBHB) for the removal of ciprofloxacin (CIP) from aqueous solutions. From an initial concentration of 160 mg/L, it was discovered that the highest sorption capacity was >76 mg/g. The adsorption capacity dropped from 34.90 mg/g to 15.77 mg/g in the presence of 0.01 N Na_3PO_4. By eliminating CIP through electron donor–acceptor contact, hydrogen bonding, and hydrophobic interaction, CBHB demonstrated a multimodal mechanism. The adsorption capacity of CBHB was reduced to 32.69 mg/g and 29.29 mg/g by reforming it with methanol and ethanol instead of water. After the regeneration process, the capacity was slightly dropped >64% (25.73 mg/g) after six regeneration cycles. The effectiveness of CBHB for CIP removal demonstrated its value as a cost-effective and long-lasting adsorbent.

Bombuwala Dewage et al.[18] presented a new CMBC made from pinewood, assessed its batch and column sorption capability, and clarified the sorption pathways for Pb^{2+}. The process of lead removal showed pH-dependent and endothermic behavior at pH 5 and 318 K. The adsorption studies that a fixed-bed column revealed a 5.8 mg/g capacity of Pb^{2+}. Based on FTIR and XPS data, the coordination between chitosan amine groups and Pb^{2+} ions was the primarily responsible for controlling the Pb^{2+} adsorption mechanism on the CMBC biochar.

Heavy metals and other contaminants can be effectively removed from aqueous solutions using chitosan–biochar composites. Due to their exceptional qualities, chitosan–biochar composites have become standard sustainable materials for environmental clean-up.

1.4.3   CHITOSAN-BASED NANOCOMPOSITES

Recently, chitosan-based nanocomposites have been widely developed by combining chitosan polymer and nanomaterials. In addition to enhancing chitosan's physical, mechanical, and thermal durability, the scattered nano-materials in the chitosan polymer matrix provide unique intrinsic qualities to the chitosan, such as high surface area with exceptional physicochemical properties. Chitosan nanocomposites were produced via physical or chemical interaction between the chitosan and nanomaterials.[64]

1.4.3.1   CHITOSAN–GRAPHENE OXIDE NANOCOMPOSITE

Graphite is oxidized to produce graphene oxide (GO) using the well-known modified Hummers' process. As the graphite is oxidized, the interlayer gap widens, resulting in GO sheets that may be exfoliated using a relatively straightforward liquid-phase process. Chitosan–GO composites are easily made by combining the GO sheets with chitosan because they are stable in colloidal solutions (da Silva Alves et al., 2021).

The use of chitosan-grafted GO (chitosan–GO) nanocomposite was examined by Samuel et al.[103] for the adsorption of hexavalent chromium (Cr(VI)). With a capacity of 104.16 mg/g that succeed at pH 2.0 in the contact time of 420 min, chitosan–GO demonstrated strong monolayer adsorption. According to the results of the desorption and regeneration investigations, the chitosan–GO material can be recycled up to 10 times with no loss in adsorption capacity. Kamal et al.[56] produced chitosan–GO nanocomposites to remove Congo red (CR) from aqueous solutions. According to the find-ings, the chitosan–GO membrane has a removal efficiency of >90% for the

CR dye. The greatest CR loading capacity on the chitosan–GO membrane was 175.9 mg/g.

1.4.3.2 CHITOSAN–CARBON NANOTUBE NANOCOMPOSITE

Carbon nanotube (CNT) has very high surface area and stability, therefore great interest in creating adsorbent materials by mixing CNT with chitosan. There are two types of CNTs, that is, single-walled (SWCNT) and multi-walled (MWCNT), which can be synthesized. CNT can be evenly distributed throughout chitosan to increase the solubility and thus reduce agglomeration (da Silva Alves et al., 2021).[64]

Chatterjee et al.[25] studied the adsorption capacity of chitosan hydrogel impregnated with MWCNT for the removal of CR as an anionic dye. According to the adsorption experiments, 0.01% CNT impregnation was the optimum content of CNT in the chitosan for the highest adsorption capacity. The Langmuir model estimated a maximum adsorption capacity of 450.4 mg/g for the chitosan–CNT beads. Therefore, CNT impregnation might be a useful technique to improve the mechanical strength and mass transfer rate of chitosan hydrogel beads. Masheane et al.[84] prepared antimicrobial chitosan–alumina/functionalized-multiwalled carbon nanotube (f-MWCNT) nanocomposites through a simple phase inversion method. The produced chitosan–alumina/f-MWCNT nanocomposites demonstrated the suppression of the twelve tested bacterial strains. Hence, the nanocomposites demonstrate a potential for usage as a biocide in water treatment for the elimination of bacteria under various environmental conditions.

1.4.3.3 CHITOSAN–CLAY NANOCOMPOSITE

A subset of clay minerals is the phyllosilicate or sheet silicate family of minerals, which are distinguished by their layered structures made up of polymeric sheets of silica (SiO_4) tetrahedra connected with octahedra sheets [including aluminum (Al), magnesium (Mg), and iron (Fe)]. The chemical weathering of other silicate minerals found on the Earth's surface results in the formation of these layered-type aluminosilicate minerals.[14,48] In addition to being inexpensive and readily available, clay minerals also have additional catalytic capabilities. They are naturally occurring substance that is utilized for water treatment.

Clay minerals have been utilized as a coagulant to help in the removal of harmful substances, such as pesticides, herbicides, heavy metals, and color.[113] However, compared to zeolites and activated carbon, neat clay minerals are less effective in removing micropollutants from water, which limits their application. Smaller surface areas, difficulty in recovering after adsorption, weak affinity for organic compounds due to hydration of the clay mineral surface, and a significant reduction in adsorption capability after regeneration rarely fulfill the criteria of wastewater treatment systems. Therefore, nanocomposites of chitosan and clay have recently gained popularity due to their high availability, simplicity of manufacturing, and effectiveness as adsorbents.[16]

Nassar et al.[93] investigated the adsorptive removal of Mn(II) ions from contaminated aqueous medium utilizing native Egyptian glauconite clay (G) and its nanocomposites with modified chitosan. The removal of Mn(II) ions from the aqueous medium was demonstrated by the adsorption experiments, and it was found that the surfactant-modified chitosan/glauconite nanocomposite exhibited the highest adsorption capacity 27.74 mg/g. Azzam et al.[10] investigated the Cu(II) ion adsorption from an aqueous solution using chitosan/metal nanoparticles/clay composite. The chitosan/silver nanoparticles/clay composite was found to have a maximum Cu(II) ion absorption of 181.5 mg/g. The adsorption effectiveness of Cu(II) ions by produced chitosan/silver nanoparticles/clay and chitosan/gold nanoparticles/clay is more than that of the individual chitosan/clay nanocomposite, demonstrating the importance of metal nanoparticles in enhancing the adsorption performance of the nanocomposites.

Chitosan–clay composites were used to successfully adsorb selenium from water, as reported by Bleiman and Mishael.[17] Chitosan–montmorillonite composites had the best adsorption efficiency. Chitosan composites containing Al oxide and Fe oxide showed good agreement with the Langmuir model on the adsorption of selenate ions, giving adsorption capacities of 18.4, 17.2, and 8.2 mg/g, respectively. In addition, Auta and Hameed[8] prepared a modified ball clay–chitosan composite (MBC–CH) adsorbent for the adsorption of methylene blue. The adsorption capacity of MB on MBC–CH was 159% greater than that of MBC. After five regeneration cycles, the MBC–CH demonstrated more resilience than the MBC in terms of adsorption efficiency. This demonstrated that MB in an aqueous solution can be successfully absorbed utilizing MBC–CH several times before being discarded.

Chitosan-based nanocomposite can be formed in various combinations, such as chitosan–GO, chitosan–CNT, and chitosan–clay for removing different pollutants for wastewater treatment applications. The chitosan polymer matrix contains scattered nanoparticles that not only increase the physical, mechanical, and thermal stability of chitosan but also provide the composite its inherent qualities, such as high surface area and remarkable physicochemical properties.

1.5 CONCLUSION

Chitosan-based adsorbents have gained tremendous interest and attention for the application in wastewater treatment, due to its excellent properties, including high porosity, high surface area, good adsorption capacity, renewability, availability, and reusability. In this chapter, the modification of chitosan-based adsorbents for the adsorption of various types of pollutants (heavy metal ions, dye, organic compounds, oil–water separation, and desalination) had been covered. We have highlighted the potential of chitosan-based adsorbents and summarized many modification approaches (grafting, polymerization, embedded with nanomaterials) in optimizing and maximizing the adsorption capacity, mechanical properties as well as its reusability to produce an economic feasible adsorbent.

KEYWORDS

- water
- chitosan
- adsorption
- wastewater
- regeneration

REFERENCES

1. Afzal, M. Z.; Sun, X. F.; Liu, J.; Song, C.; Wang, S. G.; Javed, A. Enhancement of Ciprofloxacin Sorption on Chitosan/Biochar Hydrogel Beads. *Sci. Total Environ.* **2018,** *639,* 560–569. DOI: 10.1016/j.scitotenv.2018.05.129.
2. Ahamad, T.; Ruksana; Chaudhary, A. A.; Naushad, M.; Alshehri, S. M. Fabrication of $MnFe_2O_4$ Nanoparticles Embedded Chitosan-Diphenylureaformaldehyde Resin for the

Removal of Tetracycline from Aqueous Solution. *Int. J. Biol. Macromol.* **2019,** *134,* 180–188. DOI: 10.1016/j.ijbiomac.2019.04.204.

3. Ahmad, M.; Manzoor, K.; Ikram, S. Versatile Nature of Hetero-Chitosan Based Derivatives as Biodegradable Adsorbent for Heavy Metal Ions: A Review. *Int. J. Biol. Macromol.* **2017,** *105* (1), 190–203. DOI: 10.1016/j.ijbiomac.2017.07.008.

4. Ahmad, M.; Manzoor, K.; Ikram, S. Chitosan Based Nanocomposites for Drug, Gene Delivery, and Bioimaging Applications. *Appl. Nanocomposite Mater. Drug Deliv.* **2018,** 27–38. DOI: 10.1016/B978-0-12-813741-3.00002-9.

5. Akter, M.; Hirase, N.; Sikder, M. T.; Rahman, M. M.; Hosokawa, T.; Saito, T.; Kurasaki, M. Pb(II) Remediation from Aqueous Environment Using Chitosan-Activated Carbon-Polyvinyl Alcohol Composite Beads. *Water Air Soil Pollut.* **2021,** *232* (7), 272. DOI: 10.1007/s11270-021-05243-8.

6. Ali, I.; Gupta, V. K. Advances in Water Treatment by Adsorption Technology. *Nat. Protoc.* **2006,** *1* (6), 2661–2667. DOI: 10.1038/nprot.2006.370.

7. An, B.; Choi, J. W. An Experimental Application of Four Types of Chitosan Bead for Removal of Cationic and Anionic Pollutants. *Water Air Soil Pollut.* **2019,** *230* (12). DOI: 10.1007/s11270-019-4365-9.

8. Auta, M.; Hameed, B. H. Chitosan-Clay Composite as Highly Effective and Low-Cost Adsorbent for Batch and Fixed-Bed Adsorption of Methylene Blue. *Chem. Eng. J.* **2014,** *237,* 352–361. DOI: 10.1016/j.cej.2013.09.066.

9. Azeem, M. K.; Rizwan, M.; Islam, A.; Rasool, A.; Khan, S. M.; Khan, R. U.; Rasheed, T.; Bilal, M.; Iqbal, H. M. N. In-House Fabrication of Macro-porous Biopolymeric Hydrogel and Its Deployment for Adsorptive Remediation of Lead and Cadmium from Water Matrices. *Environ. Res.* **2022,** *214* (2), 113790. DOI: 10.1016/j.envres.2022.113790.

10. Azzam, E. M. S.; Eshaq, G.; Rabie, A. M.; Bakr, A. A.; Abd-Elaal, A. A.; El Metwally, A. E.; Tawfik, S. M. Preparation and Characterization of Chitosan-Clay Nanocomposites for the Removal of Cu(II) from Aqueous Solution. *Int. J. Biol. Macromol.* **2016,** *89,* 507–517. DOI: 10.1016/j.ijbiomac.2016.05.004.

11. Bailey, S. E.; Olin, T. J.; Bricka, R. M.; Adrian, D. D. A Review of Potentially Low-Cost Sorbents for Heavy Metals. *Water Res.* **1999,** *33* (11), 2469–2479. DOI: 10.1016/S0043-1354 (98)00475-8.

12. Bejan, A.; Doroftei, F.; Cheng, X.; Marin, L. Phenothiazine-Chitosan Based Eco-adsorbents: A Special Design for Mercury Removal and Fast Naked Eye Detection. *Int. J. Biol. Macromol.* **2020,** *162,* 1839–1848. DOI: 10.1016/j.ijbiomac.2020.07.232.

13. Bhangi, B. K.; Ray, S. K. Synthesis of Cu Nanoparticles in a Chitosan Entrapped Copolymer Matrix for Photocatalytic Reduction of Textile Dye and Column Adsorption of Heavy Metal Ions from Water. *Polym. Eng. Sci.* **2022,** *62* (5), 1399–1415. DOI: 10.1002/pen.25930.

14. Bibi, I.; Icenhower, J.; Niazi, N. K.; Naz, T.; Shahid, M.; Bashir, S. Clay Minerals: Structure, Chemistry, and Significance in Contaminated Environments and Geological CO_2 Sequestration. *Environmental Materials and Waste: Resource Recovery and Pollution Prevention.* Elsevier Inc.: Amsterdam, 2016; pp 543–567. DOI: 10.1016/B978-0-12-803837-6.00021-4.

15. Binaeian, E.; Babaee Zadvarzi, S.; Yuan, D. Anionic Dye Uptake via Composite Using Chitosan-Polyacrylamide Hydrogel as Matrix Containing TiO_2 Nanoparticles: Comprehensive Adsorption Studies. *Int. J. Biol. Macromol.* **2020,** *162,* 150–162. DOI: 10.1016/j.ijbiomac.2020.06.158.

16. Biswas, S.; Fatema, J.; Debnath, T.; Rashid, T. U. Chitosan–Clay Composites for Wastewater Treatment: A State-of-the-Art Review. *ACS EST Water* **2021,** *1* (5), 1055–1085. DOI: 10.1021/acsestwater.0c00207.

17. Bleiman, N.; Mishael, Y. G. Selenium Removal from Drinking Water by Adsorption to Chitosan-Clay Composites and Oxides: Batch and Columns Tests. *J. Hazard. Mater.* **2010,** *183* (1–3), 590–595. DOI: 10.1016/j.jhazmat.2010.07.065.

18. Bombuwala Dewage, N.; Fowler, R. E.; Pittman, C. U.; Mohan, D.; Mlsna, T. Lead (Pb^{2+}) Sorptive Removal Using Chitosan-Modified Biochar: Batch and Fixed-Bed Studies. *RSC Adv.* **2018,** *8* (45), 25368–25377. DOI: 10.1039/c8ra04600j.

19. Brião, G. de V.; de Andrade, J. R.; da Silva, M. G. C.; Vieira, M. G. A. Removal of Toxic Metals from Water Using Chitosan-Based Magnetic Adsorbents: A Review. *Environ. Chem. Lett.* **2020,** *18* (4), 1145–1168. DOI: 10.1007/s10311-020-01003-y.

20. Briffa, J.; Sinagra, E.; Blundell, R. Heavy Metal Pollution in the Environment and Their Toxicological Effects on Humans. *Heliyon* **2020,** *6* (9), e04691–e04716. DOI: 10.1016/j.heliyon.2020.e04691.

21. Burakov, A. E.; Galunin, E. V.; Burakova, I. V.; Kucherova, A. E.; Agarwal, S.; Tkachev, A. G.; Gupta, V. K. Adsorption of Heavy Metals on Conventional and Nanostructured Materials for Wastewater Treatment Purposes: A Review. *Ecotoxicol. Environ. Saf.* **2018,** *148*, 702–712. DOI: 10.1016/j.ecoenv.2017.11.034.

22. Chatterjee, S.; Lee, D. S.; Lee, M. W.; Woo, S. H. Congo Red Adsorption from Aqueous Solutions by Using Chitosan Hydrogel Beads Impregnated with Nonionic or Anionic Surfactant. *Bioresour. Technol.* **2009,** *100* (17), 3862–3868. DOI: 10.1016/j.biortech. 2009.03.023.

23. Chatterjee, S.; Lee, D. S.; Lee, M. W.; Woo, S. H. Enhanced Molar Sorption Ratio for Naphthalene through the Impregnation of Surfactant into Chitosan Hydrogel Beads. *Bioresour. Technol.* **2010,** *101* (12), 4315–4321. DOI: 10.1016/j.biortech.2010.01.062.

24. Chatterjee, S.; Lee, M. W.; Woo, S. H. Influence of Impregnation of Chitosan Beads with Cetyl Trimethyl Ammonium Bromide on Their Structure and Adsorption of Congo Red from Aqueous Solutions. *Chem. Eng. J.* **2009,** *155* (1–2), 254–259. DOI: 10.1016/j. cej.2009.07.051.

25. Chatterjee, S.; Lee, M. W.; Woo, S. H. Adsorption of Congo Red by Chitosan Hydrogel Beads Impregnated with Carbon Nanotubes. *Bioresour. Technol.* **2010,** *101* (6), 1800–1806. DOI: 10.1016/j.biortech.2009.10.051.

26. Chen, C.; Kuang, Y.; Hu, L. Challenges and Opportunities for Solar Evaporation. *Joule* **2019,** *3* (3), 683–718. https://doi.org/https. DOI: 10.1016/j.joule.2018.12.023.

27. Chen, M.; Shen, Y.; Xu, L.; Xiang, G.; Ni, Z. Synthesis of a Super-absorbent Nanocomposite Hydrogel Based on Vinyl Hybrid Silica Nanospheres and Its Properties. *RSC Adv.* **2020,** *10* (67), 41022–41031. DOI: 10.1039/d0ra07074b.

28. Cheng, M.; Gao, X.; Wang, Y.; Chen, H.; He, B.; Xu, H.; Li, Y.; Han, J.; Zhang, Z. Synthesis of Glycyrrhetinic Acid-Modified Chitosan 5-Fluorouracil Nanoparticles and Its Inhibition of Liver Cancer Characteristics In Vitro and In Vivo. *Mar. Drugs* **2013,** *11* (9), 3517–3536. DOI: 10.3390/md11093517.

29. Clark, G. L.; Smith, A. F. X-ray Diffraction Studies of Chitin, Chitosan, and Derivatives. *J. Phys. Chem.* **1936,** *40* (7), 863–879. DOI: 10.1021/j150376a001.

30. Dabrowski, A.; Hubicki, Z.; Podkościelny, P.; Robens, E. Selective Removal of the Heavy Metal Ions from Waters and Industrial Wastewaters by Ion-Exchange Method. *Chemosphere* **2004,** *56* (2), 91–106. DOI: 10.1016/j.chemosphere.2004.03.006.

31. Dhillon, G. S.; Kaur, S.; Brar, S. K.; Verma, M. Green Synthesis Approach: Extraction of Chitosan from Fungus Mycelia. *Crit. Rev. Biotechnol.* **2013,** *33* (4), 379–403. DOI: 10.3109/07388551.2012.717217.
32. Du, X.; Zhang, H.; Hao, X.; Guan, G.; Abudula, A. Facile Preparation of Ion-Imprinted Composite Film for Selective Electrochemical Removal of Nickel(II) Ions. *A.C.S. Appl. Mater. Interfaces* **2014,** *6* (12), 9543–9549. DOI: 10.1021/am501926u.
33. Du, X.; Kishima, C.; Zhang, H.; Miyamoto, N.; Kano, N. Removal of Chromium(VI) by Chitosan Beads Modified with Sodium Dodecyl Sulfate (SDS). *Appl. Sci.* **2020,** *10* (14). DOI: 10.3390/app10144745.
34. Feng, Q.; Zhan, Y.; Yang, W.; Dong, H.; Sun, A.; Liu, Y.; Wen, X.; Chiao, Y. H.; Zhang, S. Layer-by-Layer Construction of Super-Hydrophilic and Self-Healing Polyvinylidene Fluoride Composite Membrane for Efficient Oil/Water Emulsion Separation. *Colloids Surf. A Physicochem. Eng. Aspects* **2021,** *629* (June), 127462. DOI: 10.1016/j.colsurfa.2021.127462.
35. Feng, Z.; Danjo, T.; Odelius, K.; Hakkarainen, M.; Iwata, T.; Albertsson, A. C. Recyclable Fully Biobased Chitosan Adsorbents Spray-Dried in One Pot to Microscopic Size and Enhanced Adsorption Capacity. *Biomacromolecules* **2019,** *20* (5), 1956–1964. DOI: 10.1021/acs.biomac.9b00186.
36. Fraser, J. R.; Laurent, T. C.; Engström-Laurent, A.; Laurent, U. G. Elimination of Hyaluronic Acid from the Blood Stream in the Human. *Clin. Exp. Pharmacol. Physiol.* **1984,** *11* (1), 17–25. DOI: 10.1111/j.1440-1681.1984.tb00235.x.
37. Garg, M.; Bhullar, N.; Bajaj, B.; Sud, D. Terephthalaldehyde as a Good Crosslinking Agent in Crosslinked Chitosan Hydrogel for the Selective Removal of Anionic Dyes. *N. J. Chem.* **2021,** *45* (11), 4938–4949. DOI: 10.1039/D0NJ05758D.
38. Gonçalves, J. O.; da Silva, K. A.; Rios, E. C.; Crispim, M. M.; Dotto, G. L.; Pinto, L. A. Single and Binary Adsorption of Food Dyes on Chitosan/Activated Carbon Hydrogels. *Chem. Eng. Technol.* **2019,** *42* (2), 454–464. DOI: 10.1002/ceat.201800367.
39. Grishkewich, N.; Mohammed, N.; Tang, J.; Tam, K. C. Recent Advances in the Application of Cellulose Nanocrystals. *Curr. Opin. Colloid Interface Sci.* **2017,** *29*, 32–45. https://doi.org/https. DOI: 10.1016/j.cocis.2017.01.005.
40. Guo, J.; Chen, S.; Liu, L.; Li, B.; Yang, P.; Zhang, L.; Feng, Y. Adsorption of Dye from Wastewater Using Chitosan–CTAB Modified Bentonites. *J. Colloid Interface Sci.* **2012,** *382* (1), 61–66. DOI: 10.1016/j.jcis.2012.05.044.
41. Guo, Y.; Zhao, F.; Zhou, X.; Chen, Z.; Yu, G. Tailoring Nanoscale Surface Topography of Hydrogel for Efficient Solar Vapor Generation. *Nano Lett.* **2019,** *19* (4), 2530–2536. DOI: 10.1021/acs.nanolett.9b00252.
42. Habiba, U.; Afifi, A. M.; Salleh, A.; Ang, B. C. Chitosan/(Polyvinyl Alcohol)/Zeolite Electrospun Composite Nanofibrous Membrane for Adsorption of Cr^{6+}, Fe^{3+} and Ni^{2+}. *J. Hazard. Mater.* **2017,** *322* (A), 182–194. DOI: 10.1016/j.jhazmat.2016.06.028.
43. Han, D.; Zhao, H.; Gao, L.; Qin, Z.; Ma, J.; Han, Y.; Jiao, T. Preparation of Carboxymethyl Chitosan/Phytic Acid Composite Hydrogels for Rapid Dye Adsorption in Wastewater Treatment. *Colloids Surf. A Physicochem. Eng. Aspects* **2021,** *628* (July), 127355. DOI: 10.1016/j.colsurfa.2021.127355.
44. Han, F.; Kambala, V. S. R.; Srinivasan, M.; Rajarathnam, D.; Naidu, R. Tailored Titanium Dioxide Photocatalysts for the Degradation of Organic Dyes in Wastewater Treatment: A Review. *Appl. Cat. A* **2009,** *359* (1–2), 25–40. DOI: 10.1016/j.apcata.2009.02.043.

45. He, G.; Wang, C.; Cao, J.; Fan, L.; Zhao, S.; Chai, Y. Carboxymethyl Chitosan-Kaolinite Composite Hydrogel for Efficient Copper Ions Trapping. *J. Environ. Chem. Eng.* **2019**, *7* (2), 102953. DOI: 10.1016/j.jece.2019.102953.

46. He, S.; Chen, C.; Kuang, Y.; Mi, R.; Liu, Y.; Pei, Y.; Kong, W.; Gan, W.; Xie, H.; Hitz, E.; Jia, C.; Chen, X.; Gong, A.; Liao, J.; Li, J.; Ren, Z. J.; Yang, B.; Das, S.; Hu, L. Nature-Inspired Salt Resistant Bimodal Porous Solar Evaporator for Efficient and Stable Water Desalination. *Energy Environ. Sci.* **2019**, *12* (5), 1558–1567. DOI: 10.1039/C9EE00945K.

47. Hu, Q.; Bin, L.; Li, P.; Fu, F.; Guan, G.; Hao, X.; Tang, B. Highly Efficient Removal of Dyes from Wastewater over a Wide Range of pH Value by a Self-Adaption Adsorbent. *J. Mol. Liq.* **2021**, *331*, 115719–115727. DOI: 10.1016/j.molliq.2021.115719.

48. Huggett, J. M. Clay Minerals. *Encyclopedia of Geology*, 2004; Issue 1978, pp 358–365. DOI: 10.1016/B0-12-369396-9/00273-2.

49. Ivetic, D. Z.; Sciban, M. B.; Vasic, V. M.; Kukic, D. V.; Prodanovic, J. M.; Antov, M. G. Evaluation of Possibility of Textile Dye Removal from Wastewater by Aqueous Two-Phase Extraction. *Desalin. Water Treat.* **2013**, *51* (7–9), 1603–1608. DOI: 10.1080/19443994.2012.714650.

50. Jamali, M.; Akbari, A. Facile Fabrication of Magnetic Chitosan Hydrogel Beads and Modified by Interfacial Polymerization Method and Study of Adsorption of Cationic/Anionic Dyes from Aqueous Solution. *J. Environ. Chem. Eng.* **2021**, *9* (3), 105175. DOI: 10.1016/j.jece.2021.105175.

51. Jiang, C.; Wang, X.; Wang, G.; Hao, C.; Li, X.; Li, T. Adsorption Performance of a Polysaccharide Composite Hydrogel Based on Crosslinked Glucan/Chitosan for Heavy Metal Ions. *Compos. B Eng.* **2019**, *169*, 45–54. DOI: 10.1016/j.compositesb.2019.03.082.

52. Joseph, L.; Jun, B. M.; Flora, J. R. V.; Park, C. M.; Yoon, Y. Removal of Heavy Metals from Water Sources in the Developing World Using Low-Cost Materials: A Review. *Chemosphere* **2019**, *229*, 142–159. DOI: 10.1016/j.chemosphere.2019.04.198.

53. Jóźwiak, T.; Filipkowska, U. Sorption Kinetics and Isotherm Studies of a Reactive Black 5 Dye on Chitosan Hydrogel Beads Modified with Various Ionic and Covalent Cross-Linking Agents. *J. Environ. Chem. Eng.* **2020**, *8* (2), 103564. DOI: 10.1016/j.jece.2019.103564.

54. Jóźwiak, T.; Filipkowska, U. The Use of Air-Lift Adsorber with a Floating Filling from a Cross-Linked Chitosan Hydrogels for Reactive Black 5 Removal. *Sci. Rep.* **2021**, *11* (1), 1–11. DOI: 10.1038/s41598-021-92856-y.

55. Junaid, M.; Liu, S.; Liao, H.; Liu, X.; Wu, Y.; Wang, J. Wastewater Plastisphere Enhances Antibiotic Resistant Elements, Bacterial Pathogens, and Toxicological Impacts in the Environment. *Sci. Total Environ.* **2022**, *841*, 156805. DOI: 10.1016/j.scitotenv.2022.156805.

56. Kamal, M. A.; Bibi, S.; Bokhari, S. W.; Siddique, A. H.; Yasin, T. Synthesis and Adsorptive Characteristics of Novel Chitosan/Graphene Oxide Nanocomposite for Dye Uptake. *React. Funct. Polym.* **2017**, *110*, 21–29. DOI: 10.1016/j.reactfunctpolym.2016.11.002.

57. Kanmani, P.; Aravind, J.; Kamaraj, M.; Sureshbabu, P.; Karthikeyan, S. Environmental Applications of Chitosan and Cellulosic Biopolymers: A Comprehensive Outlook. *Bioresour. Technol.* **2017**, *242*, 295–303. DOI: 10.1016/j.biortech.2017.03.119.

58. Katare, Y.; Singh, P.; Sankhla, M. S.; Singhal, M.; Jadhav, E. B.; Parihar, K.; Nikalje, B. T.; Trpathi, A.; Bhardwaj, L. Microplastics in Aquatic Environments: Sources, Ecotoxicity,

Detection & Remediation. *Biointerface Res. Appl. Chem.* **2022,** *12* (3), 3407–3428. DOI: 10.33263/BRIAC123.34073428.

59. Khan, M.; Lo, I. M. C. A Holistic Review of Hydrogel Applications in the Adsorptive Removal of Aqueous Pollutants: Recent Progress, Challenges, and Perspectives. *Water Res.* **2016,** *106*, 259–271. DOI: 10.1016/j.watres.2016.10.008.

60. Kharissova, O. V.; Dias, H. V. R.; Kharisov, B. I. Magnetic Adsorbents on the Basis of Micro- and Nanostructurized Materials. *RSC Adv.* **2015,** *5* (9), 6695–6719. DOI: 10.1039/C4RA11423J.

61. Kordjazi, S.; Kamyab, K.; Hemmatinejad, N. Super-Hydrophilic/Oleophobic Chitosan/ Acrylamide Hydrogel: An Efficient Water/Oil Separation Filter. *Adv. Compos. Hybrid Mater.* **2020,** *3* (2), 167–176. DOI: 10.1007/s42114-020-00150-8.

62. Kumar, A.; Pal, D. Antibiotic Resistance and Wastewater: Correlation, Impact and Critical Human Health Challenges. *J. Environ. Chem. Eng.* **2018,** *6* (1), 52–58. DOI: 10.1016/j.jece.2017.11.059.

63. Kumar, S. G.; Rao, K. S. R. K. Comparison of Modification Strategies toward Enhanced Charge Carrier Separation and Photocatalytic Degradation Activity of Metal Oxide Semiconductors (TiO$_2$, WO$_3$ and ZnO). *Appl. Surf. Sci.* **2017,** *391*, 124–148. DOI: 10.1016/j.apsusc.2016.07.081.

64. Kumar, S.; Ye, F.; Dobretsov, S.; Dutta, J. Chitosan Nanocomposite Coatings for Food, Paints, and Water Treatment Applications. *Appl. Sci.* **2019,** *9* (12), 2409–2435. DOI: 10.3390/app9122409.

65. Kumar, V.; Dwivedi, S. K. A Review on Accessible Techniques for Removal of Hexavalent Chromium and Divalent Nickel from Industrial Wastewater: Recent Research and Future Outlook. *J. Cleaner Prod.* **2021,** *295*, 126229–126250. DOI: 10.1016/j.jclepro.2021.126229.

66. Kurczewska, J. Chitosan–Montmorillonite Hydrogel Beads for Effective Dye Adsorption. *J. Water Process Eng.* **2022,** *48*, 102928. DOI: 10.1016/j.jwpe.2022.102928.

67. Li, C.; Yan, Y.; Zhang, Q.; Zhang, Z.; Huang, L.; Zhang, J.; Xiong, Y.; Tan, S. Adsorption of CD^{2+} and Ni^{2+} from Aqueous Single-Metal Solutions on Graphene Oxide-Chitosan-Poly(Vinyl Alcohol) Hydrogels. *Langmuir* **2019,** *35* (13), 4481–4490. DOI: 10.1021/acs.langmuir.8b04189.

68. Li, D.; Zhang, X.; Simon, G. P.; Wang, H. Forward Osmosis Desalination Using Polymer Hydrogels as a Draw Agent: Influence of Draw Agent, Feed Solution and Membrane on Process Performance. *Water Res.* **2013,** *47* (1), 209–215. DOI: 10.1016/j.watres.2012.09.049.

69. Li, F.; Miao, G.; Gao, Z.; Xu, T.; Zhu, X.; Miao, X.; Song, Y.; Ren, G.; Li, X. A Versatile Hydrogel Platform for Oil/Water Separation, Dye Adsorption, and Wastewater Purification. *Cellulose* **2022,** *29* (8), 4427–4438. DOI: 10.1007/s10570-022-04535-4.

70. Li, Q.; Zhao, X.; Li, L.; Hu, T.; Yang, Y.; Zhang, J. Facile Preparation of Polydimethyl-siloxane/Carbon Nanotubes Modified Melamine Solar Evaporators for Efficient Steam Generation and Desalination. *J. Colloid Interface Sci.* **2021,** *584*, 602–609. DOI: 10.1016/j.jcis.2020.10.002.

71. Li, R.; An, Q.-D.; Xiao, Z.-Y.; Zhai, B.; Zhai, S.-R.; Shi, Z. Preparation of PEI/CS Aerogel Beads with a High Density of Reactive Sites for Efficient Cr(VI) Sorption: Batch and Column Studies. *RSC Adv.* **2017,** *7* (64), 40227–40236. DOI: 10.1039/C7RA06914F.

72. Li, Y.; Zhang, H.; Ma, C.; Yin, H.; Gong, L.; Duh, Y.; Feng, R. Durable, Cost-Effective and Superhydrophilic Chitosan–Alginate Hydrogel-Coated Mesh for Efficient Oil/

Water Separation. *Carbohydr. Polym.* **2019,** *226* (August), 115279. DOI: 10.1016/j. carbpol.2019.115279.

73. Liang, W.; Li, M.; Zhang, Z.; Jiang, Y.; Awasthi, M. K.; Jiang, S.; Li, R. Decontamination of Hg(II) from Aqueous Solution Using Polyamine-co-Thiourea Inarched Chitosan Gel Derivatives. *Int. J. Biol. Macromol.* **2018,** *113,* 106–115. DOI: 10.1016/j.ijbiomac. 2018.02.101.

74. Lima, V. V. C.; Dalla Nora, F. B.; Peres, E. C.; Reis, G. S.; Lima, É. C.; Oliveira, M. L. S.; Dotto, G. L. Synthesis and Characterization of Biopolymers Functionalized with APTES (3-Aminopropyltriethoxysilane) for the Adsorption of Sunset Yellow Dye. *J. Environ. Chem. Eng.* **2019,** *7* (5), 103410. https://doi.org/https. DOI: 10.1016/j.jece. 2019.103410.

75. Limchoowong, N.; Sricharoen, P.; Chanthai, S. A Novel Bead Synthesis of the Chiron-Sodium Dodecyl Sulfate Hydrogel and Its Kinetics-Thermodynamics Study of Superb Adsorption of Alizarin Red S from Aqueous Solution. *J. Polym. Res.* **2019,** *26* (12). DOI: 10.1007/s10965-019-1944-9.

76. Liu, Q.; Yu, H.; Zeng, F.; Li, X.; Sun, J.; Li, C.; Lin, H.; Su, Z. HKUST-1 Modified Ultrastability Cellulose/Chitosan Composite Aerogel for Highly Efficient Removal of Methylene Blue. *Carbohydr. Polym.* **2021,** *255* (September), 117402. DOI: 10.1016/j. carbpol.2020.117402.

77. Liu, Y.; Hu, L.; Yao, Y.; Su, Z.; Hu, S. Construction of Composite Chitosan–Glucose Hydrogel for Adsorption of Co^{2+} Ions. *Int. J. Biol. Macromol.* **2019,** *139,* 213–220. DOI: 10.1016/j.ijbiomac.2019.07.202.

78. Liu, Y.; Su, M.; Fu, Y.; Zhao, P.; Xia, M.; Zhang, Y.; He, B.; He, P. Corrosive Environments Tolerant, Ductile and Self-Healing Hydrogel for Highly Efficient Oil/Water Separation. *Chem. Eng. J.* **2018,** *354,* 1185–1196. DOI: 10.1016/j.cej.2018.08.071.

79. Lü, T.; Ma, R.; Ke, K.; Zhang, D.; Qi, D.; Zhao, H. Synthesis of Gallic Acid Functionalized Magnetic Hydrogel Beads for Enhanced Synergistic Reduction and Adsorption of Aqueous Chromium. *Chem. Eng. J.* **2021,** *408* (September). DOI: 10.1016/j.cej.2020.127327.

80. Lu, T.; Zhu, Y.; Qi, Y.; Wang, W.; Wang, A. Magnetic Chitosan–Based Adsorbent Prepared via Pickering High Internal Phase Emulsion for High-Efficient Removal of Antibiotics. *Int. J. Biol. Macromol.* **2018,** *106,* 870–877. DOI: 10.1016/j.ijbiomac.2017.08.092.

81. Luo, Q.; Huang, X.; Luo, Y.; Yuan, H.; Ren, T.; Li, X.; Xu, D.; Guo, X.; Wu, Y. Fluorescent Chitosan-Based Hydrogel Incorporating Titanate and Cellulose Nanofibers Modified with Carbon Dots for Adsorption and Detection of Cr(VI). *Chem. Eng. J.* **2021,** *407* (VI), 127050. DOI: 10.1016/j.cej.2020.127050.

82. Ma, Z. W.; Zhang, K. N.; Zou, Z. J.; Lü, Q. F. High Specific Area Activated Carbon Derived from Chitosan Hydrogel Coated Tea Saponin: One-Step Preparation and Efficient Removal of Methylene Blue. *J. Environ. Chem. Eng.* **2021,** *9* (3), 105251. DOI: 10.1016/j.jece.2021.105251.

83. Maity, S.; Naskar, N.; Jana, B.; Lahiri, S.; Ganguly, J. Fabrication of Thiophene-Chitosan Hydrogel-Trap for Efficient Immobilization of Mercury(II) from Aqueous Environs. *Carbohydr. Polym.* **2021,** *251* (May 2020), 116999. DOI: 10.1016/j.carbpol.2020.116999.

84. Masheane, M.; Nthunya, L.; Malinga, S.; Nxumalo, E.; Barnard, T.; Mhlanga, S. Anti-microbial Properties of Chitosan-Alumina/f-MWCNT Nanocomposites. *J. Nanotechnol.* **2016,** *2016,* 1687–9503. https://doi.org/10.1155/2016/5404529

85. Maskawat Marjub, M.; Rahman, N.; Dafader, N. C.; Sultana Tuhen, F.; Sultana, S.; Tasneem Ahmed, F. Acrylic Acid-Chitosan Blend Hydrogel: A Novel Polymer Adsorbent

for Adsorption of Lead(II) and Copper(II) Ions from Wastewater. *J. Polym. Eng.* **2019,** *39* (10), 883–891. DOI: 10.1515/polyeng-2019-0139.

86. Midya, L.; Das, R.; Bhaumik, M.; Sarkar, T.; Maity, A.; Pal, S. Removal of Toxic Pollutants from Aqueous Media Using Poly(Vinyl Imidazole) Crosslinked Chitosan Synthesised through Microwave Assisted Technique. *J. Colloid Interface Sci.* **2019,** *542,* 187–197. DOI: 10.1016/j.jcis.2019.01.121.

87. Mittal, H.; Al Alili, A.; Morajkar, P. P.; Alhassan, S. M. GO Crosslinked Hydrogel Nanocomposites of Chitosan/Carboxymethyl Cellulose—A Versatile Adsorbent for the Treatment of Dyes Contaminated Wastewater. *Int. J. Biol. Macromol.* **2021,** *167,* 1248–1261. DOI: 10.1016/j.ijbiomac.2020.11.079.

88. Mohamed, H. G.; Aboud, A. A.; Abd El-Salam, H. M. Synthesis and Characterization of Chitosan/Polyacrylamide Hydrogel Grafted Poly(N-Methylaniline) for Methyl Red Removal. *Int. J. Biol. Macromol.* **2021,** *187* (July), 240–250. DOI: 10.1016/j.ijbiomac. 2021.07.124.

89. Mohamed, N. A.; Al-Harby, N. F.; Almarshed, M. S. Enhancement of Adsorption of Congo Red Dye onto Novel Antimicrobial Trimellitic Anhydride Isothiocyanate-Cross-Linked Chitosan Hydrogels. *Polym. Bull.* **2020,** *77* (12), 6135–6160. DOI: 10.1007/ s00289-019-03058-6.

90. Mohammadzadeh Pakdel, P.; Peighambardoust, S. J. Review on Recent Progress in Chitosan-Based Hydrogels for Wastewater Treatment Application. *Carbohydr. Polym.* **2018,** *201,* 264–279. DOI: 10.1016/j.carbpol.2018.08.070.

91. Mokhtar, A.; Abdelkrim, S.; Djelad, A.; Sardi, A.; Boukoussa, B.; Sassi, M.; Bengueddach, A. Adsorption Behavior of Cationic and Anionic Dyes on Magadiite-Chitosan Composite Beads. *Carbohydr. Polym.* **2020,** *229,* 115399. DOI: 10.1016/j.carbpol.2019.115399.

92. Muñoz, I.; Rodríguez, C.; Gillet, D.; Moerschbacher, M. Life Cycle Assessment of Chitosan Production in India and Europe. *Int. J. Life Cycle Assess.* **2018,** *23* (5), 1151–1160. DOI: 10.1007/s11367-017-1290-2.

93. Nassar, M. Y.; El-Shahat, M. F.; Osman, A.; Sobeih, M. M.; Zaid, M. A. Adsorptive Removal of Manganese Ions from Polluted Aqueous Media by Glauconite Clay-Functionalized Chitosan Nanocomposites. *J. Inorg. Organomet. Polym. Mater.* **2021,** *31* (10), 4050–4064. DOI: 10.1007/s10904-021-02028-8.

94. Nešović, K.; Janković, A.; Kojić, V.; Vukašinović-Sekulić, M.; Perić-Grujić, A.; Rhee, K. Y.; Mišković-Stanković, V. Silver/Poly(Vinyl Alcohol)/Chitosan/Graphene Hydrogels— Synthesis, Biological and Physicochemical Properties and Silver Release Kinetics. *Compos. B Eng.* **2018,** *154,* 175–185. DOI: 10.1016/j.compositesb.2018.08.005.

95. Ohemeng-Boahen, G.; Sewu, D. D.; Tran, H. N.; Woo, S. H. Enhanced Adsorption of Congo Red from Aqueous Solution Using Chitosan/Hematite Nanocomposite Hydrogel Capsule Fabricated via Anionic Surfactant Gelation. *Colloids Surf. A Physicochem. Eng. Aspects* **2021,** *625* (April), 126911. DOI: 10.1016/j.colsurfa.2021.126911.

96. Pavithra, K. G.; P.; S. K.; V. Removal of Colorants from Wastewater: A Review on Sources and Treatment Strategies. *J. Ind. Eng. Chem.* **2019,** *75,* 1–19. DOI: 10.1016/j. jiec.2019.02.011.

97. Pérez-Calderón, J.; Santos, M. V.; Zaritzky, N. Synthesis, Characterization and Application of Cross-Linked Chitosan/Oxalic Acid Hydrogels to Improve Azo Dye (Reactive Red 195) Adsorption. *React. Funct. Polym.* **2020,** *155,* 104699. DOI: 10.1016/ j.reactfunctpolym.2020.104699.

98. Perumal, S.; Atchudan, R.; Yoon, D. H.; Joo, J.; Cheong, I. W. Graphene Oxide-Embedded Chitosan/Gelatin Hydrogel Particles for the Adsorptions of Multiple Heavy Metal Ions. *J. Mater. Sci.* **2020**, *55* (22), 9354–9363. DOI: 10.1007/s10853-020-04651-1.

99. Qiu, B.; Tao, X.; Wang, H.; Li, W.; Ding, X.; Chu, H. Biochar as a Low-Cost Adsorbent for Aqueous Heavy Metal Removal: A Review. *J. Anal. Appl. Pyrol.* **2021**, *155*, 105081. DOI: 10.1016/j.jaap.2021.105081.

100. Quiroga-Flores, R.; Noshad, A.; Wallenberg, R.; Önnby, L. Adsorption of Cadmium by a High-Capacity Adsorbent Composed of Silicate-Titanate Nanotubes Embedded in Hydrogel Chitosan Beads. *Environ. Technol.* **2020**, *41* (23), 3043–3054. DOI: 10.1080/09593330.2019.1596167.

101. Ren, J.; Yang, S.; Hu, Z.; Wang, H. Self-Propelled Aerogel Solar Evaporators for Efficient Solar Seawater Purification. *Langmuir* **2021**, *37* (31), 9532–9539. DOI: 10.1021/acs.langmuir.1c01387.

102. Rezanejade Bardajee, G.; Sadat Hosseini, S.; Vancaeyzeele, C. Graphene Oxide Nano-composite Hydrogel Based on Poly(Acrylic Acid) Grafted onto Salep: An Adsorbent for the Removal of Noxious Dyes from Water. *New J. Chem.* **2019**, *43* (8), 3572–3582. DOI: 10.1039/C8NJ05800H.

103. Samuel, M. S.; Bhattacharya, J.; Raj, S.; Santhanam, N.; Singh, H.; Pradeep Singh, N. D. Efficient Removal of Chromium(VI) from Aqueous Solution Using Chitosan Grafted Graphene Oxide (CS-GO) Nanocomposite. *Int. J. Biol. Macromol.* **2019**, *121*, 285–292. DOI: 10.1016/j.ijbiomac.2018.09.170.

104. Saruchi, K. V. Separation of Crude Oil from Water Using Chitosan Based Hydrogel. *Cellulose* **2019**, *26* (10), 6229–6239. DOI: 10.1007/s10570-019-02539-1.

105. Saruchi, K. V.; Vikas, P.; Kumar, R.; Kumar, B.; Kaur, M. *J. Petrol. Sci. Eng.* **2016**. Low cost natural polysaccharide and vinyl monomer based IPN for the removal of crude oil from water, *141*, 1–8. https://doi.org/https. DOI: 10.1016/j.petrol.2016.01.007.

106. Sathya, K.; Nagarajan, K.; Carlin Geor Malar, G.; Rajalakshmi, S.; Raja Lakshmi, P. A Comprehensive Review on Comparison Among Effluent Treatment Methods and Modern Methods of Treatment of Industrial Wastewater Effluent from Different Sources. *Appl. Water Sci.* **2022**, *12* (4), 70. DOI: 10.1007/s13201-022-01594-7.

107. Shah, K. M.; Billinge, I. H.; Chen, X.; Fan, H.; Huang, Y.; Winton, R. K.; Yip, N. Y. Drivers, Challenges, and Emerging Technologies for Desalination of High-Salinity Brines: A Critical Review. *Desalination* **2022**, *538*, 115827. DOI: 10.1016/j.desal.2022.115827.

108. Shannon, M. A.; Bohn, P. W.; Elimelech, M.; Georgiadis, J. G.; Mariñas, B. J.; Mayes, A. M. Science and Technology for Water Purification in the Coming Decades. *Nature* **2008**, *452* (7185), 301–310. DOI: 10.1038/nature06599.

109. Shi, Y.; Hu, H.; Ren, H. Dissolved Organic Matter (DOM) Removal from Biotreated Coking Wastewater by Chitosan-Modified Biochar: Adsorption Fractions and Mechanisms. *Bioresour. Technol.* **2020**, *297*, 122281. DOI: 10.1016/j.biortech.2019.122281.

110. Singh, N.; Riyajuddin, S.; Ghosh, K.; Mehta, S. K.; Dan, A. Chitosan-Graphene Oxide Hydrogels with Embedded Magnetic Iron Oxide Nanoparticles for Dye Removal. *ACS Appl. Nano Mater.* **2019**, *2* (11), 7379–7392. DOI: 10.1021/acsanm.9b01909.

111. Sorour, M. H.; Hani, H. A.; Shaalan, H. F.; El Sayed, M. M.; El-Sayed, M. M. H. Softening of Seawater and Desalination Brines Using Grafted Polysaccharide Hydrogels. *Desalin. Water Treat.* **2015**, *55* (9), 2389–2397. DOI: 10.1080/19443994.2014.947783.

112. Su, M.; Liu, Y.; Li, S.; Fang, Z.; He, B.; Zhang, Y.; Li, Y.; He, P. A Rubber-Like, Underwater Superoleophobic Hydrogel for Efficient Oil/Water Separation. *Chem. Eng. J.* **2019**, *361*, 364–372. DOI: 10.1016/j.cej.2018.12.082.

113. Sultana, S.; Karmaker, B.; Saifullah, A. S. M.; Galal Uddin, M.; Moniruzzaman, M. Environment-Friendly Clay Coagulant Aid for Wastewater Treatment. *Appl. Water Sci.* **2022**, *12* (6), 1–10. DOI: 10.1007/s13201-021-01540-z.

114. Sun, J.; Yan, X.; Lv, K.; Sun, S.; Deng, K.; Du, D. Photocatalytic Degradation Pathway for Azo Dye in TiO$_2$/UV/O$_3$ System: Hydroxyl Radical Versus Hole. *J. Mol. Cat. A: Chem.* **2013**, *367*, 31–37. DOI: 10.1016/j.molcata.2012.10.020.

115. Tamjidi, S.; Esmaeili, H.; Kamyab Moghadas, B. Application of Magnetic Adsorbents for Removal of Heavy Metals from Wastewater: A Review Study. *Mater. Res. Express* **2019**, *6* (10), 102004–102020. DOI: 10.1088/2053-1591/ab3ffb.

116. Tang, S.; Yang, J.; Lin, L.; Peng, K.; Chen, Y.; Jin, S.; Yao, W. Construction of Physically Crosslinked Chitosan/Sodium Alginate/Calcium Ion Double-Network Hydrogel and Its Application to Heavy Metal Ions Removal. *Chem. Eng. J.* **2020**, *393*, 124728. DOI: 10.1016/j.cej.2020.124728.

117. Tang, Y. Z.; Gin, K. Y. H.; Aziz, M. A. The Relationship Between pH and Heavy Metal Ion Sorption by Algal Biomass. *Adsorpt. Sci. Technol.* **2003**, *21* (6), 525–537. DOI: 10.1260/026361703771953587.

118. Tanhaei, B.; Ayati, A.; Iakovleva, E.; Sillanpää, M. Efficient Carbon Interlayed Magnetic Chitosan Adsorbent for Anionic Dye Removal: Synthesis, Characterization and Adsorption Study. *Int. J. Biol. Macromol.* **2020**, *164*, 3621–3631. DOI: 10.1016/j.ijbiomac.2020.08.207.

119. Tripodo, G.; Trapani, A.; Rosato, A.; Di Franco, C.; Tamma, R.; Trapani, G.; Ribatti, D.; Mandracchia, D. Hydrogels for Biomedical Applications from Glycol Chitosan and PEG Diglycidyl Ether Exhibit Pro-angiogenic and Antibacterial Activity. *Carbohydr. Polym.* **2018**, *198*, 124–130. DOI: 10.1016/j.carbpol.2018.06.061.

120. Vakili, M.; Amouzgar, P.; Cagnetta, G.; Wang, B.; Guo, X.; Mojiri, A.; Zeimaran, E.; Salamatinia, B. Ultrasound-Assisted Preparation of Chitosan/Nano-activated Carbon Composite Beads Aminated with (3-Aminopropyl)Triethoxysilane for Adsorption of Acetaminophen from Aqueous Solutions. *Polymers* **2019,** *11* (10). DOI: 10.3390/polym11101701.

121. Vakili, M.; Zwain, H. M.; Mojiri, A.; Wang, W.; Gholami, F.; Gholami, Z.; Giwa, A. S.; Wang, B.; Cagnetta, G.; Salamatinia, B. Effective Adsorption of Reactive Black 5 onto Hybrid Hexadecylamine Impregnated Chitosan-Powdered Activated Carbon Beads. *Water* **2020**, *12* (8). DOI: 10.3390/w12082242.

122. Venkata Mohan, S.; Chandrasekhar Rao, N.; Karthikeyan, J. Adsorptive Removal of Direct Azo Dye from Aqueous Phase onto Coal Based Sorbents: A Kinetic and Mechanistic Study. *J. Hazard. Mater.* **2002**, *90* (2), 189–204. DOI: 10.1016/S0304-3894(01)00348-X.

123. Verma, A.; Thakur, S.; Mamba, G.; Prateek, G.; Gupta, R. K.; Thakur, P.; Thakur, V. K. Graphite Modified Sodium Alginate Hydrogel Composite for Efficient Removal of Malachite Green Dye. *Int. J. Biol. Macromol.* **2020**, *148*, 1130–1139. DOI: 10.1016/j.ijbiomac.2020.01.142.

124. Wang, C.; Yediler, A.; Lienert, D.; Wang, Z.; Kettrup, A. Ozonation of an Azo Dye C.I. Remazol Black 5 and Toxicological Assessment of Its Oxidation Products. *Chemosphere* **2003**, *52* (7), 1225–1232. DOI: 10.1016/S0045-6535(03)00331-X.

125. Wang, M.; Payne, K. A.; Tong, S.; Ergas, S. J. Hybrid Algal Photosynthesis and Ion Exchange (HAPIX) Process for High Ammonium Strength Wastewater Treatment. *Water Res.* **2018**, *142*, 65–74. DOI: 10.1016/j.watres.2018.05.043.

126. Wang, S.; Lu, A.; Zhang, L. Recent Advances in Regenerated Cellulose Materials. *Prog. Polym. Sci.* **2016,** *53,* 169–206. DOI: 10.1016/j.progpolymsci.2015.07.003.

127. Wang, W.; Hu, J.; Zhang, R.; Yan, C.; Cui, L.; Zhu, J. A pH-Responsive Carboxymethyl Cellulose/Chitosan Hydrogel for Adsorption and Desorption of Anionic and Cationic Dyes. *Cellulose* **2021,** *28* (2), 897–909. DOI: 10.1007/s10570-020-03561-4.

128. Wang, W.; Ni, J.; Chen, L.; Ai, Z.; Zhao, Y.; Song, S. Synthesis of Carboxymethyl Cellulose–Chitosan–Montmorillonite Nanosheets Composite Hydrogel for Dye Effluent Remediation. In *Int. J. Biol. Macromol.* **2020,** *165* (A). Elsevier B.V, 1–10. DOI: 10.1016/j.ijbiomac.2020.09.154.

129. Xu, T.; Xu, Y.; Wang, J.; Lu, H.; Liu, W.; Wang, J. Sustainable Self-Cleaning Evaporator for Long-Term Solar Desalination Using Gradient Structure Tailored Hydrogel. *Chem. Eng. J.* **2021,** *415,* 128893. DOI: 10.1016/j.cej.2021.128893.

130. Yang, J.; Li, M.; Wang, Y.; Wu, H.; Ji, N.; Dai, L.; Li, Y.; Xiong, L.; Shi, R.; Sun, Q. High-Strength Physically Multi-cross-Linked Chitosan Hydrogels and Aerogels for Removing Heavy-Metal Ions. *J. Agric. Food Chem.* **2019,** *67* (49), 13648–13657. DOI: 10.1021/acs.jafc.9b05063.

131. Yang, J.; Han, Y.; Sun, Z.; Zhao, X.; Chen, F.; Wu, T.; Jiang, Y. PEG/Sodium Tripoly-phosphate-Modified Chitosan/Activated Carbon Membrane for Rhodamine B Removal. *A.C.S. Omega* **2021,** *6* (24), 15885–15891. DOI: 10.1021/acsomega.1c01444.

132. Ye, Z.; Yin, X.; Chen, L.; He, X.; Lin, Z.; Liu, C.; Ning, S.; Wang, X.; Wei, Y. An Integrated Process for Removal and Recovery of Cr(VI) from Electroplating Wastewater by Ion Exchange and Reduction–Precipitation Based on a Silica-Supported Pyridine Resin. *J. Cleaner Prod.* **2019,** *236,* 117631. DOI: 10.1016/j.jclepro.2019.117631.

133. Yu, F.; Chen, Z.; Guo, Z.; Irshad, M. S.; Yu, L.; Qian, J.; Mei, T.; Wang, X. Molybdenum Carbide/Carbon-Based Chitosan Hydrogel as an Effective Solar Water Evapora-tion Accelerator. *ACS Sustain. Chem. Eng.* **2020,** *8* (18), 7139–7149. DOI: 10.1021/acssuschemeng.0c01499.

134. Zhang, A.; Li, X.; Xing, J.; Xu, G. Adsorption of Potentially Toxic Elements in Water by Modified Biochar: A Review. *J. Environ. Chem. Eng.* **2020,** *8* (4). DOI: 10.1016/j.jece.2020.104196.

135. Zhang, H.; Omer, A. M.; Hu, Z.; Yang, L. Y.; Ji, C.; Ouyang, Kun, X. Fabrication of Magnetic Bentonite/Carboxymethyl Chitosan/Sodium Alginate Hydrogel Beads for Cu(II) Adsorption. *Int. J. Biol. Macromol.* **2019,** *135* (II), 490–500. DOI: 10.1016/j.ijbiomac.2019.05.185.

136. Zhang, L.; Zeng, Y.; Cheng, Z. Removal of Heavy Metal Ions Using Chitosan and Modified Chitosan: A Review. *J. Mol. Liq.* **2016,** *214,* 175–191. DOI: 10.1016/j.molliq.2015.12.013.

137. Zhang, S.; Dong, Y.; Yang, Z.; Yang, W.; Wu, J.; Dong, C. Adsorption of Pharmaceuticals on Chitosan-Based Magnetic Composite Particles with Core-Brush Topology. *Chem. Eng. J.* **2016,** *304,* 325–334. DOI: 10.1016/j.cej.2016.06.087.

138. Zhang, W.; Hu, L.; Hu, S.; Liu, Y. Optimized Synthesis of Novel Hydrogel for the Adsorption of Copper and Cobalt Ions in Wastewater. *RSC Adv.* **2019,** *9* (28), 16058–16068. DOI: 10.1039/c9ra00227h.

139. Zhang, Y.; Cao, Z.; Luo, Z.; Li, W.; Fu, T.; Qiu, W.; Lai, Z.; Cheng, J.; Yang, H.; Ma, W.; Liu, C. Facile Fabrication of Underwater Superoleophobic Membrane Based on Polyacrylamide/Chitosan Hydrogel Modified Metal Mesh for Oil–Water Separation. *J. Polym. Sci.* **2022,** *60* (15), 2329–2342. DOI: 10.1002/pol.20210923.

140. Zhou, L.; Liu, Z.; Liu, J.; Huang, Q. Adsorption of Hg(II) from Aqueous Solution by Ethylenediamine-Modified Magnetic Crosslinking Chitosan Microspheres. *Desalination* **2010**, *258* (1–3), 41–47. DOI: 10.1016/j.desal.2010.03.051.

141. Zhu, Y.; Zheng, Y.; Wang, F.; Wang, A. Monolithic Supermacroporous Hydrogel Prepared from High Internal Phase Emulsions (HIPEs) for Fast Removal of Cu^{2+} and Pb^{2+}. *Chem. Eng. J.* **2016**, *284*, 422–430. DOI: 10.1016/j.cej.2015.08.157.

142. Zhuang, L.; Li, Q.; Chen, J.; Ma, B.; Chen, S. Carbothermal Preparation of Porous Carbon-Encapsulated Iron Composite for the Removal of Trace Hexavalent Chromium. *Chem. Eng. J.* **2014**, *253*, 24–33. https://doi.org/https. DOI: 10.1016/j.cej.2014.05.038.

143. Zinadini, S.; Zinatizadeh, A. A.; Rahimi, M.; Vatanpour, V.; Zangeneh, H.; Beygzadeh, M. Novel High Flux Antifouling Nanofiltration Membranes for Dye Removal Containing Carboxymethyl Chitosan Coated Fe_3O_4 Nanoparticles. *Desalination* **2014**, *349*, 145–154. DOI: 10.1016/j.desal.2014.07.007.

CHAPTER 2

Wastewater Treatment by Porous Composites

CINTIL JOSE[1], BIJU PETER[1], and SABU THOMAS[2]

[1]Newman College, Thodupuzha, Kerala, India

[2]School of Energy Materials, Mahatma Gandhi University, Kottayam, Kerala, India

ABSTRACT

Wastewater is a mixture of liquid or water-borne waste removed from homes, institutions, commercial and industrial organizations, as well as groundwater, surface water, and storms. Water is necessary for human survival and well-being; hence, appropriate water supplies are crucial. Controlling pollution's negative consequences and enhancing human living conditions are thus critical. Wastewater contains heavy metals, inorganic chemicals, organic pollutants, and a range of other complicated molecules. Water contamination is one of the most pressing issues today, as a result of industrialization and human activity. Nanofibers are becoming more widely used as a result of their unique qualities, such as their high surface-to-volume ratio, surface modification capabilities, and conversion to small mesh mats. Because of their appealing features, such as small diameters (at the micro- and nanoscale), substantial porosity, and a huge surface area available for chemical functionalization, polymer nanofibers have been proven to be extremely appropriate for this purpose. In this regard, the current chapter looks at different types of composite nanofibers, as well as their adsorption performance and applications for heavy metals, dyes, and other pollutants.

Mechanics and Physics of Porous Materials: Novel Processing Technologies and Emerging Applications.
Chin Hua Chia, Tamara Tatrishvili, Ann Rose Abraham, & A. K. Haghi (Eds.)
© 2024 Apple Academic Press, Inc. Co-published with CRC Press (Taylor & Francis)

2.1 INTRODUCTION

The modern world is dealing with a number of environmental pollution issues, and water pollution is one of the biggest threats to both aquatic life and human health. As garbage from these sources is typically dumped through pipes into rivers, reservoirs, or streams, factories, power plants, and waste processing facilities are the main contributors of water pollution. Typically, these pollutants include metals, dyes, certain oils, pharmaceutical wastes, and others. Heavy metals are naturally occurring track materials in the fluvial environment, but due to anthropogenic activities, including industrial waste, agriculture, and mining, their amounts are rising. One of the key factors that eat up the majority of water, poison waterways, and harm the environment is the textile industry. Different types of organic and inorganic dyes are released during the wet processing of textiles, and these dyes quickly reduce the BOD and COD levels of water. Oils and pharmaceutical wastes are also to blame for polluting the water. Therefore, to ensure the safety of their use and to maintain a clean and tidy environment, these contaminants should be removed from water bodies.[1-4]

Many techniques for treating wastewater have been developed throughout the years, including ion exchange, biological treatment, chemical precipitation, membrane separation, reverse osmosis, adsorption, coagulation, and flocculation filtration, among others. Adsorption was discovered to be an economical and environmentally favorable way for removing contaminants from wastewater. Nanofibers created using electrospinning procedures are crucial in the treatment of wastewater. They may be transformed into mats or membranes to remove contaminants from aqueous solutions and have a high specific surface area, high porosity, and superior functional capability.[5-7] While conventional adsorption has numerous advantages, it also has drawbacks, including low efficiency, higher energy consumption, and a difficult recycling process. However, by using nanofiber technology, these issues have been eliminated. Nanofibers created using the electrospinning technique, which spins fiber and uses an electric field to draw charged threads out of polymer solutions.[8]

Nanofibers are good supporting materials for loading functional materials or being modified with various chemical groups in water treatment applications due to their high surface area to volume ratio, great flexibility, porous structure, reusability, nontoxicity, and low cost. The nanofibers used as filtering medium give high contact between adsorbent and aqueous media, improving adsorption capacity with the ease of recovery and recycling. This is made possible by their increased surface area and pore volume. The separation capacity can be significantly increased by adding different functional groups (carboxylate, amino, acid, and hydroxyl groups) or integrating adsorbents like metal oxides, graphene, graphene oxide (GO), and metal–organic frameworks in the nanofibers.[9,10]

2.2 SOME TYPES OF WATER POLLUTANTS

2.2.1 *ORGANIC POLLUTANTS*

The household sewage system, agricultural, food, paper, and leather industries can all release organic contaminants into water, including colors, phenols, detergents, and medications. Because the oxidative degradation of such pollutants requires a lot of oxygen, the amount of dissolved oxygen in the water diminishes, endangering the survival of aquatic creatures and ecosystems. The majority of these dyes have hydrophobic qualities, collect in the water, and can permeate the tissues of some aquatic species and people when they are introduced to surface water, where they can form degradable byproducts that may have carcinogenic effects.[11–14] Additionally, dyes are highly harmful pollutants that are released into the environment and water supplies by a variety of businesses, including the textile, rubber, and plastics industries. Humans suffer severe injury as a result of the discharge of such materials into the water. The dyes also hinder light from permeating water, which limits photosynthesis in aquatic plants, one of their more severe impacts. It is crucial to remove dyes from colored water. This is due to the fact that dye in water will block sunlight from penetrating the water, even at low concentrations. As a result, aquatic species' lives will be in danger. Additionally, the colors in drinking water may have cancer-causing effects.[15,16]

2.2.2 *INORGANIC POLLUTANTS*

The discharge of effluents loaded with poisonous heavy metal ions into water bodies is one of the riskiest consequences of many sectors' explosive growth, including electroplating, tanneries, battery production, electroplating, and so on. All living things, including people, animals, and plants, are affected by the presence of these toxins in the aquasphere, even at extremely low quantities. Heavy metals can have harmful effects on the environment and people's health due to their carcinogenic and poisonous properties. Human kidneys and central nervous systems can be harmed by some metal cations.[17–20] It is thought that exposure to these metal ions, even at low amounts, poses dangers to aquatic, terrestrial, and human organisms. In addition to the growing issues with water shortages, eliminating these toxins from water is a critical necessity and continues to be significant for scientists.

2.2.3   OIL POLLUTION

The primary source of crude oil and its byproducts is the processing of natural oil. Oil usage and transportation-related leakage and spill incidents have contaminated the water and land, harming living things. Oil and water can be separated using a variety of processes, such as in-situ burning, filtration, centrifugation, neutralization, and so on. However, these methods have some drawbacks, including inefficient operation, high prices, long processing times, and secondary contamination. Additionally, the oil that failed to properly separate from the water still poses threats to both aquatic life and human health.[21–23] Therefore, governments and researchers must look into effective and environmentally friendly materials and treatment methods right away. Scientists are currently researching and producing many types of innovative polymeric nanomaterials that have good physicochemical qualities and may be employed in diverse applications, utilizing cutting-edge methodologies. As a result, new technologies have been established.[24,25]

2.3   ELECTROSPUN NANOFIBERS FOR WATER TREATMENT

One of the many well-established methods for creating fibers at the micro- or nanoscale is electrospinning. Excellent control over the fiber diameters, nanostructures, and morphology is made possible by the adaptability of electrospinning, which also improves the materials' catalytic, mechanical, electrical, biomedical, optical, and adsorptive capabilities.[26] The electrospinning process may produce nanofibers into a variety of intriguing forms for a variety of applications since it has access to a large selection of polymers and the tools for additive incorporation. Recent developments in the electrospinning process have made it possible to create intriguing nanostructures that can be used to improve pollutant removal by drawing inspiration from natural items. The branch and trunk fibers make up the tree-like structure. The thin branches serve as connections, reduce membrane pore size, and increase surface area, while the trunk fibers, which serve as supports, can enhance mechanical properties. The nanofibers formed the spider web-like pattern. Hierarchical bio-inspired composite nanofibers made of PVA, PAA, GO–COOH, and polydopamine showed an effective adsorption capacity for removing dye while using an eco-friendly and controllable manufacturing technique.[27]

2.4 ELECTROSPINNING BIO-BASED POLYMERS FOR WATER TREATMENT

Bio-based polymers like cellulose, chitosan, zein, collagen, silk, hyaluronic acid, alginate, and DNA have been extensively employed to create nanofibrous composite membranes as a result of sustainability and environmental concerns. The advantages of biocompatibility, biodegradability, safety, and nontoxicity have led to a growth in the commercialization of these polymers' applications in water filtering over the past few years.[28] The presence of diverse functional groups in bio-based polymers, which can be used for pollution collection, is one of their distinctive characteristics. Chemisorption or physisorption are two possible dye adsorption methods onto polymers. The former is irreversible and frequently associated with strong bonds (like covalent or ionic bonds). The latter is better because it can be reversed. Van der Waals forces, hydrogen bonds, hydrophobic interaction, and electrostatic attraction control the physisorption.[29]

2.5 CARBON NANOFIBERS AS SUPPORTING MATERIALS

Applications for catalysis, the environment, and energy have all made use of carbon nanofibers with distinctive and adaptable morphologies. Nanofibers made of polyacrylonitrile (PAN) have been widely used as a productive precursor in the production of CNF. Catalytic and bioactive compounds have been loaded onto CNFs in an effective manner. For applications involving dye removal, a variety of metal oxide nanoparticles have been immobilized in/onto CNFs. Electrodeposition, chemical synthesis, and dry synthesis are the principal methods used to adorn CNFs with active chemicals. It is interesting to note that the CNFs also have an adsorption capacity, which results from the aromatic rings of the CNFs and the adsorbate binding together via a stacking relationship. The excellent conductivity and chemical inertness of CNFs improve their use in photocatalytic dye degradation and support reusability. Before being broken down by photocatalysts placed on the surface of the nanofibers, the dye molecules have a high probability of being drawn to CNFs with a negative conjugative structure.[30,31]

2.6 ELECTROSPUN NANOFIBERS FOR WATER TREATMENT

Demand for high-quality water has rapidly increased recently all throughout the world. Decontaminating wastewater that has been released into rivers, reservoirs, or streams from diverse sources has thus been the focus of numerous works. As a result, researchers are looking for and creating new, affordable, sustainable technologies to treat or recycle wastewater. The primary class of ions that significantly affects ecosystems and human health is the class of heavy metal ions. Furthermore, organic contaminants have severe toxicity impacts on both human health and the environment at low concentrations. As a result, the need for water treatment has increased. As a result, there is an urgent need for the development of effective and cutting-edge filtration and separation techniques to address the problems associated with water purification. The fabricated electrospun nanofiber mats have been regarded as suitable platforms to combine filtration and adsorption techniques—with significant purification efficiency—to remove organic pollutants and heavy metal ions from contaminated water because of their special qualities and potential benefits. However, in order for nanofiber membranes to be the best possible material for eliminating such contaminants, they need to have a large number of interaction sites. As a result, researchers have given the manufacturing of innovative nanofiber architectures a lot of attention. The main pollutant-removal method is anticipated to be the adsorption process, via electrostatic interactions, ion exchange, or chelation, when electrospun nanofibers are used as ultrafiltration membranes in water purification operations. Furthermore, size-based rejection must be taken into account if electrospun nanofibers are employed as NF/RO membranes.[32]

2.6.1   REMOVAL OF DYES AND ORGANIC POLLUTANTS

It has been evaluated whether several innovative and promising nanofibers and their prospective alterations could be used to remove various organic contaminants from wastewater. Tetracycline (TC) removal from wastewater has been studied using self-sustained electrospun PAN nanofiber-supported polyamide (PA) thin-film composite (PA/PAN-eTFC) membranes. Initially, electrospinning was used to create the PAN nanofiber's support (Figure 2.1). After that, a paper laminator was used to laminate the resulting PAN nanofiber mat. With a water contact angle of 32.3 1.3, a stress of 13 0.77 MPa, and a strain of 68 0.28 percent, respectively, the laminated nanofiber support demonstrated hydrophilicity and mechanical characteristics. When using 2.0-M sodium chloride as the draw solution, the PA composite mat that was created on top of it obtained >57 LMH water flux and displayed low structural characteristics (S = 168 m) and increased perm selectivity (A = 1.47 LMH bar, B = 0.278 LMH). The manufactured PA/PAN-eTFC membranes had a high TC rejection rate (>99.9%), and the FO-MD hybrid technique allowed for a 15–20% water recovery after 7 h.[33]

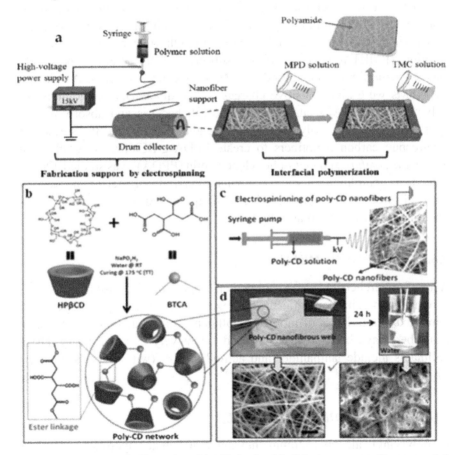

FIGURE 2.1 Schematic illustration of the fabrication (a) of electrospun nanofiber-supported TFC membrane and (b–d) of the cross-linked poly-CD nanofibrous web.[33]

Source: Reprinted/Modified with permission from Ref. [33]. Copyright 2016. Elsevier B.V.

In a study by Zhao et al., a facile strategy was used to prepare positively charged PES nanofibrous membranes (NFMs) containing different amounts of quaternary ammonium salt (QAS) polymer (PMETAC). The detailed procedures for the preparation of electrospinning solution and fabrication of positively charged NFMs are illustrated in Figure 2.2.[34] The electrospun nanofiber membrane of poly(L-lactic acid) (PLLA) covered with *p*-toluenesulfonic acid-doped polyaniline (pTSA-PANI), which is used to remove MO, is another significant example of such modified materials. The interactions between dye–benzene rings and pTSA-PANI rings and the superior wettability features of pTSA-PANI/PLLA membranes as compared to PLLA membranes provide evidence that the loading of PLLA membranes with pTSA-PANI increases the removal efficiency of the membranes. Additionally, by combining electrospinning with the in-situ cross-link polymerization of poly[2-(methacryloyloxy)-ethyl] trimethyl ammonium chloride (PMETAC) in poly(ether sulfone) (PES) solution, positively charged nanofibrous membranes (NFMs) have been created. As shown, the NFMs were given positive charges by the quaternary ammonium salt polymer of PMETAC, enabling them to kill bacteria and absorb anionic dyes. By oxidatively polymerizing *o*-, *p*-, and *m*-PDA onto the surface of electrospun carbon nanofibers to create PPDA-g-ECNFs, the grafting of poly(*para*-, *ortho*-, and *meta*-phenylenediamin (PPDA)) was studied. On the other hand, employing a vanillin biomass-derived polymer as the starting material through an electrospinning technique, aldehyde-containing nanofiber membranes were created.[34]

2.6.2   REMOVAL OF METALS

A lot of researchers have concentrated on employing electrospun nanofibers to remove inorganic contaminants from wastewater over the past 10 years. To remove different kinds of inorganic pollutants, multifunctional electrospun nanofibers have been created and used extensively in recent years. These pollutants include lanthanides and actinides, Pb^{2+} (with other metal ions) Cu^{2+}, Cd^{2+}, Ni^{2+}, Cd^{2+}, Zn^{2+}, and so on.[35,36]

2.6.3   LANTHANIDE AND ACTINIDE

A poly(vinyl alcohol)/sodium hexametaphosphate hydrogel nanofiber (PVA/SHMP HENF) with a 3D structure was created for lanthanide and actinide using an electrospinning process. Citric acid (CA) served as

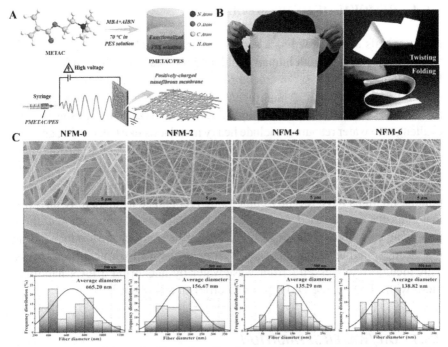

FIGURE 2.2 (a) Schematic illustration for the fabrication of electrospinning solution and fabrication of positively charged PES NFMs; (b) representative optical images of the as-prepared large-sized NFM, and twisted and folded NFMs; and (c) SEM images and the corresponding nanofiber diameter distributions curves for the different NFMs.[34]

Source: Reprinted/Adapted with permission from Ref. [34]. Copyright 2019. Elsevier Inc.

the cross-linking agent during the heat processing used to carry out the cross-linking reaction. The created PVA/SHMP HENF showed impressive removal for several lanthanide metals, including La^{3+}, Tb^{3+}, and Nd^{3+}, as well as good swelling characteristics and high pH stability. Additionally, it has an eight-cycle regeneration and reuse lifespan. To increase its effectiveness against lanthanide ions like Eu^{3+} and Y^{3+}, PAN-272 nanofiber membranes were also created by combining PAN electrospun nanofiber with a commercial organic extractant, Cyanex272. The collected findings demonstrate the fabrication of a nanofiber membrane as a suitable material for lanthanide metal ion recovery. The application of polymer (PAN) nanofibers with lower pressure-reactive filtration to adsorb uranium(VI) (U^{6+}) was achieved, however, through chemical modification. The binding agents were hexadecylphosphonic acid, which included phosphorus, or nitrogen-based Aliquat®336(Aq) (HPDA).[37,38]

2.6.4 REMOVAL OF HEAVY METALS

Heavy metals are regarded as a serious environmental pollutant and provide a significant threat to both human health and the health of other living things. Electrospun nanofibers serve crucial roles in decreasing and removing heavy metals from wastewater. A significant class of inorganic contaminants that pollute water and have hazardous physiological effects as well as posing serious challenges to water resources include heavy metal ions like Cd, Cu, Pb, and Hg. They also collect in the bodies of organisms and are difficult to disintegrate. The amounts of these metal ions in natural drinking water frequently exceed permissible limits. Different nanofibrous nanomaterials created by electrospinning methods have recently come to be recognized as incredibly effective materials that may be used to remove heavy metal ions. An illustration of these promising nanofibers is the $-Fe_2O_3$/PAN nanofibers, which are produced using the electrospinning technology and a straightforward hydrothermal process. This method removes Pb^{2+} from wastewater by anchoring a PAN nanofiber mesh with $-Fe_2O_3$ ($-Fe_2O_3$/PAN) as an efficient adsorbent.[39,40]

2.6.5 OIL/WATER SEPARATION

Large amounts of oily sewage are released into the environment on a yearly basis as a result of oil spill accidents and other oil-related operations, including those of the food, textile, tannery, and metallurgy sectors, which have led to major water pollution issues. Therefore, to address these issues, scientists focus on the treatment and separation of oil–water mixtures using various technologies.[41–43] To detoxify and separate oil–water mixtures, a variety of nanomaterials, nanofibers, and methods have been created. Due to their special qualities and benefits, such as their higher porosity with a high internal area, low density, reusability, and controllable porosity size, electrospun nanofibrous materials play a crucial role and have attracted significant interest as oil absorbents used to collect oil spills from water surfaces. The literature reports that nanofiber membranes are successful at removing oil spills from the water's surface on a wide scale, with promising outcomes.[44–47]

2.7 CONCLUSIONS

Nanofibers are regarded as a promising new class of nanomaterials with a variety of uses, including water filtration, because of their distinctive properties.

Submicron polymeric nanofibers have been produced using a variety of techniques, including nanolithography, self-assembly, multicomponent spinning, flash spinning, electrospinning, and so on. Compared to other conventional techniques, it is clear that electrospinning processes are extraordinarily versatile in the fabrication of a wide variety of polymeric nanofiber nanomaterials, and they can be viewed as a reliable and multifunctional method of water filtration without the need for significant infrastructure. Many constraints still need to be resolved despite the enormous efforts made to create and enhance the distinctive structures and capabilities of the nanofibers utilized for wastewater treatment and O/W separation. The chemical and physical characteristics of the media as well as the environment have a considerable impact on the characteristics of polymer, composite, and nanofiber membranes. The availability, long-term durability, and cost-effectiveness of the manufactured nanofibers, as well as their eco-friendliness, degradability, renewability, lack of toxicity, mechanical properties, and thermal properties, should all be taken into account and further optimized to overcome the other drawbacks. The development and synthesis of novel nanofiber materials with particular properties and self-cleaning capabilities are required to address other significant challenges facing wastewater treatment, such as the adhesion of pollutants, microbial contaminations, biofilm formation, and eventual fouling during treatment processes. These challenges must be addressed using various multifunctional electrospun nanofiber materials. To lessen the fouling, additional efforts should be made to enhance the surface-coating and impregnation processes.

KEYWORDS

- **wastewater treatment**
- **composites**
- **nanofibers**
- **electrospinning**
- **pollution**

REFERENCES

1. Sanaeepur, H.; Ebadi Amooghin, A. E.; Shirazi, M. M. A.; Pishnamazi, M.; Shirazian, S. Water Desalination and Ion Removal Using Mixed Matrix Electrospun Nanofibrous

Membranes: A Critical Review. *Desalination* **2022,** *521,* 115350. DOI: 10.1016/j.desal. 2021.115350.

2. Peydayesh, M.; Mezzenga, R. Protein Nanofibrils for Next Generation Sustainable Water Purification. *Nat. Commun.* **2021,** *12* (1), 3248. DOI: 10.1038/s41467-021-23388-2.

3. Adam, M. R.; Othman, M. H. D.; Kurniawan, T. A.; Puteh, M. H.; Ismail, A. F.; Khongnakorn, W.; Rahman, M. A.; Jaafar, J. Advances in Adsorptive Membrane Technology for Water Treatment and Resource Recovery Applications: A Critical Review. *J. Environ. Chem. Eng.* **2022,** *10* (3), 107633. DOI: 10.1016/j.jece.2022.107633.

4. Sultana, M.; Rownok, M. H.; Sabrin, M.; Rahaman, M. H.; Nur Alam, S. M. A Review on Experimental Chemically Modified Activated Carbon to Enhance Dye and Heavy Metals Adsorption. Clean. *Eng. Technol.* **2022,** *6,* 100382.

5. Patel, K. D.; Kim, H.-W.; Knowles, J. C.; Poma, A. Molecularly Imprinted Polymers and Electrospinning: Manufacturing Convergence for Next-Level Applications. *Adv. Funct. Mater.* **2020,** *30* (32), 2001955. DOI: 10.1002/adfm.202001955.

6. Qureshi, U. A.; Khatri, Z.; Ahmed, F.; Ibupoto, A. S.; Khatri, M.; Mahar, F. K.; Brohi, R. Z.; Kim, I. S. Highly Efficient and Robust Electrospun Nanofibers for Selective Removal of Acid Dye. *J. Mol. Liq.* **2017,** *244,* 478–488. DOI: 10.1016/j.molliq.2017.08.129.

7. Gopiraman, M.; Bang, H.; Yuan, G.; Yin, C.; Song, K. H.; Lee, J. S.; Chung, I. M.; Karvembu, R.; Kim, I. S. Noble Metal/Functionalized Cellulose Nanofiber Composites for Catalytic Applications. *Carbohydr. Polym.* **2015,** *132,* 554–564. DOI: 10.1016/j. carbpol.2015.06.051.

8. Khatri, M.; Ahmed, F.; Shaikh, I.; Phan, D. N.; Khan, Q.; Khatri, Z.; Lee, H.; Kim, I. S. Dyeing and Characterization of Regenerated Cellulose Nanofibers with Vat Dyes. *Carbohydr. Polym.* **2017,** *174,* 443–449. DOI: 10.1016/j.carbpol.2017.06.125.

9. Abdolmaleki, A. Y.; Zilouei, H.; Khorasani, S. N.; Zargoosh, K. Adsorption of Tetracycline from Water Using Glutaraldehyde-Crosslinked Electrospun Nanofibers of Chitosan/Poly(Vinyl Alcohol). *Water Sci. Technol.* **2018,** *77* (5–6), 1324–1335. DOI: 10.2166/wst.2018.010.

10. Nabeela Nasreen, S. A. A.; Sundarrajan, S.; Syed Nizar, S. A.; Ramakrishna, S. Nanomaterials: Solutions to Water-Concomitant Challenges. *Membranes* **2019,** *9* (3), 40. DOI: 10.3390/membranes9030040.

11. Subrahmanya, T. M.; Arshad, A. B.; Lin, P. T.; Widakdo, J.; H K, M.; Austria, H. F. M.; Hu, C. C.; Lai, J. Y.; Hung, W. S. A Review of Recent Progress in Polymeric Electrospun Nanofiber Membranes in Addressing Safe Water Global Issues. *RSC Adv.* **2021,** *11* (16), 9638–9663. DOI: 10.1039/d1ra00060h.

12. Zia, Q.; Tabassum, M.; Lu, Z.; Khawar, M. T.; Song, J.; Gong, H.; Meng, J.; Li, Z.; Li, J. Porous Poly(L-Lactic Acid)/Chitosan Nanofibres for Copper Ion Adsorption. *Carbohydr. Polym.* **2020,** *227,* 115343. DOI: 10.1016/j.carbpol.2019.115343.

13. Zhang, K.; Li, Z.; Deng, N.; Ju, J.; Li, Y.; Cheng, B.; Kang, W.; Yan, J. Tree-Like Cellulose Nanofiber Membranes Modified by Citric Acid for Heavy Metal Ion (Cu^{2+}) Removal. *Cellulose* **2019,** *26* (2), 945–958. DOI: 10.1007/s10570-018-2138-z

14. Pervez, M. N.; Balakrishnan, M.; Hasan, S. W.; Choo, K.-H.; Zhao, Y.; Cai, Y.; Zarra, T.; Belgiorno, V.; Naddeo, V. A Critical Review on Nanomaterials Membrane Bioreactor (NMs-MBR) for Wastewater Treatment. *NPJ Clean Water* **2020,** *3* (1), 43. DOI: 10.1038/ s41545-020-00090-2.

15. Fahimirad, S.; Fahimirad, Z.; Sillanpää, M. Efficient Removal of Water Bacteria and Viruses Using Electrospun Nanofibers. *Sci. Total Environ.* **2021,** *751,* 141673. DOI: 10.1016/j.scitotenv.2020.141673.

16. Lee, H.; Kim, M.; Sohn, D.; Kim, S. H.; Oh, S.-G.; Im, S. S.; Kim, I. S. Electrospun Tungsten Trioxide Nanofibers Decorated with Palladium Oxide Nanoparticles Exhibiting Enhanced Photocatalytic Activity. *RSC Adv.* **2017**, *7* (10), 6108–6113. DOI: 10.1039/C6RA24935C

17. Khatri, Z.; Nakashima, R.; Mayakrishnan, G.; Lee, K.-H.; Park, Y.-H.; Wei, K.; Kim, I. Preparation and Characterization of Electrospun Poly(ε-Caprolactone)-Poly(L-Lactic Acid) Nanofiber Tubes. *J. Mater. Sci.* **2013**, *48* (10), 3659–3664. DOI: 10.1007/s10853-013-7161-8.

18. Lee, H.; Nishino, M.; Sohn, D.; Lee, J. S.; Kim, I. S. Control of the Morphology of Cellulose Acetate Nanofibers Via Electrospinning. *Cellulose* **2018**, *25* (5), 2829–2837. DOI: 10.1007/s10570-018-1744-0.

19. Zhang, Z.; Wu, X.; Kou, Z.; Song, N.; Nie, G.; Wang, C.; Verpoort, F.; Mu, S. Rational Design of Electrospun Nanofiber-Typed Electrocatalysts for Water Splitting: A Review. *Chem. Eng. J.* **2022**, *428*, 131133. DOI: 10.1016/j.cej.2021.131133.

20. Forghani, S.; Almasi, H.; Moradi, M. Electrospun Nanofibers as Food Freshness and Time-Temperature Indicators: A New Approach in Food Intelligent Packaging. *Innov. Food Sci. Emerg. Technol.* **2021**, *73*, 102804. DOI: 10.1016/j.ifset.2021.102804.

21. Liu, J.; Zhang, F.; Hou, L.; Li, S.; Gao, Y.; Xin, Z.; Li, Q.; Xie, S.; Wang, N.; Zhao, Y. Synergistic Engineering of 1D Electrospun Nanofibers and 2D Nanosheets for Sustainable Applications. *Sustain. Mater. Technol.* **2020**, *26*, e00214. DOI: 10.1016/j.susmat.2020.e00214.

22. Háková, M.; Havlíková, L. C.; Švec, F.; Solich, P.; Šatínský, D. Nanofibers as Advanced Sorbents for On-Line Solid Phase Extraction in Liquid Chromatography: A Tutorial. *Anal. Chim. Acta* **2020**, *1121*, 83–96. DOI: 10.1016/j.aca.2020.04.045.

23. Wang, Y.; Yokota, T.; Someya, T. Electrospun Nanofiber-Based Soft Electronics. *N.P.G. Asia Mater.* **2021**, *13* (1), 22. DOI: 10.1038/s41427-020-00267-8.

24. Wen, D. L.; Sun, D. H.; Huang, P.; Huang, W.; Su, M.; Wang, Y.; Han, M. D.; Kim, B.; Brugger, J.; Zhang, H. X.; Zhang, X. S. Recent Progress in Silk Fibroin-Based Flexible Electronics. *Microsyst. Nanoeng.* **2021**, *7*, 35. DOI: 10.1038/s41378-021-00261-2.

25. Hamano, F.; Seki, H.; Ke, M.; Gopiraman, M.; Lim, C. T.; Kim, I. S. Cellulose Acetate Nanofiber Mat with Honeycomb-Like Surface Structure. *Mater. Lett.* **2016**, *169*, 33–36. DOI: 10.1016/j.matlet.2015.11.069.

26. Kharaghani, D.; Tajbakhsh, Z.; Nam, P. D.; Kim, I. S. *Application of Nanowires for Retinal Regeneration*; IntechOpen, 2019. DOI: 10.5772/intechopen.90149.

27. Zhu, Q.; Tang, X.; Feng, S.; Zhong, Z.; Yao, J.; Yao, Z. ZIF-8@SiO$_2$ Composite Nanofiber Membrane with Bioinspired Spider Web-Like Structure for Efficient Air Pollution Control. *J. Membr. Sci.* **2019**, *581*, 252–261. DOI: 10.1016/j.memsci.2019.03.075.

28. Xing, R.; Wang, W.; Jiao, T.; Ma, K.; Zhang, Q.; Hong, W.; Qiu, H.; Zhou, J.; Zhang, L.; Peng, Q. Bioinspired Polydopamine Sheathed Nanofibers Containing Carboxylate Graphene Oxide Nanosheet for High-Efficient Dyes Scavenger. *ACS Sustainable Chem. Eng.* **2017**, *5* (6), 4948–4956. DOI: 10.1021/acssuschemeng.7b00343.

29. Alharbi, H. F.; Haddad, M. Y.; Aijaz, M. O. *Membranes Incorporated with Metal Oxide Nanoparticles for Heavy Metal Ion Adsorption. Coatings* **2020**, *10*(3), 285. DOI: 10.3390/coatings10030285.

30. Bahalkeh, F.; Habibi Juybari, M.; Zafar Mehrabian, R.; Ebadi, M. Removal of Brilliant Red Dye (Brilliant Red E-4BA) from Wastewater Using Novel Chitosan/SBA-15 Nanofiber. *Int. J. Biol. Macromol.* **2020**, *164*, 818–825. DOI: 10.1016/j.ijbiomac.2020.07.035.

31. Bahmani, E.; Koushkbaghi, S.; Darabi, M.; ZabihiSahebi, A.; Askari, A.; Irani, M. Fabrication of Novel Chitosan-g-PNVCL/ZIF-8 Composite Nanofibers for Adsorption of Cr(VI), As(V) and Phenol in a Single and Ternary Systems. *Carbohydr. Polym.* **2019,** *224* (June), article 115148. DOI: 10.1016/j.carbpol.2019.115148.

32. Balagangadharan, K.; Dhivya, S.; Selvamurugan, N. Chitosan Based Nanofibers in Bone Tissue Engineering. *Int. J. Biol. Macromol.* **2017,** *104* (B), 1372–1382. DOI: 10.1016/j.ijbiomac.2016.12.046.

33. Pan, S.-F.; Dong, Y.; Zheng, Y.-M.; Zhong, L.-B.; Yuan, Z.-H. Self-Sustained Hydrophilic Nanofiber Thin Film Composite Forward Osmosis Membranes: Preparation, Characterization and Application for Simulated Antibiotic Wastewater Treatment. *J. Membr. Sci.* **2017,** *523,* 205–215. DOI: 10.1016/j.memsci.2016.09.045.

34. Lv, C.; Chen, S.; Xie, Y.; Wei, Z.; Chen, L.; Bao, Jianxu; He, C.; Zhao, W.; Sun, S.; Zhao, C. Positively-Charged Polyethersulfone Nanofibrous Membranes for Bacteria and Anionic Dyes Removal. *J. Colloid Interface Sci.* **2019,** *556,* 492–502. DOI: 10.1016/j. jcis.2019.08.062.

35. Nie, G.; Zhao, X.; Luan, Y.; Jiang, J.; Kou, Z.; Wang, J. Key Issues Facing Electrospun Carbon Nanofibers in Energy Applications: On-Going Approaches and Challenges. *Nanoscale* **2020,** *12* (25), 13225–13248. DOI: 10.1039/d0nr03425h.

36. Göktaş, R. K.; MacLeod, M. Remoteness from Sources of Persistent Organic Pollutants in the Multi-Media Global Environment. *Environ. Pollut.* **2016,** *217,* 33–41. DOI: 10.1016/j.envpol.2015.12.058.

37. Iwuozor, K. O.; Emenike, E. C.; Aniagor, C. O.; Iwuchukwu, F. U.; Ibitogbe, E. M.; Okikiola, T. B.; Omuku, P. E.; Adeniyi, A. G. Removal of Pollutants from Aqueous Media Using Cow Dung-Based Adsorbents. *Curr. Res. Green Sustain. Chem.* **2022,** *5,* 100300.

38. Isaeva, V. I.; Vedenyapina, M. D.; Kurmysheva, A. Y.; Weichgrebe, D.; Nair, R. R.; Nguyen, N. P. T.; Kustov, L. M. Modern Carbon-Based Materials for Adsorptive Removal of Organic and Inorganic Pollutants from Water and Wastewater. *Molecules* **2021,** *26* (21), 6628. DOI: 10.3390/molecules26216628.

39. Agasti, N. Decontamination of Heavy Metal Ions from Water by Composites Prepared from Waste. *Curr. Res. Green Sustain. Chem.* **2021,** *4,* 100088.

40. Ge, M.; Cao, C.; Huang, J.; Zhang, X.; Tang, Y.; Zhou, X.; Zhang, K.; Chen, Z.; Lai, Y. Rational Design of Materials Interface at Nanoscale Towards Intelligent Oil–Water Separation. *Nanoscale Horiz.* **2018,** *3* (3), 235–260. DOI: 10.1039/c7nh00185a.

41. Cui, J.; Li, F.; Wang, Y.; Zhang, Q.; Ma, W.; Huang, C. Electrospun Nanofiber Membranes for Wastewater Treatment Applications. *Sep. Purif. Technol.* **2020,** *250,* 117116. DOI: 10.1016/j.seppur.2020.117116.

42. Li, Y.; Zhu, J.; Cheng, H.; Li, G.; Cho, H.; Jiang, M.; Gao, Q.; Zhang, X. Developments of Advanced Electrospinning Techniques: A Critical Review. *Adv. Mater. Technol.* **2021,** *6* (11), 2100410. DOI: 10.1002/admt.202100410.

43. Wang, C.; Wang, J.; Zeng, L.; Qiao, Z.; Liu, X.; Liu, H.; Zhang, J.; Ding, J. Fabrication of Electrospun Polymer Nanofibers with Diverse Morphologies. *Molecules* **2019,** *24* (5), 834. DOI: 10.3390/molecules24050834.

44. Sakib, M. N.; Mallik, A. K.; Rahman, M. M. Update on Chitosan-Based Electrospun Nanofibers for Wastewater Treatment: A Review. *Carbohydr. Polym. Technol. Appl.* **2021,** *2,* 100064. DOI: 10.1016/j.carpta.2021.100064.

45. El-Aswar, E. I.; Ramadan, H.; Elkik, H.; Taha, A. G. A Comprehensive Review on Preparation, Functionalization and Recent Applications of Nanofiber Membranes in

Wastewater Treatment. *J. Environ. Manage.* **2022,** *301*, 113908. DOI: 10.1016/j.jenvman. 2021.113908.

46. Zamel, D.; Khan, A. U. New Trends in Nanofibers Functionalization and Recent Applications in Wastewater Treatment. *Polym. Adv. Technol.* **2021,** *32* (12), 4587–4597. DOI: 10.1002/pat.5471.

47. Zhang, J.; Liu, L.; Si, Y.; Yu, J.; Ding, B. Electrospun Nanofibrous Membranes: An Effective Arsenal for the Purification of Emulsified Oily Wastewater. *Adv. Funct. Mater.* **2020,** *30* (25), 2002192. DOI: 10.1002/adfm.202002192.

CHAPTER 3

Porous Nanocomposite Polymers for Water Treatment and Remediation

BINDU M. and PRADEEPAN PERIYAT

Department of Environmental Studies, Kannur University, Mangattuparamba Campus, Kannur, Kerala, India

ABSTRACT

Polymer nanocomposites have outstanding features that their components individually lack. The properties, namely, shape, size, surface area, and optical and catalytic activity, can be fine-tuned for desired applications. Their light weight and flexibility make them suitable for environmental and industrial applications. One of the major environmental issues, we are facing today, is water pollution. Proper water management and reclamation are very important. The best method to conserve water is wastewater recycling. The use of polymer nanocomposites for water treatment and remediation offers a solution to recycling. This chapter highlights various aspects of water treatment by polymer nanocomposites.

3.1 INTRODUCTION

In the present scenario, we are facing an alarming increase in environmental pollution, owing to widespread industrialization and agricultural activities. Nowadays, nanotechnology has emerged as a solution to minimize the effects of pollutants, and remedial measures based on this strategy were highly successful in preventing further damages to the environment. In a quest for maximizing water resources, nanotechnology has gained special attention

Mechanics and Physics of Porous Materials: Novel Processing Technologies and Emerging Applications.
Chin Hua Chia, Tamara Tatrishvili, Ann Rose Abraham, & A. K. Haghi (Eds.)

for the treatment and remediation of wastewater. Due to the anthropogenic activities, freshwater sources undergo continuous fall in both quantity and quality.[1]

Polymer nanocomposites have emerged as a promising candidate for wastewater remediation and treatment. The selectivity, remediation capacity, and reusability are influenced by polymer nanocomposite modification strategies. Nanocomposites with a polymer matrix and a nanofiller exhibit certain characteristics such as improved antifouling property, permeability, excellent mechanical strength, thermal stability, adsorption, good pore formation features, and easiness in functionalization.[2] Particle size, morphology, interaction with the polymer matrix, and concentration of the nanofillers have crucial role in determining the overall performance of the polymer nanocomposites.[3,4] In the following section, various aspects of water treatment, namely, dye removal, removal of metal ions, and microorganism from water, by using polymer nanocomposites, have been discussed in detail.

3.2 POLYMER NANOCOMPOSITES AS ADSORBENT MATERIAL

Adsorption is the basic process employed in wastewater treatment strategies. The process has been found to be very effective for the removal of organic/inorganic components from wastewater. Polymer nanocomposites have been proven to be very effective for the adsorption process, owing to their high specific area, tunable surface chemistry, and active sites of adsorption.[5] In addition to that, they are highly stable to various treatment conditions. Literature reports reveal that a variety of polymer nanocomposites have been developed for removing pollutants from water/wastewater. They have been employed for the removal of organic dyes, heavy metals, pharmaceuticals, oils, greases, and some other organic and inorganic constituents.[6–10] Adsorption process is controlled by various factors such as pH of the solution, temperature, contact time, speed of agitation, and concentration of adsorbent.

The metal ion extraction from solution is pH dependent, and it influences metal ion solubility, degree of adsorbate ionization, and concentration of the counter-ions on the polymer nanocomposites. The functional groups of the polymer nanocomposites or any groups grafted onto it have crucial role in the removal of components. Various processes/interactions take place in the adsorption including complexation reactions, electrostatic interactions, hydrogen bonding, reduction, and hydrophobic effect.[11–13] Hence, a combination of two or more processes has been involved for the adsorption to attain equilibrium. So, the adsorption process has more than one mechanism for maximum efficient removal. Removal is very efficient if the process is spontaneous and fast. While implementing a lab-scale process to an industrial scale, the process time is always a challenging factor. The reaction kinetics and equilibrium conditions should be optimized during the removal process. The desired characteristics of an efficient removal include the following: the process should be spontaneous, less contact time, excellent percentage removal, and reuse capability.

3.3 DYE REMOVAL BY USING POLYMER NANOCOMPOSITES

Industries such as textiles, paper, coating, and cosmetics are major contributors of organic synthetic dyes. They make use of various dyes to impart color to the product and the discharged wastewater from these industries creates severe environmental issues. Water polluted with organic dyes is toxic and carcinogenic. Hence, it is very important to remove these dyes and make wastewater reusable for industrial purpose, to prevent environmental impacts. Extensive research work has been going on in this regard by employing polymer nanocomposites.[14–16] The dye adsorbing material should possess large surface area, number of active sites, and thermal and chemical stability.[17]

A recent work by Srivastava et al. demonstrated the development of chitosan/CuO nanocomposite beads for the removal of dyes, namely, Congo red and eriochrome black T.[18] Ninety-seven percent dye removal efficiency has been achieved within 2 h. An amount of 120 mg/g of Congo red and 236 mg/g eriochrome black T was adsorbed onto the polymer nanocomposite beads. The dye adsorption has been influenced by certain features such as surface properties of the adsorbent, hydrophilicity/hydrophobicity of both the adsorbent and adsorbate, charge of the dye molecule, electrostatic interaction, hydrogen bonding, and van der Waals interactions.[19] The possible interaction between chitosan/CuO nanocomposite beads and the dye molecules has been depicted in Figure 3.1.

There occurs an electrostatic interaction between the negatively charged dye molecules and positively charged adsorbent, chitosan. In addition to this, strong hydrogen bonding exists between N of the dye molecules and hydroxyl group of chitosan. These interactions make favorable adsorption of dyes onto the chitosan-based polymer nanocomposites. Chitosan/CuO nanocomposite beads exhibit higher dye adsorption rate and percentage dye removal in comparison with virgin chitosan. The CuO nanoparticles in the nanocomposite act as a supporting medium and facilitate the settling of dyes onto the adsorbent. Highly electronegative oxygen atom of CuO offers further possibility of hydrogen bonding with the hydroxyl and amine group of dye molecules.

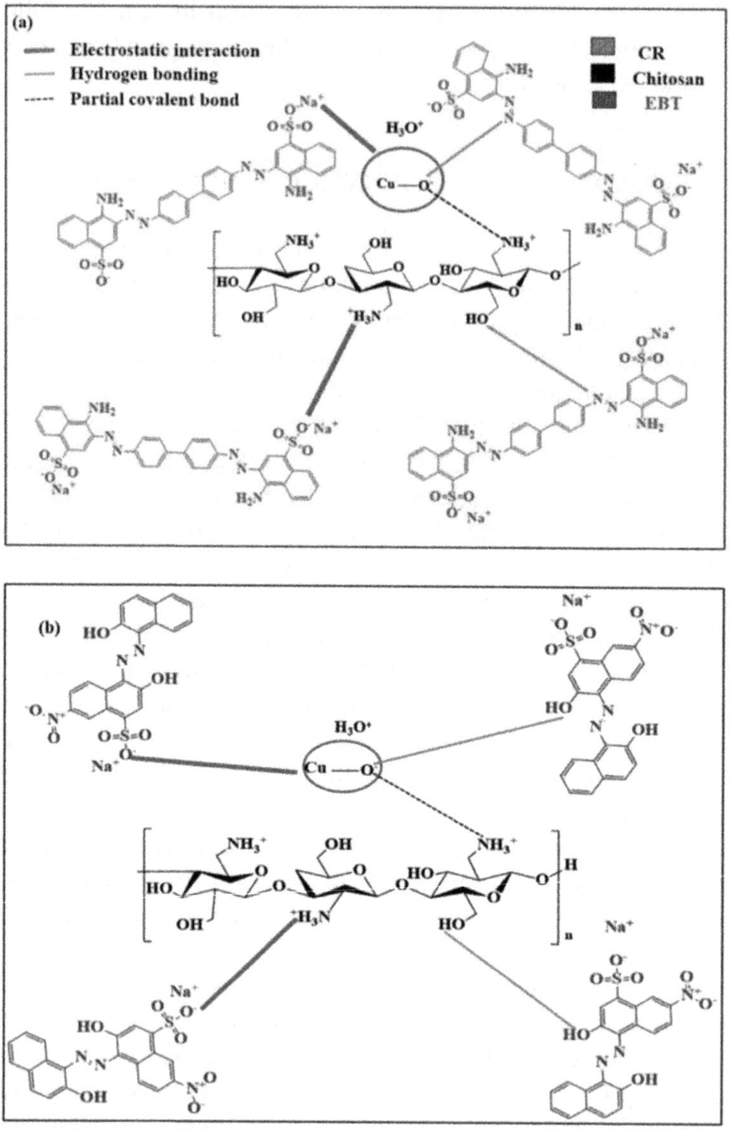

FIGURE 3.1 Mechanism of (a) Congo red and (b) eriochrome black T dye adsorption onto chitosan/CuO nanocomposite beads.

Source: Reprinted with permission from Ref. [18]; copyright © Elsevier.

Table 3.1 shows the effect of initial dye concentration on the percentage removal. It has been observed that the percentage dye removal is inversely related to their initial concentration. At lower dye concentrations, the active sites on the adsorbent are vacant, and as the dye concentration increases, more and more active sites are occupied.[20] Owing to the higher concentration gradient, the dye molecules get diffused into the inner pores of the beads and operate the dye molecules to the activated adsorption sites.

TABLE 3.1   Percentage Dye Removal as a Function of Dye Concentration (Reprinted with Permission from Ref. [18]; copyright © Elsevier).

Sl. no.	Concentration of dyes (mg/L)	Concentration of the adsorbent (g)		% Dye removal			
				Chitosan beads		Chitosan–CuO nanocomposite beads	
		CR	EBT	CR	EBT	CR	EBT
1	10	0.05	1.0	17.62	33.22	69.86	98.01
2	20	0.05	1.0	14.06	32.72	49.88	98.88
3	30	0.05	1.0	12.66	29.09	42.78	98.83
4	40	0.05	1.0	10.94	28.93	39.78	98.11
5	50	0.05	1.0	9.11	28.69	34.23	97.56

CR: Congo red; EBT: eriochrome black T.

Chitosan–TiO_2 nanocomposites have been developed for the removal of Rose Bengal dye from industrial water.[21] The nanocomposites possess an average surface area of 95 m^2/g and uniform mesoporous structure allows excellent dye adsorption. Kinetics of dye removal has been investigated by pseudo first order, pseudo second order, and Weber–Morris models and found that the results, well matched with the pseudo-second-order model, indicate the monolayer adsorption of the dye molecules. Figure 3.2a shows a comparison of the percentage of dye removal by chitosan and chitosan/TiO_2 nanocomposites. Compared to pure chitosan, chitosan/TiO_2 nanocomposites exhibited a higher percentage of dye removal. Fifty-three percent of Rose Bengal dye removal by chitosan is attributed to the electrostatic interactions among positively charged chitosan and negatively charged dye molecules. A maximum dye adsorption of 98.8% has been observed for the chitosan/TiO_2 nanocomposites, owing to the attraction of dye molecules into the mesoporous cavities of the nanocomposites.

After each cycle of dye removal, its adsorption onto the nanocomposites get decreased till 63%; the catalyst can be reused for four cycles as indicated by Figure 3.2b. Selective removal of Rose Bengal from a mixture of alizarine red and methylene blue is also possible. To assure the complete removal of the dyes, the samples were further stirred under UV irradiation for 60 min to decompose methylene blue. This leads to 10% increase in dye adsorption efficiency as shown in Figure 3.2c.

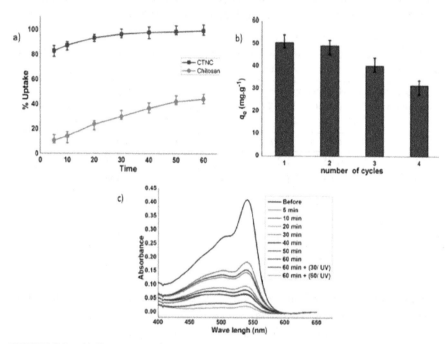

FIGURE 3.2   (a) Percentage adsorption of Rose Bengal dye on chitosan and chitosan/TiO₂ nanocomposites, (b) regeneration cycle of the nanocomposites, and (c) removal efficiency of different dyes by chitosan/TiO₂ nanocomposites

Source: Reprinted with permission from Ref. [21]; copyright © Elsevier.

Chitosan/ZnO nanocomposites have been prepared for the removal of methylene blue dye from industrial water.[22] The optimum conditions for the methylene blue degradation by chitosan/ZnO nanocomposites were found to be 20 mg/L of the dye and 60 mg/L nanocomposites at pH 9 for 1 h. Chitosan/ZnO nanocomposites exhibit a dye removal of 96.5%, while pure chitosan shows 81% efficiency. Enhanced dye removal by the nanocomposites has been attributed to the combined effect of adsorption and photocatalytic degradation of methylene blue by ZnO nanoparticles. A complexation reaction occurs between chitosan and surface Zn^{2+} ions of ZnO, namely, ligand substitution reaction. The coordinated H_2O molecule on Zn^{2+} ions is being substituted by the functional group in chitosan as shown in Figure 3.3.

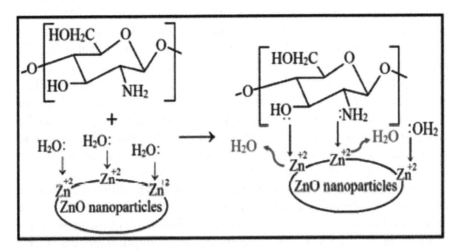

FIGURE 3.3 Complexation reaction between chitosan and Zn^{2+} ions of ZnO nanoparticles

Source: Reprinted with permission from Ref. [22]; copyright © Elsevier.

Cellulose/Graphene oxide nanocomposites have been reported to be a promising candidate for azo dye removal. The nanocomposite was developed by modified Hummer's method.[23] Graphite is oxidized to graphene oxide and the negatively charged functional groups present on the graphene oxide make hydrogen bonding with cellulose. As the concentration of the cellulose/graphene oxide nanocomposite increases from 20 to 30 mg/L, there is a significant enhancement in the dye removal efficiency. Maximum efficiency (91%) has been obtained for 30 mg/L of the nanocomposites. This may be due to the increase in number of active sites available for dye adsorption. However, when the concentration of the nanocomposites reaches to 45 mg/L, there is no considerable enhancement in the efficiency of dye removal. At higher concentration, there are chances for agglomeration of chitosan/graphene oxide particles, and active sites for dye adsorption get saturated and lead to a reduction in the dye removal. Alkaline pH has been found to be more favorable for the process. Temperature also influences dye adsorption; both are directly related. The optimum temperature has been found to be 20–40°C.

Chinthalapudi et al. reported the synthesis of cellulose fiber/Ag nanocomposites for malachite green removal.[24] The effect of contact time, concentration of cellulose/Ag nanocomposites, pH, temperature, and initial concentration of the malachite green on the dye removal has been investigated. The equilibrium time for the dye removal has been found to be 100 min. As the concentration of the dye increases from 20 to 100 mg/L, removal efficiency decreases. The optimum pH has been observed to be 8. Depending on the structure of the cellulose adsorbent and malachite green dye, three possible interactions of malachite green with cellulose have been put forward. These include ion–dipole interaction, hydrogen bonding, and electrostatic interactions. The hydrogen bonding operates between hydroxyl groups of cellulose, hydroxyl layer around positively charged Ag nanoparticles, and lone pair of electrons in nitrogen (tail end) of the malachite green.[25] Oxygen atom of the hydroxyl group and positively charged nitrogen atoms of malachite green create ion–dipole interactions.

The kinetics of dye degradation has been evaluated by fitting with pseudo-first-order and pseudo-second-order equations. Correlation coefficient of pseudo-first-order model ($R^2 = 0.998$) is higher than those of pseudo-second-order model ($R = 0.987$). This suggests that the adsorption process is well fitted with pseudo-first-order model. Table 3.2 shows a comparison of the rate constant and correlation coefficient calculated with both models. Figure 3.4a shows $\log(q_e - q_t)$ versus time plot (corresponding to pseudo-first-order kinetics) and 4 (b) t/q_t versus time plot (corresponding to pseudo-second-order kinetics).

TABLE 3.2 Rate Constant and Correlation Coefficient Calculated with Kinetic Models (Reprinted with Permission from Ref. [24]; Copyright © Elsevier).

Order	Equation	Rate constant	R^2
Pseudo first order	$\log(q_e - q_t) = \log q_e - \dfrac{K_1}{2.303}t$	0.0532 min^{-1}	0.998
Pseudo second order	$\dfrac{t}{q_t} = 0.7068 + 0.00273t$	1.23 mol/L min	0.987

q_e and q_t are the amounts of dye adsorbed at equilibrium and at any time "t," respectively.

Cellulose/Acid-activated clay nanocomposites were employed for rhodamine B degradation from industrial water.[26] It was found that the pH, temperature, concentration of the nanocomposites, and time influence the dye degradation. Due to the anionic carboxylic form of rhodamine B, sorption was more favorable at acidic pH as reported earlier.[27] When the pH is greater than 4, there is a possibility of zwitterion formation and photocatalytic activity getting decreased. The effect of pH on the sorption of rhodamine dye has been presented in Figure 3.5a. Sorption of dye molecule and sorbent dose are inversely related. The effect sorbent dose on the photocatalytic activity has been shown in Figure 3.5b. Maximum activity was observed at lower dose. As the sorbent dose increases, aggregation happens and causes a reduction in the photocatalytic activity.[28] Figure 3.5c shows the sorption capacity versus temperature plot. The color of rhodamine B before sorption was shown in Figure 3.5d and after sorption in Figure 3.5e.

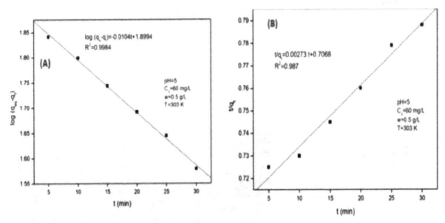

FIGURE 3.4 (a) $\log(q_e - q_t)$ versus time plot and (b) t/q_t versus time plot.

Source: Reprinted with permission from Ref. [24]; copyright © Elsevier.

Photocatalyst such as nano-TiO_2, ZnO, and CuO degrades the organic dye molecules by generating hydroxyl ions. So, when polymer nanocomposites are used for dye removal, most suitable approach is the adsorption of the dye molecules onto the active site of polymer nanocomposites followed by photocatalytic degradation of the dye molecule. Poly(1-naphthylamine)/TiO_2 nanocomposites were employed for the degradation of methylene blue dye, under visible light.[29] The nanocomposites exhibited a degradation efficiency of 60%. After 60 min, most of the dye molecules undergo decomposition on PNA/TiO_2 nanocomposites, after that the degradation will be slow till 160 min, having a maximum value of 60%. PNA, having conjugated π electron systems, gets excited under visible light irradiation. Then, the π* electrons from PNA are transferred to the conduction band of TiO_2. On the other hand, hole (h^+) is formed by the transfer of valence band electrons of TiO_2 to PNA

π orbital for regeneration of electrons. The electrons in the conduction band react with oxygen molecules to form superoxide (O_2^-) radicals. Simultaneously, the holes in the valance band of titania react with H_2O molecules to form hydroxyl radical. PNA involves charge transfer and facilitates the photocatalytic activity with excellent charge separation between electron–hole pairs. The generation of highly reactive oxyradicals and hydroxyl radical degrades the dye molecules.[30]

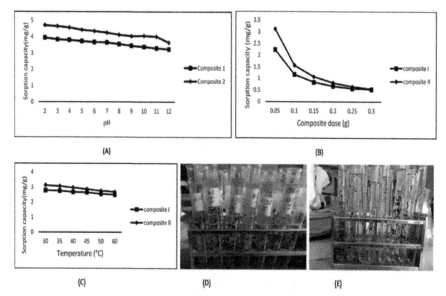

FIGURE 3.5   (a) Effect of pH on sorption capacity, (b) sorption capacity with composite dose, (c) sorption capacity versus temperature plot, (d) color of rhodamine B before sorption, and (e) rhodamine B after sorption. Composites 1 and 2 contain different compositions of clay.

Source: Reprinted with permission from Ref. [26]; copyright © Elsevier.

Polylactic acid (PLA)/TiO_2 nanocomposites were developed by *in-situ* polymerization of lactic acid in the presence of TiO_2.[31] Lactic acid coordinates with the TiO_2 nanoparticles via its carboxylic group as shown in Figure 3.6. Photocatalytic activity of the developed nanocomposites has been investigated by using model dyes, namely, malachite green and methyl orange under UV and visible light. It was found that under UV irradiation, complete decolorization of methyl orange happens in 30 min and malachite green in 15 min. However, under sunlight, the same dyes undergo complete decolorization by 20 and 8 min, respectively.

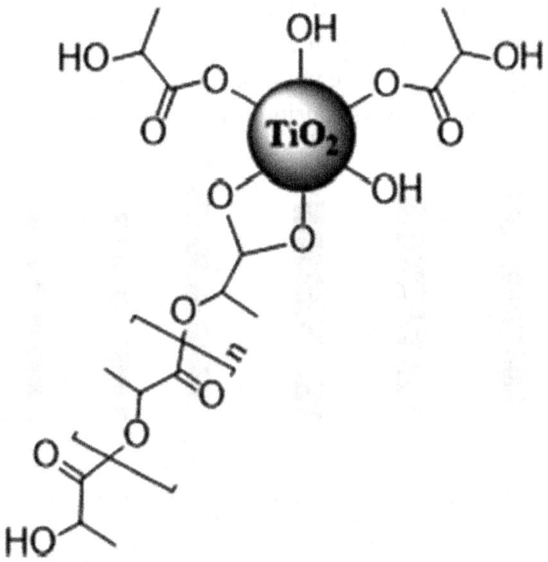

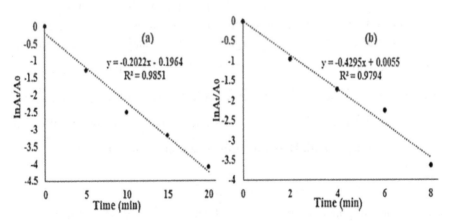

FIGURE 3.6 TiO$_2$ nanoparticles grafting onto PLA.

Source: Reprinted with permission from Ref. [31]; copyright © Wiley online library.

Kinetics of degradation has been found to follow first-order kinetics and Figure 3.7 shows the $\log \frac{A_t}{A_0}$ versus "*t*" plot. A straight line with negative slope has been observed in both cases.

FIGURE 3.7 Kinetics of degradation: (a) methyl orange and (b) malachite green [dye concentration: 10^{-4} M (10 mL); PLA/TiO$_2$ nanocomposite (50 mg)].

Source: Reprinted with permission from Ref. [31]; copyright © Wiley online library.

Figure 3.8 shows the recyclability results of PLA/TiO$_2$ nanocomposites. Degradation percentage of both the dyes by the photocatalyst after four cycles was almost similar to those with fresh catalyst. Only a slight reduction occurs due to the loss of catalyst during separation process.

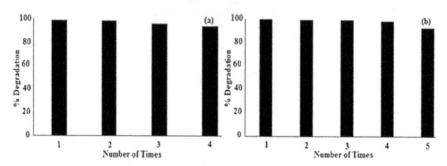

FIGURE 3.8   Recyclability results of the PLA/TiO$_2$ nanocomposites: (a) methyl orange and (b) malachite green.

Source: Reprinted with permission from Ref. [31]; copyright © Wiley online library.

Karagoz et al. reported the fabrication of polycaprolactone/Ag/TiO$_2$ nanocomposites (PCL/Ag/TiO$_2$) for photocatalytic degradation of methylene blue.[32] Figure 3.9 shows the result of photocatalytic activity studies. Figure 3.9a corresponds to the UV visible spectra of methylene blue dye over PCL/Ag/TiO$_2$ nanocomposites at different time intervals. As time goes on, the intensity of characteristic peak of methylene blue gets decreased. The solution becomes completely decolorized after 160 min. From Figure 3.9b, it was clear that the degradation efficiency increases with an increase in Ag concentration, due to the synergy between Ag and TiO$_2$ nanoparticles. Figure 3.9c shows the $\log \dfrac{C_0}{C}$ versus "*t*" plot and kinetics of degradation has been found to be pseudo first order. Rate constant of the photocatalytic reaction has been calculated from the slope of $\log \dfrac{C_0}{C}$ versus "*t*" plot and presented in Figure 3.9d. Half-life of the reaction is calculated as[33]:

$$t1/2 = \frac{0.693}{K} \tag{3.1}$$

Polyaniline/TiO$_2$ nanocomposites (PANI/TiO$_2$) were developed for the electrocatalytic decomposition of Congo red in industrial water.[34] TiO$_2$/PANI in 2:1 ratio showed 60% decolorization after 90 min. Kinetics studies of the dye removal revealed that the dye degradation follows pseudo-second-order reaction. Figure 3.10 represents the Congo red decolorization by TiO_2/PANI nanocomposites in different ratio of TiO_2 and PANI, namely, 2:1, 4:1, and 6:1, respectively. Dye removal efficiency of all the ratios was higher for the initial 10 min of the reaction; 2:1 and 4:1 TiO_2/PANI nanocomposites gradually loss their activity after 10 min. However, 6:1 system exhibits higher activity throughout the reaction.

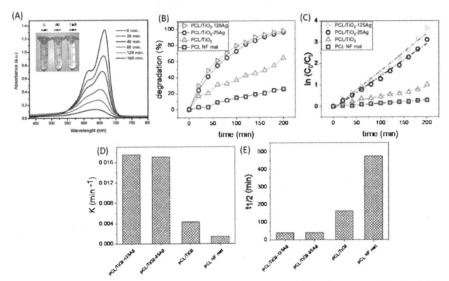

FIGURE 3.9 (a) UV–visible spectrum of methylene blue dye over PCL/Ag/TiO_2 nanocomposites at different time intervals; (b) percentage photodegradation of methylene blue over P CL nanofiber mat, PCL/TiO_2 nanocomposites, and PCL/Ag/TiO_2 nanocomposites; (c) $\log \frac{C_0}{C}$ versus "t" plot (d) pseudo-first-order rate constant; and (e) half-life of photodegradation reaction.

Source: Reprinted with permission from Ref. [32]; copyright © Elsevier.

An article by Reddy et al. reported the photocatalytic activity evaluation of TiO_2/PANI nanocomposites prepared via oxidative polymerization of aniline by oxidant ammonium persulfate, in the presence of nano-TiO_2.[35] The dispersion of TiO_2 nanoparticles in PANI matrix has been evaluated by TEM analysis. Methylene blue, rhodamine B, and phenol were used as model pollutants. Photocatalytic activity studies show that TiO_2/PANI nanocomposites (20 wt% TiO_2) exhibit higher dye degradation efficiency than pure TiO_2 nanoparticles. The reason for enhanced photoactivity of TiO_2/PANI nanocomposite is as follows: (1) the coupling between conjugated polymer PANI and inorganic semiconductor metal oxide TiO_2 facilitates effective charge separation between photogenerated holes and electrons; (2) uniform dispersion of TiO_2 nanoparticles into PANI matrix enhances the dye adsorption onto the nanocomposite surface; and (3) optical properties of TiO_2/PANI system are entirely different from those of pure TiO_2. The nanocomposites absorb near IR and visible light in addition to UV absorption. This result indicates that PANI acts as a photosensitizer for TiO_2. (4) TiO_2 band gap has been reduced by PANI and hence photocatalytic activity gets increased.

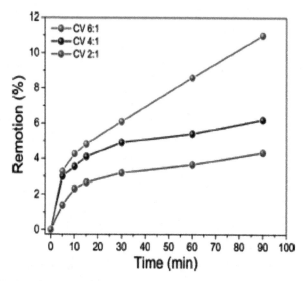

FIGURE 3.10   Decolorization of Congo red by 2:1, 4:1, and 6:1 TiO_2/PANI nanocomposites developed by cyclic voltammetry.

Source: Reprinted with permission from Ref. [34]; copyright © Elsevier.

Table 3.3 gives a detailed literature analysis of polymer nanocomposites employed for organic dye degradation from industrial wastewater.

3.4   POLYMER NANOCOMPOSITES FOR METAL ION REMOVAL

Advancements in industrialization and urbanization are the main cause of heavy metal ion pollution. Various anthropogenic activities such as mining, smelting, use of fertilizers and pesticides, and industrial effluents and chemicals lead to the contamination of heavy metal ions such as Pb^{2+}, Cd^{2+}, Cu^{2+}, Ni^{2+}, and Cr^+.[54] Presence of toxic metal ions has negative impact on water bodies. They are non-biodegradable and carcinogenic.[55,56] Cd^{2+} and Pb^{2+} have been reported to be most toxic and threat to the environment.

TABLE 3.3 Polymer Nanocomposites for Organic Dye Degradation.

Polymer nanocomposites	Dyes used	Inference	Referees
Polyhydroxy butyrate/TiO_2	Methylene blue	96% of methylene blue degradation with 1 h of visible light illumination	[36]
Polydimethylsiloxane (PDMS)/TiO_2	Methylene blue	Methylene blue undergoes degradation when PDMS content is lower than 60%, under UV light illumination	[37]
Poly(3-hexylthiophene)/$Ti_0 2$	Methyl orange	88% dye degradation within 10 h, under visible light illumination [TiO_2/ poly(3-hexylthiophene ratio: 75:1]	[38]
Polyethylene (PE)/TiO_2	Methylene blue	Visible light photocatalytic activity. Degradation of 0.3 mg of dye in 1 h	[39]
Chitosan/CdS	Congo red	85% degradation of the dye within 180 min at pH 6	[40]
Cellulose/ZnO	Methylene blue Malachite green	93.5% degradation of methylene blue and 99% degradation of malachite green within 5 min	[41]
PMMA/ZnO	Methylene blue	100% methylene blue degradation by 80 min	[42]
PMMA/CNT	Methyl green	An adsorption capacity of 6.8 mmol/g at room temperature	[43]
Polypyrrole/zeolite	Reactive blue Reactive red	86% of reactive blue and 88% of reactive red undergoes degradation	[44]
Polylactic acid/Graphene oxide	Crystal violet	98% of crystal violet has been removed	[45]
Polyacrylic acid/Fe_3O_4/starch	Congo red Methylene violet	93% Congo red degradation and 99% methylene violet degradation	[46]
Polyacrylic acid/Fe_3O_4/ carboxylated cellulose	Methylene blue	An adsorption capacity of 330 mg/g	[47]
Polystyrrene/N-doped TiO_2	Methylene blue	30% degradation of methylene blue under visible light illumination for 180 min	[48]
Polyacrylamide/ZnS	Congo red Methyl orange	75% degradation of Congo red within 2 h and 69% degradation of methyl orange within 4 h	[49]

TABLE 3.3 *(Continued)*

Polymer nanocomposites	Dyes used	Inference	Refereces
Polypyrrole/TiO_2	Rhodamine B	85% rhodamine B degradation with 120 min UV irradiation. Photocatalytic reaction follows pseudo-second-order kinetics	[50]
PVA/TiO_2/Graphene-multiwalled carbon nanotube	Methylene blue	70 % methylene blue degradation under UV irradiation	[51]
Polyether sulfone/TiO_2	Methyl orange	100% methylene blue degradation at acidic pH. Follows pseudo-first-order kinetics	[52]
Polyaniline/Fe	Methylene blue, malachite green, and methyl orange	Complete decolorization of all the dyes within 90 min UV irradiation	[53]

Cu^{2+} ions are essential trance elements for drinking water; however, their excess concentration causes iron disorder and damage to cell membranes.[57,58] Table 3.4 gives a brief idea about some heavy metal ions in wastewater, their sources, health effects, and allowed quantity in drinking water.

Removal of heavy metals from water can be achieved *via* various processes such as adsorption, precipitation, ion exchange, solvent extraction, reverse osmosis, and electrochemical methods. Adsorption is relatively simple, cost-effective, and excellent removal potential and can be implemented easily.[59] Adsorption mechanism has been defined by the physicochemical feature of both the adsorbent and heavy metals and the operating conditions such as pH of the solution, temperature, contact time, concentration of the adsorbent, and initial concentration of the adsorbate. Figure 3.11 shows a schematic representation of an adsorption process.[59,60] Metal ions of wastewater get adhered to the surface of high surface area porous adsorbent. The adsorption process is highly selective, that is, one or more metal ions can be selectively removed from a mixture of metal ions.

Polymer nanocomposites were reported to be very effective for metal ion removal, and the basic mechanism is the adsorption of these metal ions onto the nanocomposite surface. Compared to neat polymers, polymer nanocomposites exhibit higher adsorption capacity, owing to the presence of a greater number of active surface groups for interactions. An article by Saad et al. reported the development of ZnO@chitosan core–shell-structured nanocomposites for the removal of heavy metal ion from wastewater.[64] The synthesis strategy has been presented in Figure 3.12. Chitosan, the natural polymer has hydroxyl (–OH) and amino groups ($–NH_2$) in their structure, which acts as active sites for the dye adsorption.[65]

The adsorption studies were performed by means of batch experiments with Cu^{2+}, Cd^{2+}, and Pb^{2+} ions. The rate of removal has been assessed in terms of various parameters such as pH, time of contact, dose of adsorbent, and initial concentration of the metal ions, and the results have been presented in Figure 3.13.

For Pb^{2+} ions, 90% removal has been achieved at pH 6. However, for Cd^{2+} and Cu^{2+} ions, the pH for obtaining maximum percentage removal is 6.5 and 4, respectively, as indicated by Figure 3.13a. For Cu^{2+} ions, after pH 6 formation of precipitate occurs. In the case of Pb^{2+} and Cd^{2+}, adsorption increases when pH increases from 3 to 6, after that it shows a decreasing trend. At a pH less than 3, chitosan amino groups get protonated to form NH_3^+, leading to a reduction in percentage adsorption.[66] As the pH increased, the competition between the toxic ions and protons decreased. The equilibrium time was 30 min for Pb^{2+} and 60 min for both Cd^{2+} and Cu^{2+} ions (Figure 3.13b). The optimum

TABLE 3.4 Typical Heavy Metal Ions in Wastewater, Their Health Issues, and Permitted Limits in Drinking Water as Per WHO Organizations.

Metal ions	Sources[61–63]	Affected organs/systems	Permitted level (μg)[61–63]
Pb (lead)	Batteries, alloys, solder, and rust inhibitors	Kidney, bone, hematological systems, cardiovascular, and reproductive systems	10
As (arsenic)	Electronic components and production of glass	Lungs, skin, brain, cardiovascular, and immunological systems	10
Cu (copper)	Cable and electronic industry, plumbing products (corroded)	Liver, cornea, gastrointestinal systems, lungs, immunological, and hematological systems	2000
Ni (nickel)	Production of nickel alloys and stainless steel	Gastrointestinal diseases, lungs, kidneys, pulmonary fibrosis, and skin	70
Hg (mercury)	Agricultural and landfills runoffs, production of chlorine and caustic soda, refineries, laboratory apparatus, and electrical instruments	Liver, kidney, brain, immunological, cardiovascular, endocrine, and reproductive systems	6
Cr (chromium)	Steel and pulp mills	Kidney, lungs, brain, skin, pancreas, reproductive, and gastrointestinal systems	50
Cd (cadmium)	Plastic industries, steel industries, metal refineries, paints, batteries, and galvanized pipes	Liver, kidney, lungs, testes, bones, immunological, and cardiovascular systems	3
Zn (zinc)	Cosmetics, deodorants, rubber products, and brass coatings	Vomiting, nausea, anemia, stomach cramps, and skin irritations	3000

removal was observed at lower adsorbent dosage. Maximum removal of Pb^{2+} is at 0.1 g/L and both Cu^{2+} and Cd^{2+} are at 0.5 g/L of the adsorbent as indicated by Figure 3.13c. Increasing the initial metal ion concentration causes a reduction in the efficiency of removal, owing to the decrease in number of active sites for adsorption in the nanocomposites (Figure 3.13d).

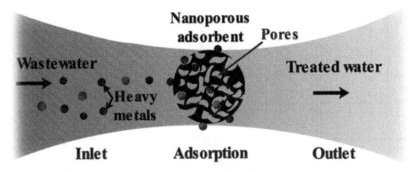

FIGURE 3.11   Schematic representation of an adsorption process.

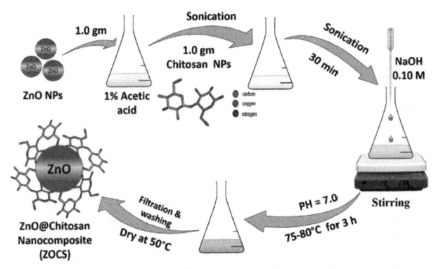

FIGURE 3.12   Development of ZnO@chitosan core–shell-structured nanocomposites.

Source: Reprinted with permission from Ref. [64]; copyright © Elsevier.

An interesting article by Mohammadnezhad et al. reported the fabrication of PMMA/MCM-41 silica and polystyrene (PS)/MCM-41 silica nanocomposites for the removal of Cd^{2+} metal ions.[67] Initially, the MCM-41 silica particles were treated with a silane coupling agent APTS (3-animopropyl triethoxy silane) to anchor PMMA and PS. A schematic representation of the synthesis strategy has been depicted in Figure 3.14.

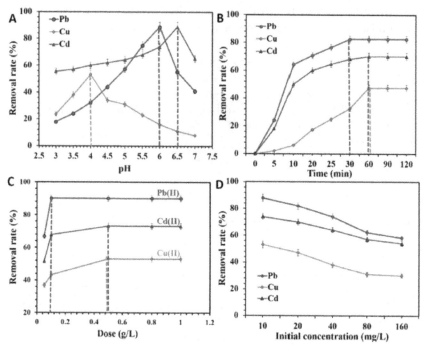

FIGURE 3.13   Effect of (a) pH, (b) time, (c) dose of adsorbent, and (d) initial concentration of heavy metals on rate of removal by ZnO@chitosan nanocomposites.

Source: Reprinted with permission from Ref. [64]; copyright © Elsevier.

Effect of pH, initial concentration of the metal ion, and contact time on adsorption efficiency have been investigated. At lower pH, adsorption is not efficient, owing to the greater interaction of H⁺ ions with Cd²⁺ ions. The optimum pH for maximum adsorption was found to be 5, with 240 min. The maximum adsorption capacity of PMMA and PS nanocomposites has been observed to be 10.5 and 26.8 mg/g, respectively. Also, it was found that the mechanical characteristics of the nanocomposites were significantly improved compared to neat PMMA and PS.

In another study, a series of polymer nanocomposites, namely, APTS-MCM-41/PMMA, APTS-MCM-41/PS, APTS-MCM-PVA, and APTS-MCM-Nylon 6, have been developed for the removal of Cr⁶⁺ ions.[68] Figure 3.15 shows the effect of pH on the percentage removal of the nanocomposites.

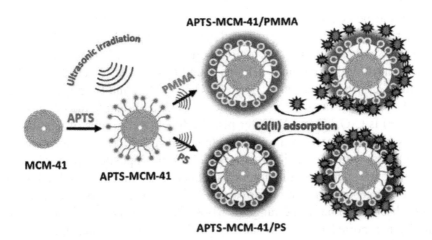

FIGURE 3.14 Development of APTS-MCM-41/PMMA and APTS-MCM-41/PS nanocomposites for Cd^{2+} ion removal.

Source: Reprinted with permission from Ref. [67]; copyright © Elsevier.

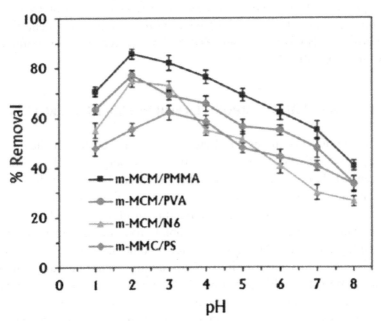

FIGURE 3.15 pH versus percentage removal plot of nanocomposites based on PMMA, PVA, nylon 6, and PS (m: APTS).

Source: Reprinted with permission from Ref. [68]; copyright © American Chemical Society.

pH and percentage removal exhibit an inverse relationship. A sudden rise in removal (%) has been observed in the pH range 1–2. Further increase in pH causes a reduction in the removal efficiency, for all the nanocomposites. Therefore, the optimum pH was found to be 2. At lower pH, that is, less than 2, Cr(6+) exists in the form $Cr_2O_7^{2-}$, CrO_4^{2-}, and $HCrO_4^-$, whereas $HCrO_4^-$ in the pH range 2–5. When the pH is greater than 5, the main form of Cr(6+) is CrO_4^{2-} ions. At lower pH, the functional groups of nanocomposites get protonated to a greater extent and it favors the adsorption of Cr(6+) ions, owing to the interaction between the oppositely charged ions. When the pH increases, this interaction decreases, causing a reduction in the Cr(6+) adsorption. The removal efficiency of the nanocomposites has been observed to be in the range of 61–85; however, no systematic study has been reported on the reusability of the nanocomposites for Cr(6+) removal.

Nanocomposites containing zirconium phosphate nanoparticles (ZrP–Cl, ZrP–N, ZrP–S) and macroporous PS resins having different surface groups (CH_2Cl and $-CH_2N^+(CH_3)_3$) were developed for Pb^{2+} ion removal.[69] The results proved that, for better dispersion of zirconia nanoparticles, charged functional groups ($-CH_2N^+(CH_3)_3$) were more effective than neutral functional groups (CH_2Cl). ZrP–N and ZrP–S showed greater removal efficiency of Pb^{2+} ions than ZrP–N.

3.5   POLYMER NANOCOMPOSITES FOR WATER DISINFECTION

The microorganisms in the drinking water influence human health to a greater extent. Therefore, removal of these microorganisms by killing the pathogens or their deactivation is highly essential. In view of this, several polymer nanocomposites, showing antibacterial activity, have been reported to be a promising candidate for disinfecting drinking water. Cellulose/Ag nanocomposites which exhibit antimicrobial properties and excellent water permeability have been utilized for water treatment.[79] Figure 3.16 shows a comparison of antibacterial activity of the cellulose nanofibers and cellulose/Ag nanocomposites. It has been observed that after 48 h, there are no bacterial colonies in cellulose/Ag nanocomposite systems.

An interesting article by Munnawar et al. reported the fabrication of polyethersulfone antifouling membranes, by incorporating chitosan/ZnO onto polyether sulfone matrix.[80] Figure 3.17 shows the permeability of the samples with respect to chitosan/ZnO hybrid nanoparticle content. As the chitosan/ZnO content increases, the water flux also increases. The lowest water flux has been observed for MH0 sample, due to the hydrophobic nature of the polyether sulfone membranes, and the highest water flux for MH3, which contained 15 w/w% chitosan/ZnO hybrid nanoparticles. Improved pore size and hydrophilicity make better attraction of water molecules and increased permeability.

TABLE 3.5 Some Polymer Nanocomposites Employed for Removing Heavy Metals.

Polymer nanocomposites	Metal ions	Inference	References
Polyaniline/rGO (reduced graphene oxide)	Hg^{2+}	Adsorption capacity was 1000 mg/g	[70]
Chitosan/GO/ferrite	Cr^{4+}	An adsorption capacity of 270 mg/g at pH 2	[71]
Cellulose/TiO$_2$ nanocomposites	Zn^{2+} Cd^{2+} Pb^{2+}	Adsorption capacity for Zn^{2+}, Cd^{2+}, and Pb^{2+} was 102.02, 102.05, and 120.5 mg/g, respectively	[72]
Polyacrylamide/ Montmorillonite nanocomposites	Co^{2+} Ni^{2+}	98. 7% of Co^{2+} and 99.3% Ni^{2+} have been removed at pH 6	[73]
PMMA/Silica nanocomposites	Cu^{2+}	The nanocomposites exhibited a maximum adsorption capacity of 42 mg/g at pH 4 within 140 min	[74]
Polyaniline/ZnO	Cr^{6+}	Maximum adsorption capacity of 345 mg/g of Cr^{6+}	[75]
Polyacrylamide/Bentonite hydrogel nanocomposites	Cd^{2+} Pb^{2+}	Within the initial 20 min, more than 95% of Pb^{2+} and Cd^{2+} ions were removed. Maximum adsorption capacity for Pb^{2+} and Cd^{2+} was found to be 138 and 200 mg/g, respectively	[76]
Bacterial cellulose/Ti$_O$2	Pb^{2+}	90% of Pb^{2+} removal within 120 min at pH 6	[77]
Poly(acrylic acid)/Starch/ Fe$_3$O$_4$	Cu^{2+} Pb^{2+}	88% Pb^{2+} and 95% Cu^{2+} have been removed at pH 5.5 and 6, respectively	[78]

The antibacterial results have been presented in Figure 3.18. The zone of inhibition increased with an increase in nanoparticle content. Further, the diameter of zone in Gram-positive bacteria is higher than those in Gram-negative bacteria. ZnO produces reactive oxygen species; Gram-negative bacteria are more resistant because their cell walls have outer membrane and are more complex.

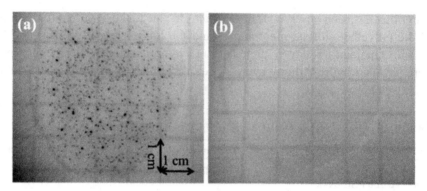

FIGURE 3.16 (a) Bacterial colonies in water, in contact with cellulose nanofibers and (b) water in contact with cellulose nanofibers/Ag nanocomposites (Ag: 3.6 wt%).

Source: Reprinted with permission from Ref. [79]; copyright © Elsevier.

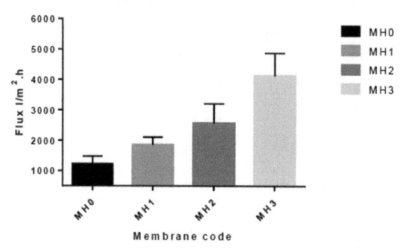

FIGURE 3.17 Water flux versus chitosan/ZnO hybrid nanoparticles loading in polyether sulfone matrix.

Source: Reprinted with permission from Ref. [80]; copyright © Elsevier. (MH0: polyether sulfones without chitosan/ZnO; MH1: 5 w/w% chitosan/ZnO; MH2: 10 w/w% chitosan/ZnO; MH3: 15 w/w % chitosan/ZnO).

In another study, polycaprolactone/ZnO nanocomposite membranes which exhibit antibacterial activity against *Staphylococcus aureus* and *Escherichia coli* have been developed.[81] Polypyrrole/carbon nanotube (CNT)/Ag nanocomposites were developed via oxidative polymerization of pyrrole with silver nitrate in presence of CNT.[82] The nanocomposites were employed for the removal of bacteria in water, and the removal efficiency was found to be 87%.

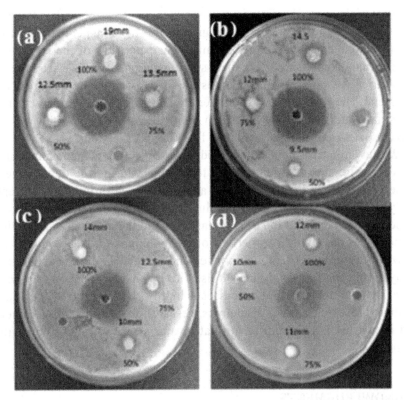

FIGURE 3.18 Antibacterial activity of chitosan/ZnO–polyether sulfone membranes: (a) *Staphylococcus aureus*, (b) *Bacillus cereus*, (c) *Escherichia coli*, and (d) *Salmonella typhi*.

Source: Reprinted with permission from Ref. [80]; copyright © Elsevier.

Even though the nanocomposites exhibit good antimicrobial properties that have been used for treating water, their large-scale production with minimal cost is highly challenging. The following things should be considered, such as (1) nanocomposites showing activity against larger number of microorganisms and (2) evaluation of the antibacterial activity by employing advanced techniques to understand the deeper idea about the mechanism of antibacterial activity.

3.6 CONCLUSIONS AND FUTURE PERSPECTIVES

This chapter highlights the present scenario of employing polymer nano-composites for wastewater treatment and remediation. Application of nanomaterials in various field increases exponentially, day by day. Polymer nanocomposites have great significance in the development of materials science. Excellent physical and chemical characteristics of the partners of polymer nanocomposites make them smart materials. We have seen the use of a variety of polymer nanocomposites for wastewater remediation and treatment. They were found to be very apt for organic dye removal from industrial water and heavy metal ion removal from water/wastewater and also can act as antifouling membrane for water purifications.

Wastewater treatment and remediation by polymer nanocomposites are very safe; special attention has to be taken during the development, reuse potential, and environmental and human health impacts. In-depth analysis has to be done on the influence of polymer chemistry on the dispersion and distribution of nanoparticles and the effect of nanoparticle immobilization on the properties of nanocomposites. The interaction between the polymer nanocomposites and targeted pollutants should be analyzed by modern analytical tools. The extension of laboratory-scale experiments into industrial processes is highly challenging. So, multidisciplinary research among academics, research institutions, government, and industries is essential to mitigate water scarcity issues and related problems.

ACKNOWLEDGMENTS

Bindu M. acknowledges KSCSTE for her Back to Lab Post-Doctoral Fellowship, Grant No. 161/2021/KSCSTE.

KEYWORDS

- **polymer nanocomposites**
- **photocatalysis**
- **dye degradation**
- **nanomaterials**
- **water remediation**
- **environmental pollution**

REFERENCES

1. Zeng, Z.; Liu, J.; Savenije, H. H. G. A Simple Approach to Assess Water Scarcity Integrating Water Quantity and Quality. *Ecol. Indic.* **2013**, *34*, 441–449.

2. Zhou, Q.; Yan, C.; Luo, W. Polypyrrole Coated Secondary Fly Ash–Iron Composites: Novel Floatable Magnetic Adsorbents for the Removal of Chromium(VI) from Wastewater. *Mater. Des.* **2016**, *92*, 701–709.

3. Wen, Y.; Yuan, J.; Ma, X.; Wang, S.; Liu, Y. Polymeric Nanocomposite Membranes for Water Treatment: A Review. *Environ. Chem. Lett.* **2019**, *17* (4), 1539–1551. DOI: 10.1007/s10311-019-00895-9.

4. Mohammed, N.; Grishkewich, N.; Tam, K. C. Cellulose Nanomaterials: Promising Sustainable Nanomaterials for Application in Water/Wastewater Treatment Processes. *Environ. Sci. Nano* **2018**, *5* (3), 623–658. DOI: 10.1039/C7EN01029J.

5. Qu, X.; Alvarez, P. J. J.; Li, Q. Applications of Nanotechnology in Water and Wastewater Treatment. *Water Res.* **2013**, *47* (12), 3931–3946. DOI: 10.1016/j.watres.2012.09.058.

6. Muliwa, A. M.; Leswifi, T. Y.; Onyango, M. S.; Maity, A. Magnetic Adsorption Separation (MAS) Process: An Alternative Method of Extracting Cr(VI) from Aqueous Solution Using Polypyrrole Coated Fe_3O_4 Nanocomposites. *Sep. Purif. Technol.* **2016**, *158*, 250–258. DOI: 10.1016/j.seppur.2015.12.021.

7. Hosseini, S.; Ekramul Mahmud, N. H. M.; Binti Yahya, R.; Ibrahim, F.; Djordjevic, I. Polypyrrole Conducting Polymer and Its Application in Removal of Copper Ions from Aqueous Solution. *Mater. Lett.* **2015**, *149*, 77–80. DOI: 10.1016/j.matlet.2015.02.113.

8. Kyzas, G. Z.; Bikiaris, D. N.; Seredych, M.; Bandosz, T. J.; Deliyanni, E. A. Removal of Dorzolamide from Biomedical Wastewaters with Adsorption onto Graphite Oxide/Poly(Acrylic Acid) Grafted Chitosan Nanocomposite. *Bioresour. Technol.* **2014**, *152*, 399–406. DOI: 10.1016/j.biortech.2013.11.046.

9. Cho, S.; Kim, N.; Lee, S.; Lee, H.; Lee, S. H.; Kim, J.; Choi, J. W. Use of Hybrid Composite Particles Prepared Using Alkoxysilane-Functionalized Amphiphilic Polymer Precursors for Simultaneous Removal of Various Pollutants from Water. *Chemosphere* **2016**, *156*, 302–311. DOI: 10.1016/j.chemosphere.2016.05.004.

10. Lu, T.; Zhang, S.; Qi, D.; Zhang, D.; Zhao, H. Thermosensitive Poly(*N*-Isopropylacrylamide)-Grafted Magnetic Nanoparticles for Efficient Treatment of Emulsified Oily Wastewater. *J. Alloys Compd.* **2016**, *688*, 513–520.

11. Ravishankar, H.; Wang, J.; Shu, L.; Jegatheesan, V. Removal of Pb(II) Ions Using Polymer Based Graphene Oxide Magnetic Nano-Sorbent. *Process Saf. Environ. Prot.* **2016**, *104*, 472–480. DOI: 10.1016/j.psep.2016.04.002.

12. Yao, W.; Ni, T.; Chen, S.; Li, H.; Lu, Y. Graphene/Fe_3O_4@polypyrrole Nanocomposites as a Synergistic Adsorbent for Cr(VI) Ion Removal. *Compos. Sci. Technol.* **2014**, *99*, 15–22. DOI: 10.1016/j.compscitech.2014.05.007.

13. Hasanzadeh, R.; Moghadam, P. N.; Bahri-Laleh, N.; Sillanpää, M. Effective Removal of Toxic Metal Ions from Aqueous Solutions: 2-Bifunctional Magnetic Nanocomposite Base on Novel Reactive PGMA-Man Copolymer@Fe_3O_4 Nanoparticles. *J. Colloid Interface Sci.* **2017**, *490*, 727–746. DOI: 10.1016/j.jcis.2016.11.098.

14. Ali Khan, M. A.; Govindasamy, R.; Ahmad, A.; Siddiqui, M. R.; Alshareef, S. A.; Hakami, A. A. H.; Rafatullah, M. Carbon Based Polymeric Nanocomposites for Dye Adsorption:

Synthesis, Characterization, and Application. *Polymers* **2021**, *13* (3), 419. DOI: 10.3390/polym13030419.

15. Peng, N.; Hu, D.; Zeng, J.; Li, Y.; Liang, L.; Chang, C. Superabsorbent Cellulose-Clay Nanocomposite Hydrogels for Highly Efficient Removal of Dye in Water. *ACS Sustain. Chem. Eng.* **2016**, *4*, 7217–7224.

16. Wang, N.; Chen, J.; Wang, J.; Feng, J.; Yan, W. Removal of Methylene Blue by Polyaniline/TiO$_2$ Hydrate: Adsorption Kinetic, Isotherm and Mechanism Studies. *Powder Technol.* **2019**, *347*, 93–102. DOI: 10.1016/j.powtec.2019.02.049.

17. Gelaw, T. B.; Sarojini, B. K.; Kodoth, A. K. Review of the Advancements on Polymer/Metal Oxide Hybrid Nanocomposite-Based Adsorption Assisted Photocatalytic Materials for Dye Removal. *Chem. Select.* **2021**, *6*, 9300–9310.

18. Srivastava, V.; Choubey, A. K. Investigation of Adsorption of Organic Dyes Present in Wastewater Using Chitosan Beads Immobilized with Biofabricated CuO Nanoparticles. *J. Mol. Struct.* **2021**, *1242*, 130749. DOI: 10.1016/j.molstruc.2021.130749.

19. Liu, L.; Gao, Z. Y.; Su, X. P.; Chen, X.; Jiang, L.; Yao, J. M. Adsorption Removal of Dyes from Single and Binary Solutions Using a Cellulose-Based Bioadsorbent. *ACS Sustainable Chem. Eng.* **2015**, *3* (3), 432–442. DOI: 10.1021/sc500848m.

20. Boukoussa, B.; Hakiki, A.; Moulai, S.; Chikh, K.; Kherroub, D. E.; Bouhadjar, L.; Guedal, D.; Messaoudi, K.; Mokhtar, F.; Hamacha, R. Adsorption Behaviours of Cationic and Anionic Dyes from Aqueous Solution on Nanocomposite Polypyrrole/SBA-15. *J. Mater. Sci.* **2018**, *53* (10), 7372–7386. DOI: 10.1007/s10853-018-2060-7.

21. Ahmed, M. A.; Abdelbar, N. M.; Mohamed, A. A. Molecular Imprinted Chitosan–TiO$_2$ Nanocomposite for the Selective Removal of Rose Bengal from Wastewater. *Int. J. Biol. Macromol.* **2018**, *107* (A), 1046–1053. DOI: 10.1016/j.ijbiomac.2017.09.082.

22. Mostafa, M. H.; Elsawy, M. A.; Darwish, M. S. A.; Hussein, L. I.; Abdaleem, A. H. Microwave-Assisted Preparation of Chitosan/ZnO Nanocomposite and Its Application in Dye Removal. *Mater. Chem. Phys.* **2020**, *248*, 122914. DOI: 10.1016/j.matchemphys.2020.122914.

23. Zaman, A.; Orasugh, J. T.; Banerjee, P.; Dutta, S.; Ali, M. S.; Das, D.; Bhattacharya, A.; Chattopadhyay, D. Facile One-Pot In-Situ Synthesis of Novel Graphene Oxide-Cellulose Nanocomposite for Enhanced Azo Dye Adsorption at Optimized Conditions. *Carbohydr. Polym.* **2020**, *246*, 116661. DOI: 10.1016/j.carbpol.2020.116661.

24. Chinthalapudi, N.; Kommaraju, V. V. D.; Kannan, M. K.; Nalluri, C. B.; Varanasi, S. Composites of Cellulose Nanofibers and Silver Nanoparticles for Malachite Green Dye Removal from Water. *Carbohydr. Polym. Technol. Appl.* **2021**, *2*, 100098. DOI: 10.1016/j.carpta.2021.100098.

25. Heidari, Z.; Pelalak, R.; Malekshah, R. E.; Pishnamazi, M.; Marjani, A.; Sarkar, S. M. Molecular Modeling Investigation on Mechanism of Cationic Dyes Removal from Aqueous Solutions by Mesoporous Materials. *J. Mol. Liq.* **2021**, *329*, 115485.

26. Kausar, A.; Shahzad, R.; Asim, S.; BiBi, S.; Iqbal, J.; Muhammad, N.; Sillanpaa, M.; Din, I. U. Experimental and Theoretical Studies of Rhodamine B Direct Dye Sorption onto Clay-Cellulose Composite. *J. Mol. Liq.* **2021**, *328*, 115165. DOI: 10.1016/j.molliq.2020.115165.

27. Zhou, C. H.; Zhang, D.; Tong, D. S.; Wu, L. M.; Yua, W. H.; Ismadji, S. Paper-Like Composites of Cellulose Acetate-Organo-Montmorillonite for Removal of Hazardous Anionic Dye in Water. *Chem. Eng. J.* **2012**, *209*, 223–234.

28. Mobarak, M.; Mohamed, E. A.; Selim, A. Q.; Eissa, M. F.; Seliem, M. K. Experimental Results and Theoretical Statistical Modeling of Malachite Green Adsorption onto MCM–41 Silica/Rice Husk Composite Modified by Beta Radiation. *J. Mol. Liq.* **2019,** *273*, 68–82. DOI: 10.1016/j.molliq.2018.09.132.

29. Ameen, S.; Akhtar, M. S.; Kim, Y. S.; Shin, H. S. Nanocomposites of Poly(1-Naphthylamine)/SiO_2 and Poly(1-Naphthylamine)/TiO_2: Comparative Photocatalytic Activity Evaluation Towards Methylene Blue Dye. *Appl. Catal. B* **2011,** *103*, 136–142.

30. Zhang, H.; Zong, R. L.; Zhao, J. A.; Zhu, Y. F. Dramatic Visible Photocatalytic Degradation Performances Due to Synergetic Effect of TiO_2 with PANI. *Environ. Sci. Technol.* **2008,** *42* (10), 3803–3807. DOI: 10.1021/es703037x.

31. Shaikh, T.; Rathore, A.; Kaur, H.; Poly(Lactic Acid) Grafting of TiO_2 Nanoparticles: A Shift in Dye Degradation Performance of TiO_2 from UV to Solar Light. *Chem. Select.* **2017,** *2*, 6901–6908.

32. Karagoz, S.; Kiremitler, N. B.; Sakir, M.; Salem, S.; Onses, M. S.; Sahmetlioglu, E.; Ceylan, A.; Yilmaz, E. Synthesis of Ag and TiO_2 Modified Polycaprolactone Electrospun Nanofibers (PCL/TiO_2-Ag NFs) as a Multifunctional Material for SERS, Photocatalysis and Antibacterial Applications. *Ecotoxicol. Environ. Saf.* **2020,** *188*, 109856. DOI: 10.1016/j.ecoenv.2019.109856.

33. Alahmad, W. R.; Alawi, M. A. Photocatalytic Degradation of Diclofenac and Ibuprofen from Simulated Wastewater Using SiO_2–TiO_2–(Ru, N) by Artificial Light. *Fresenius Environ. Bull.* **2016,** *25*, 4299–4308.

34. Larios, L. M.; Mondragón, R. M.; Orozco, R. D. M.; García, U. P.; Rivas, N. V. G.; Alamilla, R. G. Electrochemically-Assisted Fabrication of Titanium-Dioxide/Polyaniline Nanocomposite Films for the Electroremediation of Congo Red in Aqueous Effluents. *Synth. Met.* **2020,** *268*, 116464.

35. Reddy, K. R.; Karthik, K. V.; Prasad, S. B. B.; Soni, S. K.; Jeong, H. M.; Raghu, A. V. Enhanced Photocatalytic Activity of Nanostructured Titanium Dioxide/Polyaniline Hybrid Photocatalysts. *Polyhedron* **2016,** *120*, 169–174. DOI: 10.1016/j.poly.2016.08.029.

36. Yew, S. P.; Tang, H. Y.; Sudesh, K. Photocatalytic Activity and Biodegradation of Polyhydroxybutyrate Films Containing Titanium Dioxide. *Polym. Degrad. Stab.* **2006,** *91* (8), 1800–1807. DOI: 10.1016/j.polymdegradstab.2005.11.011.

37. Iketani, K.; Sun, R. D.; Toki, M.; Hirota, K.; Yamaguchi, O. Sol–Gel-Derived TiO_2/Poly(Dimethylsiloxane) Hybrid Films and Their Photocatalytic Activities. *J. Phys. Chem. Solids* **2003,** *64* (3), 507–513. DOI: 10.1016/S0022-3697(02)00357-8.

38. Wang, D. S.; Zhang, J.; Luo, Q.; Li, X. Y.; Duan, Y.; An, J. Characterization and Photocatalytic Activity of Poly(3-Hexylthiophene)-Modified TiO_2 for Degradation of Methyl Orange Under Visible Light. *J. Hazard. Mater.* **2009,** *169* (1–3), 546–550. DOI: 10.1016/j.jhazmat.2009.03.135.

39. Naskar, S.; Arumugom Pillay, S. A.; Chanda, M. Photocatalytic Degradation of Organic Dyes in Aqueous Solution with TiO_2 Nanoparticles Immobilized on Foamed Polyethylene Sheet. *J. Photochem. Photobiol. A* **1998,** *113* (3), 257–264. DOI: 10.1016/S1010-6030(97)00258-X.

40. Zhu, H.; Jiang, R.; Xiao, L.; Chang, Y.; Guan, Y.; Li, X.; Zeng, G. Photocatalytic Decolorization and Degradation of Congo Red on Innovative Crosslinked Chitosan/Nano-CdS Composite Catalyst Under Visible Light Irradiation. *J. Hazard. Mater.* **2009,** *169* (1–3), 933–940. DOI: 10.1016/j.jhazmat.2009.04.037.

41. Guan, Y.; Yu, H. Y.; Abdalkarim, S. Y. H.; Wang, C.; Tang, F.; Marek, J.; Chen, W. L.; Militky, J.; Yao, J. M. Green One-Step Synthesis of ZnO/Cellulose Nanocrystal Hybrids with Modulated Morphologies and Superfast Absorption of Cationic Dyes. *Int. J. Biol. Macromol.* **2019**, *132*, 51–62. DOI: 10.1016/j.ijbiomac.2019.03.104.

42. Mohammed, M. I.; Khafagy, R. M.; Hussien, M. S. A.; Sakr, G. B.; Ibrahim, M. A.; Yahia, I. S.; Zahran, H. Y. Enhancing the Structural, Optical, Electrical, Properties and Photocatalytic Applications of ZnO/PMMA Nanocomposite Membranes: Towards Multifunctional Membranes. *J. Mater. Sci. Mater. Electron.* **2022**, *33*, 1977–2002.

43. Hussein, M. A.; Albeladi, H. K.; Elsherbiny, A. S.; El-Shishtawy, R. M.; A. N. Al-Romaizan, Cross-Linked Poly(Methyl Methacrylate)/Multiwall Carbon Nanotube Nanocomposites for Environmental Treatment. *Adv. Polym. Technol.* **2018**, *37*, 3240–3251.

44. Senguttuvan, S.; Janaki, V.; Senthilkumar, P.; Kamala-Kannan, S. Polypyrrole/Zeolite Composite-A Nanoadsorbent for Reactive Dyes Removal from Synthetic Solution. *Chemosphere* **2022**, *287*, 132164. DOI: 10.1016/j.chemosphere.2021.132164.

45. Zhou, G.; Wang, K. P.; Liu, H. W.; Wang, L.; Xiao, X. F.; Dou, D. D.; Fan, Y. B. Three-Dimensional Polylactic Acid@Graphene Oxide/Chitosan Sponge Bionic Filter: Highly Efficient Adsorption of Crystal Violet Dye. *Int. J. Biol. Macromol.* **2018**, *113*, 792–803. DOI: 10.1016/j.ijbiomac.2018.02.017.

46. Saberi, A.; Alipour, E.; Sadeghi, M. Superabsorbent Magnetic Fe_3O_4-Based Starch-Poly(Acrylic Acid) Nanocomposite Hydrogel for Efficient Removal of Dyes and Heavy Metal Ions from Water. *J. Polym. Res.* **2019**, *26*, 271.

47. Samadder, R.; Akter, N.; Roy, A. C.; Uddin, M. M.; Hossen, M. J.; Azam, M. S. Magnetic Nanocomposite Based on Polyacrylic Acid and Carboxylated Cellulose Nanocrystal for the Removal of Cationic Dye. *RSC Adv.* **2020**, *10* (20), 11945–11956. DOI: 10.1039/d0ra00604a.

48. Vaiano, V.; Sacco, O.; Sannino, D.; Ciambelli, P.; Longo, S.; Venditto, V.; Guerra, G. N. Doped TiO_2/s-PS Aerogels for Photocatalytic Degradation of Organic Dyes in Wastewater Under Visible Light Irradiation. *J. Chem. Technol. Biotechnol.* **2014**, *89*, 1175–1181.

49. Pathania, D.; Gupta, D.; Al-Muhtaseb, A. H.; Sharma, G.; Kumar, A.; Naushad, M.; Ahamad, T.; Alshehri, S. M. Photocatalytic Degradation of Highly Toxic Dyes Using Chitosan-g-Poly(Acrylamide)/ZnS in Presence of Solar Irradiation. *J. Photochem. Photobiol. A* **2016**, *329*, 61–68. DOI: 10.1016/j.jphotochem.2016.06.019.

50. He, M. Q.; Bao, L. L.; Sun, K. Y.; Zhao, D. X.; Li, W. B.; Xia, J. X.; Li, H. M. Synthesis of Molecularly Imprinted Polypyrrole/Titanium Dioxide Nanocomposites and Its Selective Photocatalytic Degradation of Rhodamine B Under Visible Light Irradiation. *Express Polym. Lett.* Lett. **2014**, *8* (11), 850–861. DOI: 10.3144/expresspolymlett.2014.86.

51. Jung, G.; Kim, H. I. Synthesis and Photocatalytic Performance of PVA/TiO_2/Graphene-MWCNT Nanocomposites for Dye Removal. *J. Appl. Polym. Sci.* **2014**, *131* (17), n/a–n/a. DOI: 10.1002/app.40715.

52. Hir, Z. A. M.; Moradihamedani, P.; Abdullah, A. H.; Mohamed, M. A. Immobilization of TiO_2 into Polyethersulfone Matrix as Hybrid Film Photocatalyst for Effective Degradation of Methyl Orange Dye. *Mater. Sci. Semicond. Process.* **2017**, *57*, 157–165. DOI: 10.1016/j.mssp.2016.10.009.

53. Haspulat, B.; Gülce, A.; Gülce, H. Efficient Photocatalytic Decolorization of Some Textile Dyes Using Fe Ions Doped Polyaniline Film on ITO Coated Glass Substrate. *J. Hazard. Mater.* **2013**, *260*, 518–526. DOI: 10.1016/j.jhazmat.2013.06.011.

54. Chibuike, G. U.; Obiora, S. C. Heavy Metal Polluted Soils: Effect on Plants and Bioremediation Methods. *Appl. Environ. Soil Sci.* **2014**, *2014*, 1–12. DOI: 10.1155/2014/752708.

55. Taseidifar, M.; Makavipour, F.; Pashley, R. M.; Rahman, A. F. M. M. Removal of Heavy Metal Ions from Water Using Ion Flotation. *Environ. Technol. Innov.* **2017**, *8*, 182–190. DOI: 10.1016/j.eti.2017.07.002.

56. García-Niño, W. R.; Pedraza-Chaverrí, J. Protective Effect of Curcumin Against Heavy Metals-Induced Liver Damage. *Food Chem. Toxicol.* **2014**, *69*, 182–201. DOI: 10.1016/j.fct.2014.04.016.

57. Zaidi, M. I.; Asrar, A. A.; Mansoor, A. M.; Farooqui, M. A. F. Heavy Metal Concentration Along Road Site Trees of Quetta and Its Effect on Public Health. *J. Appl. Sci.* **2005**, *5* (4), 708–711. DOI: 10.3923/jas.2005.708.711.

58. Gaetke, L. M.; Chow, C. K. Copper Toxicity, Oxidative Stress, and Antioxidant Nutrients. *Toxicology* **2003**, *189* (1–2), 147–163. DOI: 10.1016/s0300-483x(03)00159-8.

59. Yang, X.; Wan, Y.; Zheng, Y.; He, F.; Yu, Z.; Huang, J.; Wang, H.; Ok, Y. S.; Jiang, Y.; Gao, B. Surface Functional Groups of Carbon-Based Adsorbents and Their Roles in the Removal of Heavy Metals from Aqueous Solutions: A Critical Review. *Chem. Eng. J.* **2019**, *366*, 608–621. DOI: 10.1016/j.cej.2019.02.119.

60. Naef, A. A. Q.; Mohammed, R. H.; Lawal, D. U. Removal of Heavy Metal Ions from Wastewater: A Comprehensive and Critical Review. *NPJ Clean Water* **2021**, *4*, 36.

61. Demiral, İ.; Samdan, C.; Demiral, H. Enrichment of the Surface Functional Groups of Activated Carbon by Modification Method. *Surf. Interfaces* **2021**, *22*, 100873. DOI: 10.1016/j.surfin.2020.100873.

62. Kumar, A. S. K.; Jiang, S. J.; Tseng, W. L. Effective Adsorption of Chromium(VI)/Cr(III) from Aqueous Solution Using Ionic Liquid Functionalized Multiwalled Carbon Nanotubes as a Super Sorbent. *J. Mater. Chem. A* **2015**, *3* (9), 7044–7057.

63. Martínez-Huitle, C. A.; Panizza, M. Electrochemical Oxidation of Organic Pollutants for Wastewater Treatment. *Curr. Opin. Electrochem.* **2018**, *11*, 62–71. DOI: 10.1016/j.coelec.2018.07.010.

64. Saad, A. H. A.; Azzam, A. M.; El-Wakeel, S. T.; Mostafa, B. B.; M. B. A. El-Latifb. *Environ. Nanotechnol. Monit. Manage.* **2018**, *9*, 67–75.

65. da Silva Alves, D. C.; Healy, B.; Pinto, L. A. A.; Cadaval, T. R. S.; Breslin, C. B. Recent Developments in Chitosan-Based Adsorbents for the Removal of Pollutants from Aqueous Environments. *Molecules* **2021**, *26* (3), 594. DOI: 10.3390/molecules26030594.

66. Liu, T.; Zhao, L.; Sun, D.; Tan, X. Entrapment of Nanoscale Zero-Valent Iron in Chitosan Beads for Hexavalent Chromium Removal from Wastewater. *J. Hazard. Mater.* **2010**, *184* (1–3), 724–730. DOI: 10.1016/j.jhazmat.2010.08.099.

67. Mohammadnezhad, G.; Abad, S.; Soltani, R.; Dinari, M. Study on Thermal, Mechanical and Adsorption Properties of Amine-Functionalized MCM-41/PMMA and MCM-41/PMMA Nanocomposites Prepared by Ultrasonic Irradiation. *Ultrason. Sonochem.* **2017**, *39*, 765–773.

68. Dinari, M.; Soltani, R.; Mohammadnezhad, G. Kinetics and Thermodynamic Study on Novel Modified-Mesoporous Silica MCM-41/Polymer Matrix Nanocomposites: Effective Adsorbents for Trace Cr(VI) Removal. *J. Chem. Eng. Data* **2017**, *62*, 2316–2329.

69. Zhang, Q.; Pan, B.; Zhang, S.; Wang, J.; Zhang, W.; Lv, L. *J. Nanopart. Res.* **2011**, *13*, 5355.

70. Li, R.; Liu, L.; Yang, F. Preparation of Polyaniline/Reduced Graphene Oxide Nanocomposite and Its Application in Adsorption of Aqueous Hg(II). *Chem. Eng. J.* **2013**, *229*, 460–468. DOI: 10.1016/j.cej.2013.05.089.

71. Samuel, M. S.; Shah, S. S.; Subramaniyan, V.; Qureshi, T.; Bhattacharya, J.; Pradeep Singh, N. D. Preparation of Graphene Oxide/ Chitosan/Ferrite Nanocomposite for

Chromium(VI) Removal from Aqueous Solution. *Int. J. Biol. Macromol.* **2018**, *119*, 540–547. DOI: 10.1016/j.ijbiomac.2018.07.052.

72. Fallah, Z.; Isfahani, H. N.; Tajbakhsh, M.; Tashakkorian, H.; Amouei, A. TiO_2-Grafted Cellulose via Click Reaction: An Efficient Heavy Metal Ions Bioadsorbent from Aqueous Solutions. *Cellulose* **2017**, *25*, 639–660.

73. Moreno-Sader, K.; García-Padilla, A.; Realpe, A.; Acevedo-Morantes, M.; Soares, J. B. P. Removal of Heavy Metal Water Pollutants (Co^{2+} and Ni^{2+}) Using Polyacrylamide/Sodium Montmorillonite (PAM/Na-MMT) Nanocomposites. *ACS Omega* **2019**, *4* (6), 10834–10844. DOI: 10.1021/acsomega.9b00981.

74. Mohammadnezhad, G.; Moshiri, P.; Dinari, M.; Steiniger, F. In Situ Synthesis of Nanocomposite Materials Based on Modified Mesoporous Silica MCM-41 and Methyl Methacrylate for Copper (II) Adsorption from Aqueous Solution. *J. Iran. Chem. Soc.* **2019**, *16*, 1491–1500.

75. Ahmad, R.; Hasan, I. Efficient Remediation of an Aquatic Environment Contaminated by Cr(VI) and 2, 4-Dinitrophenol by XG-g-polyaniline@ZnO Nanocomposite. *J. Chem. Eng. Data* **2017**, *62* (5), 1594–1607. DOI: 10.1021/acs.jced.6b00963.

76. Khan, S. A.; Siddiqui, M. F.; Khan, T. A. Ultrasonic-Assisted Synthesis of Polyacrylamide/Bentonite Hydrogel Nanocomposite for the Sequestration of Lead and Cadmium from Aqueous Phase: Equilibrium, Kinetics and Thermodynamic Studies. *Ultrason. Sonochem.* **2020**, *60*, 104761. DOI: 10.1016/j.ultsonch.2019.104761.

77. Shoukat, A.; Wahid, F.; Khan, T.; Siddique, M.; Nasreen, S.; Yang, G.; Ullah, M. W.; Khan, R. Titanium Oxide-Bacterial Cellulose Bioadsorbent for the Removal of Lead Ions from Aqueous Solution. *Int. J. Biol. Macromol.* **2019**, *129*, 965–971. DOI: 10.1016/j.ijbiomac.2019.02.032.

78. Saberi, A.; Alipour, E.; Sadeghi, M. Superabsorbent Magnetic Fe_3O_4-Based Starch-Poly(Acrylic Acid) Nanocomposite Hydrogel for Efficient Removal of Dyes and Heavy Metal Ions from Water. *J. Polym. Res.* **2019**, *26*, 271.

79. Chen, L.; Peng, X. Silver Nanoparticle Decorated Cellulose Nanofibrous Membrane with Good Antibacterial Ability and High-Water Permeability. *Appl. Mater. Today* **2017**, *9*, 130–135.

80. Munnawar, I.; Iqbal, S. S.; Anwar, M. N.; Batool, M.; Tariq, S.; Faitma, N.; Khan, A. L.; Khan, A. U.; Nazar, U.; Jamil, T.; Ahmad, N. M. Synergistic Effect of Chitosan-Zinc Oxide Hybrid Nanoparticles on Antibiofouling and Water Disinfection of Mixed Matrix Polyethersulfone Nanocomposite Membranes. *Carbohydr. Polym.* **2017**, *175*, 661–670. DOI: 10.1016/j.carbpol.2017.08.036.

81. Liu, L.; Zhang, Y.; Li, C.; Cao, J.; He, E.; Wu, X.; Wang, F.; Wang, L. Facile Preparation PCL/Modified Nano ZnO Organic–Inorganic Composite and Its Application in Antibacterial Materials. *J. Polym. Res.* **2020**, *27* (3), 78. DOI: 10.1007/s10965-020-02046-z.

82. Salam, M. A.; Obaid, A. Y.; El-Shishtawy, R. M.; Mohamed, S. A. Synthesis of Nanocomposites of Polypyrrole/Carbon Nanotubes/Silver Nano Particles and Their Application in Water Disinfection. *RSC Adv.* **2017**, *7*, 16878–16884.

CHAPTER 4

Porous Polymer-Inorganic Nanocomposites Based on Natural Alumosilicate Minerals of Ukraine for Wastewater Treatment from Toxic Metal Ions

I. SAVCHENKO[1], E. YANOVSKA[1], L. VRETIK[1], D. STERNIK[2], and O. KYCHKYRUK[3]

[1]Taras National Taras Shevchenko University of Kyiv, Kyiv, Ukraine

[2]Maria Curie-Skłodowska University, Lublin, Poland

[3]Ivan Franko Zhytomyr State University, Zhytomyr, Ukraine

ABSTRACT

This chapter presented the structure, classification, and chemical composition of various organo-mineral composite materials and a descriptive analysis of their adsorption behavior. This work is devoted to the synthesis and research of the adsorption properties of composite materials based on silica and aluminosilicate minerals of Ukraine immobilized by different methods in their surface layer with polymers of different chemical natures in relation to toxic metal ions. The main factors influencing the adsorption properties of such organo-mineral composites are considered, namely, the chemical nature of the polymer, the inorganic carrier, and the method of fixing the polymer on the surface. This chapter confirmed that modified forms of clay minerals have excellent feasibility in removing different toxic aquatic metal pollutants.

Mechanics and Physics of Porous Materials: Novel Processing Technologies and Emerging Applications.
Chin Hua Chia, Tamara Tatrishvili, Ann Rose Abraham, & A. K. Haghi (Eds.)

4.1 INTRODUCTION

The constant increase in the volume of polluted natural and wastewater due to anthropogenic influence requires the search for new safe methods of their treatment. Existing chemical and physicochemical methods of cleaning polluted water (chlorination, ozonation, osmosis, etc.), consisting of active chemical action or physical impact on water, allow to remove pollutants from it. At the same time, the physical and chemical properties of water deteriorate and the natural balance of salts dissolved in it is disturbed. The set of these factors has already begun to have a significant impact on the state of local water ecosystems and the Earth's hydrosphere as a whole.

At the same time, in nature, there are safe methods of water purification with equalization of the salt balance in it when water passes through surface and underground horizons of minerals that have great adsorption properties for anthropogenic toxic substances (clays, aluminosilicates, zeolites, etc.). Such a mechanism worked and ensured a balance between the planet's geospheres throughout their existence. Accordingly, its reasonable use should form the basis of the most modern technologies of water purification. The addition of such natural mineral adsorbents during the purification of wastewater and polluted natural waters at the settling stage allows to get rid of dangerous anthropogenic pollutants by adsorption without adding chemical reagents and to improve the structure and mineralization of water.[1]

As ecologically safe industrial adsorbents-cleaners, the use of natural aluminosilicate minerals of Ukraine, such as saponite (the largest deposits are in the Khmelnytskyi region of Ukraine), Transcarpathian clinoptilolites of the Sokyrnytskyi and Tushinsky deposits, bentonite, vermiculite, colored clays of the Luhansk region and Crimea, is promising. These minerals are widely known for their adsorption properties for a number of toxic organic substances, toxic metal ions, pathogenic bacteria, and so on.[2-8] However, their natural properties can be improved, in particular, by fixing on their surface homo- or copolymers of different chemical natures, which are able to participate in the processes of complex formation or ion exchange, resulting in composite materials with valuable sorption properties.

The main drawback of the development of new wastewater treatment technologies, which include adsorption processes, is the emergence of a new type of waste—the solid phase of the adsorbent. In our opinion, the insufficient practical use of Sokyrnytsky clinoptilolite, whose deposits have long been industrially developed, is caused by this very reason. Today, the question of the patterns and conditions of possible desorption of toxic ions and substances from the surface of this zeolite remains open, and as a result, the safety conditions of its burial and storage with adsorbed toxicants are unclear.

In Ref. [1], we showed that the granular fraction of Sokyrnytskyi clinoptilolite without additional thermal or chemical treatment can be used for further purification of wastewater from heavy metal cations that enter biological treatment facilities after treatment by physicochemical methods. This process occurs by adding clinoptilolite to water and contact in a static mode with periodic movement during one day from Mn(II), Cu(II), and Pb(II), during 5 days—from Fe(III) and Zn(II). After that, the solid phase of clinoptilolite can be used as a component of hard road surface and in technical construction, since the adsorbed ions of toxic metals are practically not washed off the surface of this zeolite by precipitation (rain and melt water).

In Ref. [8], we found that Podilsky saponite and clinoptilolites of the Sokyrnytskyi and Tushinsky deposits show adsorption activity for oxoanions Cr(VI), Mo(VI), W(VI), and As(V) and can be effectively used to clean polluted natural water bodies (lakes, ponds, small rivers) and sewage sludge.

The main methods of fixing polymers on the surface of inorganic matrices are physical and chemical fixation, sol–gel method, and *in-situ* immobilization. Each of these methods has its advantages and disadvantages.

Fixation on the surface of porous inorganic matrices of polyelectrolytes with a quaternary nitrogen atom either in the side chain (polycations) or in the main chain (polyionenes) allows to obtain effective adsorbents with ion-exchange properties for anionic forms of such anthropogenic inorganic pollutants as acid residues of mineral acids (in particular, nitrates, phosphates, arsenates) or oxoanions of multivalent metals (chromates, molybdates, tung-states, vanadates, and their polyanions).[9]

In Refs. [10–12], we showed that *in-situ* immobilization of the surface of saponite from the Tashkiv deposit (Khmelnytskyi region, Ukraine) and clinoptilolite from the Sokyrnytsky deposit (Transcarpathia, Ukraine) with polyaniline leads to an increase in the sorption capacity of these minerals for phosphate and arsenate ions in neutral aqueous environment in 2–3 times.

Immobilization of polymers with nitrogen-, sulfur-, or oxygen-containing groups in the main chain, which can be ligands, allows to obtain adsorbents with complexing properties, which effectively remove cations of transition toxic metals from wastewater.

4.2   EXPERIMENTAL METHODS AND MATERIALS

4.2.1   METHODS

4.2.1.1   FTIR SPECTRA

FTIR spectra of the samples of composite and the original mineral were recorded using an IR spectrometer with Fourier transformation (Thermo Nicolet Nexus FT-IR, USA). For this purpose, the samples were ground in an agate mortar and pressed with KBr. The FTIR spectra were recorded in the spectral range of 500–4000 cm^{-1} with 16 scans per spectrum at a resolution of 4 cm^{-1}.

4.2.1.2   THERMAL ANALYSIS

The amount of immobilized polymer on the surface of the modified saponite was evaluated by thermogravimetric analysis, which was obtained on a synchronous TG/DTA analyzer "Shimadzu DTG-60 H" (Shimadzu, Japan) in the temperature range 15–1000°C. The heating rate of the samples was 10°/min.

Differential scanning calorimetry was performed on an instrument "STA 449 Jupiter F1" (Netzsch, Germany) with a mass spectroscopic attachment "QMS 403C" (Germany).

4.2.1.3   LOW-TEMPERATURE ADSORPTION–DESORPTION OF NITROGEN

The values of the specific surface area and the average pore diameter were calculated from the isotherms of low-temperature adsorption–desorption of nitrogen using sorbometer software "ASAP 2420 V1.01" (Micromeritics, USA). Before measurement, the samples were degassed at 100°C for 24 h.

4.2.1.4   SURFACE MORPHOLOGY ANALYSIS

The surface morphology of composite was observed by using a scanning electron microscope (SEM, LEO 1430VP, Carl Zeiss, Germany).

4.2.1.5   DESCRIPTION OF ADSORPTION PROCESSES

Properties of the obtained composite to adsorb Cu(II), Pb(II), Cd(II), and Fe(III) were studied in static mode. Working solutions of nitrates of the corresponding metals were prepared in volumetric flasks of 25, 50, or 100 ml, diluting the solutions to the mark with water, and then added the required volume to flat-bottomed flasks containing 0.1 g of adsorbent. The reaction was proceeding while the flasks were shaken mechanically. Equilibrium concentrations of ions were measured using atomic absorption methods.

Working nitrate solutions of Cu(II), Pb(II), and Fe(III) are prepared with the sets of "standard sample solutions" of these salts on 1 M HNO_3 background (produced by A.V. Bogatsky FHI in Odesa) with concentrations of 1 and 10 mg/ml.

4.2.1.6 ANALYSIS OF ION CONCENTRATIONS

Sorption capacity (A) was calculated by the following formula:

$$A = (c_0 - [M])V/m \tag{4.1}$$

where c_0 is the concentration of the metal in the starting solution, $[M]$ is the concentration of the metal at equilibrium, V is the volume of the starting solution, and m is the mass of the composite.

Atomic absorption spectrophotometer "C115-M" (Ukraine) in the flame of a mixture of "acetylene–air" was used to measure the equilibrium concentration of copper(II), cadmium(II), lead(II), and iron(III). Equilibrium concentration in solutions was calculated by comparing the intensities of spectral lines with those for standard solutions. Maxima wavelengths were 324.7 nm for Cu(II), 228.8 nm for Cd(II), 283.3 nm for Pb(II), and 248.3 nm for Fe(III), aperture being 0.5 cm wide.

4.3 RESULTS AND DISCUSSION

4.3.1 SYNTHESIS, STUDY OF THE STRUCTURE, AND SORPTION PROPERTIES OF A COMPOSITE BASED ON CLINOPTILOLITE OF THE TUSHINSKY DEPOSIT WITH IN-SITU IMMOBILIZED POLY [N-(4-CARBOXYPHENYL)METHACRYLAMIDE] WITH RESPECT TO CU(II), PB(II), AND CD(II) IONS

Poly[*n*-(4-carboxyphenyl)methacrylamide]-clinoptilolite composite (hereafter clinoptilolite]–PKSF-MKA) was synthesized by radical polymerization of 4-carboxyphenylmethacrylamide, carried out in the presence of clinoptilolite from the Tushinsky deposit.[13] The scheme of *in-situ* immobilization of poly[*n*-(4-carboxyphenyl)methacrylamide] on the surface of clinoptilolite is presented in Figure 4.1.

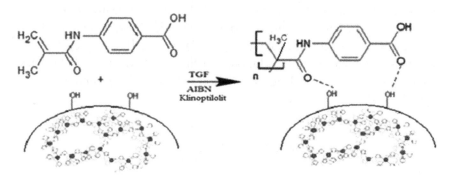

FIGURE 4.1   Reaction scheme of *in-situ* polymerization of 4-carboxyphenylmethacrylamide on the surface of clinoptilolite.

The fact of *in-situ* polymerization on the surface of clinoptilolite was established by means of a comparative analysis of the IR spectra of the original (a) and modified (b) clinoptilolite (Figure 4.2). The comparative analysis of the IR spectra proves that in the spectrum of the composite clino-ptilolite]–PKSF-MKA, on unlike the original clinoptilolite, a broad absorption band is observed in the region of 3400 cm^{-1}, which appeared as a result of the formation of a hydrogen bond between the polymer and clinoptilolite. The absorption band in the region of 2930 cm^{-1} can be attributed to valence vibrations υ(C–H) bonds. The band in the region of 1660 cm^{-1} is due to υ(C=O) bonds and in the region of 1510 cm^{-1} due to the vibrations of the amide system "2" (υ(C–N) + υ(NH). In the spectrum of the composite in the region from 1410 to 1180 cm^{-1}, there are intense absorption bands that can be attributed to the vibrations of the aromatic system of the polymer,[16] and the band around 800 cm^{-1} to the deformation vibrations of the CH– bonds of the polymer. Also, both spectra show an intense absorption band in the region of 1080 cm^{-1}—this is υ(Si–O) (please refer to Figure 4.2).

In order to determine the amount of immobilized polymer, a thermogravi-metric analysis of the original clinoptilolite and the synthesized clinoptilo-lite]–PKSF-MKA composite was carried out. The resulting thermograms are shown in Figures 4.3 and 4.4. From the given thermograms, it can be seen that the majority of the immobilized polymer decomposes in the temperature range from 100 to 500°C. At the same time, 18.6 ± 0.1 wt% of the mass of the composite is lost. Taking into account the fact that the mass loss of the original clinoptilolite in this temperature range is 8.9 ± 0.1 wt%, the mass fraction of the immobilized polymer in the synthesized composite was calculated, which is 9.7 ± 0.1 wt%.

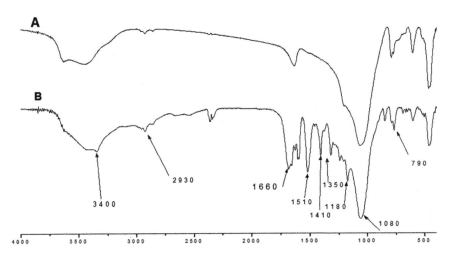

FIGURE 4.2 IR spectra of the original clinoptilolite (a) and the clinoptilolite]–PKSF-MKA composite (b).

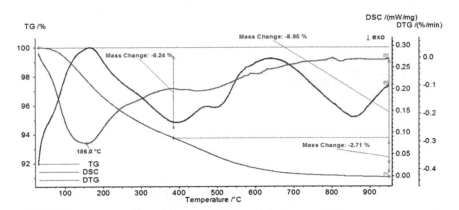

FIGURE 4.3 Thermogram of the original clinoptilolite.

Figure 4.5a shows the thermogram of the clinoptilolite]–PKSF-MKA composite combined with the mass spectrum in 3D format, and Figure 4.5b shows the mass spectrum of the synthesized composite in 2D format. Mass spectral studies show that thermal degradation of the polymer is mainly observed at a temperature higher than 400°C. At the same time, the main products with a relative mass of 28 Da are CO and (or) N_2, 44—CO_2, 18—water.

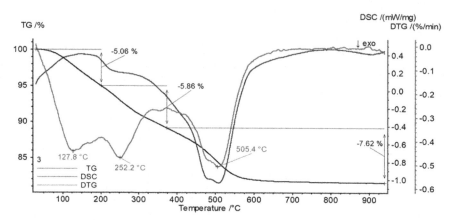

FIGURE 4.4 Thermogram of the clinoptilolite]–PKSF-MKA composite.

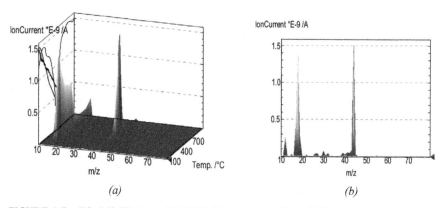

FIGURE 4.5 TG–MS-3D (a) and MS-2D (b) composite clinoptilolite]–PKSF-MCA.

The number of carboxyl groups in the composition of poly[N-(4-carboxyphenyl)methacrylamide] immobilized on the surface of clinoptilolite was determined by acid–base titration with 0.01 N NaOH. According to the results of the calculations of the obtained experimental titration data (excluding the titration data of the original clinoptilolite), it was found that the number of carboxyl groups, which is equal to the number of active ones in terms of participation in the complexation of the centers of the immobilized polymer in the composition of the composite, is 0.17 mmol/g.

To study the surface morphology of the clinoptilolite]–PKSF-MKA composite, nitrogen adsorption–desorption isotherms of the original clinoptilolite and composite were obtained. The obtained isotherms are similar and belong to type III isotherms according to the *IUPAC* classification.[14] The obtained data indicate that the immobilized polymer practically does not affect the structure of the surface layer of clinoptilolite.

A detailed comparison of the surface parameters of the original clinoptilolite and the synthesized composite (clinoptilolite]–PKSF-MKA) based on the conducted research and literature data is presented in Table 4.1. From the data in this table, it can be concluded that

- the surface area after immobilization is insignificantly reduced, which is a logical result of fixation on the surface of a nonporous polymer;
- pores with a size of 30–40 nm prevail in this composite. Therefore, after polymer modification, the surface of clinoptilolite has a macroporous character;
- polymer modification "smoothes out" the pores of the original mineral, reducing their depth and expanding them. This is evidenced by a simultaneous 2.5-fold increase in the size of pores and a halving of their volume.

TABLE 4.1 Comparison of Surface Parameters of the Original Clinoptilolite and Composites Synthesized on Its Basis.

Sample	$S_{specific\ surface}$ (m²/g)			Average pore volume (cm³/g)	Average pore size (nm)
	According to the isotherm	BJH	Langmuir		
The original clinoptilolite	13	23	14	0.04	13
Clinoptilolite]– PKSF-MKA	3	3	5	0.02	32

Changes in the surface morphology of clinoptilolite after modification with poly[*p*-(4-carboxyphenyl)methacrylamide] were monitored using scanning electron microscopy (SEM). SEM photo of the original clinoptilolite and clinoptilolite]–PKSF-MCA is shown in Figure 4.6. Analysis of these photographs confirms the above assumption about changes in the porosity of the original mineral as a result of surface modification with the selected polymer. At the same time, the method of placing the polymer on the surface can be attributed to the "film" method, which is the best option for placing the polymer on a solid surface for participation in complexation processes.

FIGURE 4.6   SEM image of the original clinoptilolite (magnification 5000×) (a) and the clinoptilolite]–PKSF-MCA composite (magnification 50,000) (b).

Sorption characteristics of the clinoptilolite]–PKSF-MKA composite with respect to Cu(II), Cd(II), Pb(II), Mn(II), and Fe(III) ions were studied in static mode according to the method described in detail in Ref. [13]. When the synthesized composite came into contact with $NaHCO_3$ solution (pH 8.1), partial washing of the immobilized polymer was observed, as a result of which the solution took on a yellow color.

The results of studies of the sorption capacity of the clinoptilolite]–PKSF-MKA composite for Cu(II), Pb(II), Mn(II), Fe(III), and Cd(II) ions at different values of acidity and chemical composition of the medium proved that the synthesized composite shows the highest sorption activity for trace amounts of all investigated metals in a neutral environment without the addition of buffer solutions. However, the maximum sorption of Fe(III) ions was recorded in a weakly acidic environment in the presence of a phthalate buffer. Therefore, further study of the sorption properties of the clinoptilolite]–PKSF-MKA composite was carried out with initial solutions of Cu(II), Pb(II), and Cd(II) salts without adding buffers.

As a result of studies of the dependence of the degree of sorption of Cu(II), Pb(II), and Cd(II) ions on the surface of the clinoptilolite]–PKSF-MKA composite on the contact time in the static mode, it was found that all studied ions are maximally sorbed within 30 min of contact. After that, the sorption equilibrium is established and the degree of sorption changes slightly. In order to establish the values of the sorption capacity of the modified clinoptilolite with respect to selected transition metal ions, their sorption isotherms were constructed and compared with the isotherms for the original clinoptilolite. The obtained Cd(II) ion sorption isotherms are presented in Figure 4.7 as an example.

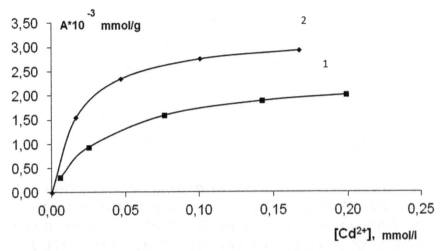

FIGURE 4.7 Cd(II) ion sorption isotherm on the original clinoptilolite (1) and the clinoptilo-lite]–PKSF-MKA composite (2).

Based on the obtained isotherms, a preliminary conclusion can be made that the sorption capacity of clinoptilolite after modification with poly[*N*-(4-carboxyphenyl) methacrylamide] increases with respect to all studied ions. This phenomenon is most likely explained by complexation processes with nitrogen or oxygen atoms of the immobilized polymer. All obtained isotherms belong to the 2L type, which allows to calculate the sorption capacity of the composite with respect to the selected metal ions.

The sorption capacity of the modified mineral with respect to each of the studied transition metal ions was calculated from the data of the sorption isotherms and presented in Table 4.2. The data in Table 4.2 allow us to assert that the sorption capacity of clinoptilolite after *in-situ* modification with poly[*N*-(4-carboxyphenyl)methacrylamide] in the molar ratio increases by 1.9 times with respect to Pb(II) ions, 1.4 times with respect to Cd(II) ions, and 1.8 times with respect to Cu(II) ions.

Comparing the data in Table 4.2 with the concentration of complexation-active centers in the composition of immobilized poly[*N*-(4-carboxyphenyl) methacrylamide] on the surface of clinoptilolite (found according to acid–base titration data (0.17 mmol/g)), it can be stated that a small part (less than 10%) of these active centers took part in complexation with the studied ions. Taking into account the rather successful ("film") arrangement of the polymer on the surface of clinoptilolite for complexation processes, the obtained results can be explained by the chemical nature of the polymer.

TABLE 4.2 Comparison of the Sorption Capacity of the Clinoptilolite]–PKSF-MKA Composite and the Original Clinoptilolite with Respect to Cu(II), Pb(II), and Cd(II) Ions.

Cation	Sorption capacity			
	Clinoptilolite		Clinoptilolite]–PKSF-MKA	
	mmol/g	mg/g	mmol/g	mg/g
Pb(II)	0.010	1.982	0.019	3.952
Cd(II)	0.022	2.463	0.030	3.345
Cu(II)	0.009	0.576	0.016	1.024

4.3.2 SYNTHESIS, STUDY OF THE STRUCTURE, AND SORPTION PROPERTIES OF COMPOSITES BASED ON CLINOPTILOLITE FROM THE TUSHINSKY DEPOSIT WITH IN-SITU MODIFIED AND ADSORBED POLY[8-OXYQUINOLINE METHACRYLATE] WITH RESPECT TO CU(II), PB(II), FE(III), AND MN(II) IONS

By radical polymerization of 8-oxyquinoline methacrylate, carried out in the presence of clinoptilolite from the Tushinsky deposit, a composite clinoptilolite-poly[8-oxyquinoline methacrylate] (hereinafter, clinoptilolite]–POKSYN-MK (*in situ*)) was synthesized.[15] Figure 4.8 shows the reaction scheme of *in-situ* polymerization of 8-oxyquinoline methacrylate on the mineral surface.

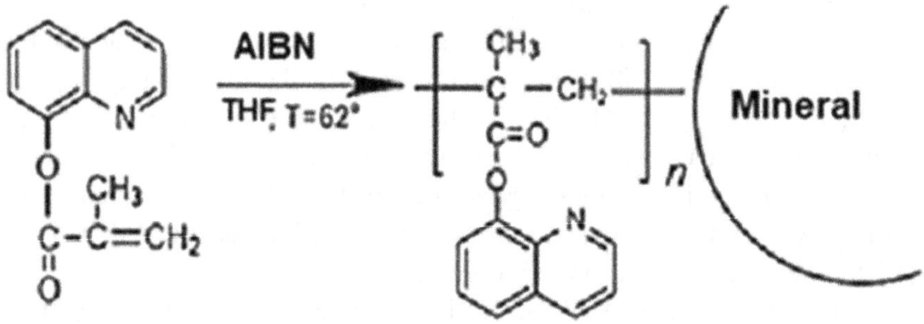

FIGURE 4.8 Reaction scheme of *in-situ* polymerization of 8-oxyquinoline methacrylate on the surface of the mineral.

By adsorption on the surface of this clinoptilolite poly[8-oxyquinoline methacrylate], clinoptilolite]–POKSYN-MK (*absorb*) was synthesized.[16]

The fact of immobilization of poly[8-oxyquinoline methacrylate] on the surface of clinoptilolite in both cases was confirmed by comparative analysis of the IR spectra of the original mineral and the synthesized composites. A comparative analysis of the IR spectra of the synthesized composites and the original clinoptilolite proves that a number of new absorption bands are observed in the spectra of the composites. The most informative for confirming the presence of poly[8-oxyquinoline methacrylate] in the surface layer of the selected mineral are a number of bands in the region of 1200–1800 cm^{-1}, which can be interpreted as follows:

- vibrations at 1590–1598 cm^{-1} correspond to valence vibrations υ(C–N) of the quinoline aromatic system;
- absorption bands at 1652 cm^{-1} and 1712 cm^{-1} correspond to valence vibrations of C=O groups;
- absorption bands in the region of 1750–1758 cm^{-1} can be attributed to valence vibrations of υ(Ar–COO–R);
- absorption bands in the range from 1390 to 1505 cm^{-1} correspond to skeletal vibrations of C–C bonds of the quinoline aromatic system.[16]

The presence of these absorption bands in the IR spectra of the synthesized composites can be evidence of the presence of poly[8-oxyquinoline methacrylate] in their composition.

In order to determine the mass fraction of the polymer in the composition of the synthesized composites, their thermogravimetric analysis was carried out. The obtained thermograms are shown in Figures 4.9 and 4.10. As follows from the data in Figure 4.9, the most intensive thermal destruction of *in-situ* immobilized poly[8-oxyquinoline methacrylate] occurs in the temperature range from 120 to 370°C. At the same time, 11.1 ± 0.1 wt% of the mass of the composite is lost. Taking into account the fact that the mass loss of the original mineral in this temperature range is 8.9 ± 0.1 wt%, respectively, it can be concluded that the mass of the *in-situ* immobilized polymer in the composition of the composite is 2.2 ± 0.1 wt%

Figure 4.10 shows that thermal destruction of adsorbed poly[8-oxyquinoline methacrylate] occurs in the temperature range from 148 to 417°C. At the same time, the mass loss of the composite clinoptilolite]–POKSYN-MK (*absorb*) is 30.3 ± 0.1 wt%, and the mass loss of the original clinoptilolite is 8.9 ± 0.1 wt%. Therefore, the mass fraction of adsorbed polymer in the composition of this composite is 21.4 ± 0.1 wt%.

In order to detail the process of thermal destruction of both synthesized composites based on clinoptilolite, their thermograms, combined with the mass spectrum, were obtained in 3D format (Figure 4.11a and 12a) and a number of mass spectra in 2D format at different temperatures, an example of which is shown in Figures 4.11b and 4.12b. The analysis of the obtained mass spectra shows that the thermal destruction of the *in-situ* immobilized polymer occurs with the formation of two main compounds, which correspond to the most intense mass peaks (see Fig. 4.11b): at mass 18, which corresponds to the formation of water, and at mass 44, which corresponds to the release of carbon dioxide.

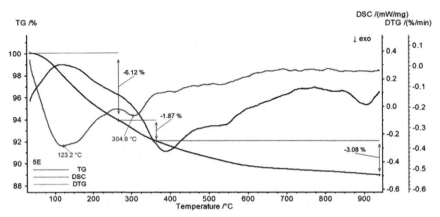

FIGURE 4.9　Thermogram of the composite clinoptilolite]–POKSYN-MK (*in situ*).

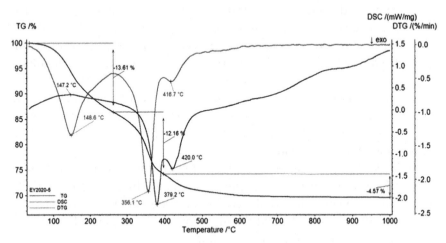

FIGURE 4.10　Thermogram of the composite clinoptilolite]–POKSYN-MK (*absorb*).

As can be seen from Figure 4.12b, there are three intense peaks in the mass spectrum of the composite clinoptilolite]–POKSYN-MK (*absorb*), which indicate that the thermal destruction of the adsorbed polymer occurs mainly with the formation of water (mass peak at 18), carbon(II) oxide (mass peak at 28), and carbon dioxide (mass peak at 44). The two peaks of low intensity present in the spectrum can be attributed to the formation of soot (mass peak at 12) and ethane (mass peak at 30).

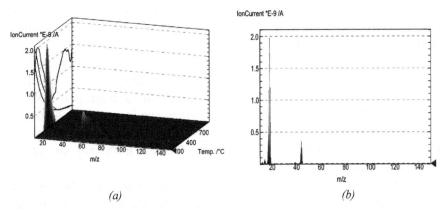

(a) *(b)*

FIGURE 4.11 TG–MS-3D (a) and MS-2D (b) of the composite clinoptilolite]–POKSYN-MK (*in situ*).

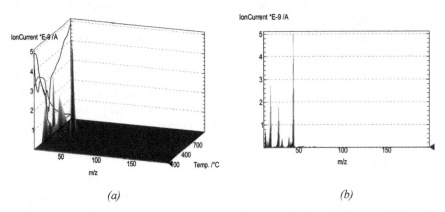

(a) *(b)*

FIGURE 4.12 TG–MS-3D (a) and MS-2D (b) of the composite clinoptilolite]–POKSYN-MK (*absorb*).

The low-temperature nitrogen adsorption–desorption method was used to study the surface parameters of clinoptilolite before and after poly[8-oxyquinoline methacrylate] immobilization. A comparison of the obtained nitrogen adsorption–desorption isotherms for the synthesized composites based on clinoptilolite and the original mineral shows that these isotherms are similar and belong to type IV isotherms according to the IUPAC classification.[14] The similarity of nitrogen adsorption–desorption isotherms indicates that the surface structure of clinoptilolite does not change as a result of polymer fixation.

The values of the surface area of the synthesized composites, the average volume, and the pore size of its surface, calculated by computer processing using various methods from the data of nitrogen adsorption–desorption isotherms, are shown in Table 4.1.

From the data in Table 4.1, it follows that as a result of *in-situ* immobilization of poly[8-oxyquinoline methacrylate] on the surface of clinoptilolite, the specific surface area decreases by 5.7 times, the average pore volume decreases by 1.5 times, and the average pore diameter increases by 2.2 times. That is, in the case of *in-situ* polymerization, the immobilization of a small mass (2.2%) of the polymer led to a slight decrease in the porosity of the mineral surface.

In the composite clinoptilolite]–POXYN-MK (*absorb*), the specific surface area and average pore volume decrease significantly, and the average pore size increases sharply compared to the parameters of the original mineral, that is, the porosity of its surface after adsorption of a large mass of poly[8-oxyquinoline methacrylate] (21.4% in the composition of the composite) sharply decreased.

Changes in the surface morphology of clinoptilolite after *in-situ* modification and adsorption of poly[8-oxyquinoline methacrylate] were monitored using SEM. SEM photos of the surfaces of the synthesized composites are presented in Figures 4.13 and 4.14. As can be seen from the photos shown in Figure 4.13, the *in-situ* immobilized polymer covers the porous surface of the mineral with a thin layer. As a result, the structure of its surface does not change but is only slightly smoothed, which is in good agreement with the data in Table 4.1.

And the photos shown in Figure 4.14 confirm the fact that a large amount of adsorbed polymer tightly covers the porous surface of the mineral, significantly reducing its porosity, which also correlates well with the data in Table 4.1.

The immobilized polymer in the synthesized composites contains oxyquinoline groups, which should easily form complexes with many transition metal ions. To confirm these considerations, we investigated the sorption of Cu(II), Pb(II), Mn(II), and Fe(III) ions, which have a high affinity for 8-oxyquinoline.[17] The study of the sorption capacity of the synthesized composites with respect to the listed metal ions included the study of the sorption rate of selected ions on the surface of the synthesized composites in a static mode from aqueous solutions of nitrates, the construction of sorption isotherms for each of the studied metal ions on the surface of the synthesized composites from aqueous solutions of nitrates, the determination of the sorption capacity of the composites with respect to of each of the investigated metal ions, and comparing these values with those for the original mineral.

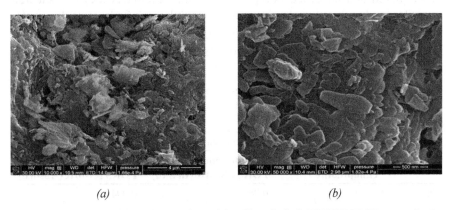

(a) *(b)*

FIGURE 4.13 SEM photo of the surface of the clinoptilolite]–POKSYN-MK composite (*in situ*): a (magnification 10,000×) and b (magnification 50,000×).

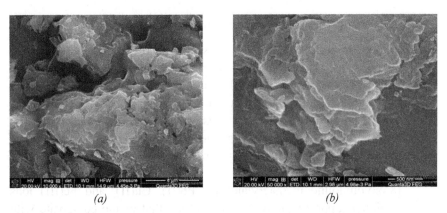

(a) *(b)*

FIGURE 4.14 SEM photo of the surface of the clinoptilolite]–POKSYN-MK composite *(absorb)*: a (magnification 10,000×) and b (magnification 50,000×).

It was experimentally found that the time of optimal sorption of Cu(II) and Pb(II) ions on the surface of a composite based on clinoptilolite with *in-situ* immobilized poly[8-oxyquinoline methacrylate] in the static mode is 1 h, and Fe(III) and Mn(II) 2 h. The established optimal time of contact of all metal ions with the surface of the composite was taken into account during the experiments of constructing sorption isotherms.

In order to establish the values of the sorption capacity of the synthesized composites with respect to selected ions of transition metals, their sorption isotherms were constructed. All obtained isotherms made it possible to unambiguously establish the sorption capacity of composites with *in-situ* immobilized and adsorbed poly[8-oxyquinoline methacrylate] for selected ions. The values of the sorption capacity of the original mineral and the synthesized composites for Pb(II), Cu(II), and Fe(III) ions, which are calculated from the sorption isotherms, are presented in Table 4.3.

TABLE 4.3   Comparison of the Sorption Capacity of the Original Clinoptilolite and Composites Based on It with *In-Situ* Immobilized and Adsorbed Poly[8-Oxyquinoline Methacrylate] for Cu(II), Pb(II), and Fe(III) Ions.

Cations	Sorption capacity (mmol/g)		
	Original clinoptilolite	**Clinoptilolite]– POKSYN-MK (In Situ)**	**Clinoptilolite]– POKSYN-MK (Absorb)**
Cu^{2+}	0.009	0.067	0.094
Fe^{3+}	0.008	0.011	0.018
Pb^{2+}	0.010	0.024	0.034

The following conclusions can be drawn from the data in Table 4.3:

- *In-situ* immobilization and adsorption of poly[8-oxyquinoline methacrylate] on the surface of clinoptilolite leads to an increase in the sorption capacity for Pb(II), Cu(II), and Fe(III) ions compared to the original mineral. No increase in sorption capacity was recorded for Mn(II) ions. Moreover, the greatest increase was recorded for Cu(II) ions, which can be explained by the greatest stability of complexes of this metal with oxyquinoline groups.
- The sorption capacity of the composite with adsorbed poly [8-oxyquinoline methacrylate] for Pb(II), Cu(II), and Fe(III) ions is higher (approximately 1.5 times) than that of the composite with *in-situ* immobilized poly[8-oxyquinoline methacrylate], which can be explained by a much larger amount of adsorbed polymer compared to the *in-situ* immobilized polymer.

Considering that the mass of poly[8-oxyquinoline methacrylate] adsorbed on the surface of clinoptilolite is 10 times greater than that of the *in-situ* immobilized polymer, and the sorption capacity of the *in-situ* immobilized polymer is only 1.5 times lower with respect to Pb(II), Cu(II), and Fe(III), we can conclude: the efficiency of active sorption centers during *in-situ* immobilization is several times higher than as a result of adsorption of the finished polymer.

A comparison of the data in Tables 4.2 and 4.3 shows that the composite based on clinoptilolite with *in-situ* immobilized poly[8-oxyquinoline methacrylate] is a significantly more effective sorbent for Cu(II) and Pb(II) ions than the composite based on clinoptilolite with *in-situ* immobilized poly[*N*-(4-carboxyphenyl)methacrylamide].

4.3.3 SYNTHESIS, STUDY OF THE STRUCTURE, AND SORPTION PROPERTIES OF A SAPONITE-BASED COMPOSITE WITH IN-SITU IMMOBILIZED POLY[N-(4-CARBOXYPHENYL)METHACRYLAMIDE]

The composite poly[*N*-(4-carboxyphenyl)methacrylamide]-saponite (here-inafter, saponite]–PKSF-MKA) was synthesized by radical polymerization of 4-carboxyphenylmethacrylamide, carried out in the presence of saponite, according to the method given in Ref. [18]. The fact of *in-situ* polymerization on the saponite surface was established by comparative analysis of the IR spectra of the original (1) and modified (2) saponite, shown in Figure 4.15.

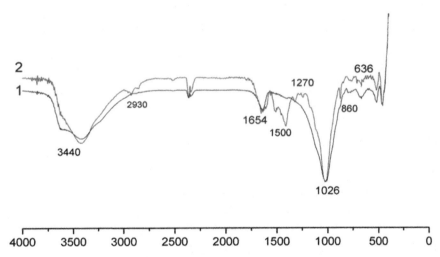

FIGURE 4.15 IR spectra of the original saponite (1) and the saponite]–PKSF-MKA composite (2).

Comparative analysis of the obtained IR spectra proves that in the spectrum of the saponite]–PKSF-MKA composite, unlike the original saponite, there is a broad absorption band in the region of 3440 cm⁻¹, which appeared due to the formation of hydrogen bonds between the polymer and the saponite. The absorption band in the region of 2930 cm⁻¹ can be attributed to valence vibrations of (C–H) bonds, the band in the region of 1654 cm⁻¹ to $\upsilon(CO)$ bonds (amide I), the band in the region of 1500 cm⁻¹ to vibrations of the amide 2 system ($\upsilon(C–N) + \upsilon(N–H)$). And the band around 860 cm⁻¹ corresponds to the deformation vibrations of CH bonds of the polymer. Also, both spectra show an intense absorption band in the region of 1020–1030 cm⁻¹, which corresponds to $\upsilon(Si–O–Si)$.[16]

To determine the concentration and limits of the thermal stability of the immobilized polymer, a thermogravimetric analysis of the original saponite and the saponite]–PKSF-MKA composite was performed. The resulting thermograms are shown in Figures 4.16 and 4.17. From the given thermograms, it is clear that the majority of the immobilized polymer decomposes in the temperature range from 150°C to 500°C. At the same time, 16.9 ± 0.1% of the mass of the composite is lost. Taking into account the fact that the mass loss of the original saponite in this temperature range is 10.5 ± 0.1 wt%, the mass fraction of the immobilized polymer in the synthesized composite was calculated, which is 6.4 ± 0.1 wt.%.

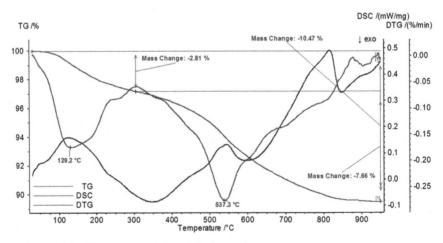

FIGURE 4.16   Thermogram of the original saponite.

Figure 4.18a shows the thermogram of the composite combined with the mass spectrum in 3D format, and Figure 4.18b shows its mass spectrum in 2D format. Mass spectral studies show that thermal degradation of the polymer is mainly observed at a temperature higher than 400°C. At the same time, the main products with a relative mass of 18 and 44 Da are water and CO_2.

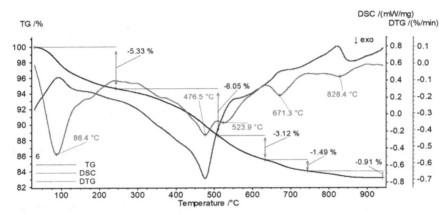

FIGURE 4.17 Thermogram of the saponite]–PKSF-MKA composite.

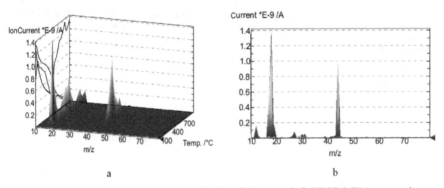

FIGURE 4.18 TG–MS-3D (a) and MS-2D (b) of the saponite]–PKSF-MKA composite.

Nitrogen adsorption–desorption isotherms of original saponite and the composite saponite]–PKSF-MKA were obtained to study the surface morphology of the composite. All of the isotherms are similar and belong to IV type of isotherms and exhibit H3-type hysteresis according to the *IUPAC* classification.[14] The shapes of the isotherms indicate that the pores of both the saponite and the composite consist of plane-parallel layers, which is in a good agreement with the literature data.[5–7] The obtained experimental data indicate that the immobilized copolymer does not significantly affect the surface layer structure of saponite.

A detailed comparison of original saponite surface parameters and composite saponite]–PKSF-MKA based on our investigations and on the literature data is presented in Table 4.4. The data indicate the following:

- the surface area after immobilization of the polymer decreases 3.5 times, which is a logical result of attachment of a nonporous polymer on the surface;
- modification increases the pore size of saponite, which before and after modification has both micro- and mesoporous character.

Analysis of Table 4.4 data allows us to conclude that polymer modification "smoothes out" the pores of original mineral, simultaneously reducing their depth and expanding them, as evidenced by a simultaneous 1.5-fold increase in the size of the pores and a halving of their volume.

TABLE 4.4   Surface Parameters Comparative Characteristics for Original Saponite and the Synthesized Composites.

Sample	S_{pit} (m²/g)			Average pore volume (cm³/g)	Average pore size (nm)
	Calculated from isotherm	BJH	Langmuir		
Original saponite	41	42	65	0.08	8
Saponite]–PKSF-MKA	12	12	18	0.04	10–12
Saponite]–PMCNM-AzoB	8	8	14	0.04	18
Saponite]–POQ-MA *(in situ)*	2	2	2.3	0.01	23
Saponite]–POQ-MA *(absorb)*	1	1	1.3	0.01	42
Saponite with *in-situ* immobilized copolymer poly(4-vinylpyridine-*co*-styrene)	3	3	6.4	0.01	1.4

Changes in saponite surface morphology after modification with poly[*N*-(4-carboxyphenyl)methacrylamide] were monitored using SEM. SEM photo of original saponite and composite saponite]–PCPh-MAA is presented in Figure 4.19.

These photos, first, confirm the above-mentioned assumption about changes in the porosity of the original mineral as a result of surface modification with the polymer. Second, it allows us to say about the "linear" placing the polymer on the saponite surface.

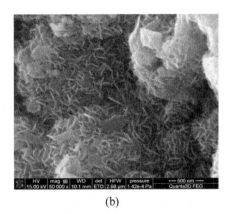

(a)   (b)

FIGURE 4.19 SEM-photo of the original saponite (a) (magnification 50,000×) and composite saponite]–PKSF-MKA (b) (magnification 50,000×).

Sorption characteristics of saponite]–PKSF-MKA toward Cu(II), Cd(II), Pb(II), and Fe(III) ions were studied in static mode according to the method described in Ref. [18]. It was found that partial washing of the immobilized polymer was observed at a contact of the synthesized composite with $NaHCO_3$ solution (pH 8.1) resulting a yellow color of solution acquired.

The sorption capacity of saponite]–PKSF-MKA for Cu(II), Pb(II), Fe(III), and Cd(II) ions was studied at different values of acidity and chemical composition of the medium. We have shown that saponite]–PKSF-MKA exhibits the highest sorption activity for microquantities of all investigated metals in a neutral medium. Taking into account all of the above, further studies of the sorption properties of modified saponite were carried out with solutions of Cu(II), Pb(II), Cd(II), and Fe(III) salts without adding buffers.

Studying of the sorption degree dependence from the contact time with the surface of saponite]–PKSF-MKA in the static mode made for Cu(II), Pb(II), Fe(III), and Cd(II) ions demonstrated that all the ions are maximally sorbed within 30 min contact, and then the sorption equilibrium is established.[18]

In order to establish the values of the sorption capacity of the composite with these ions of transition metals, their sorption isotherms were constructed and compared with the sorption isotherms of the original mineral. Examples of isotherms of Cd(II) ions are shown in Figure 4.20.

Based on the obtained isotherms, a preliminary conclusion was made that the sorption capacity of saponite after *in-situ* modification with poly[*N*-(4-carboxyphenyl)methacrylamide] increases toward Cd(II), Cu(II), and

Fe(III) ions and practically does not changes to Pb(II) ions. All obtained isotherms belong to the 2L type, which allows us to unambiguously establish the value of the sorption capacity of the synthesized composite for each of the selected ions.

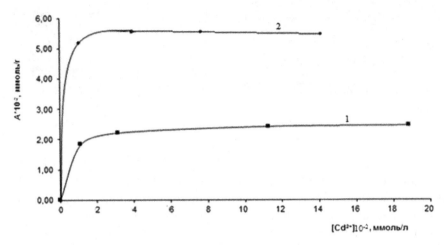

FIGURE 4.20   Sorption isotherms for Cd(II) ions on the original saponite surface (number 1) and composite saponite]–PKSF-MKA (number 2).

The sorption capacity[18] of the modified saponite calculated from the data of the sorption isotherms is summarized in Table 4.5. The data in the table allow us to state that the sorption capacity of the saponite after the modification with poly[N-(4-carboxyphenyl)methacrylamide] in the molar ratio increases for Fe(III) ions in 1.7 times, for Cd(II) and Cu(II) ions in 2 times.

TABLE 4.5   Sorption Capacity Values Comparison for Saponite]–PCPh-MAA and Original Saponite toward Cu(II), Pb(II), Cd(II), and Fe(III) ions.

Cation	Sorption capacity			
	Saponite		Saponite]–PKSF-MKA	
	mmol/g	mg/g	mmol/g	mg/g
Fe(III)	0.014	0.78	0.024	1.34
Pb(II)	0.016	3.31	0.018	3.73
Cd(II)	0.002	0.30	0.005	0.59
Cu(II)	0.017	1.08	0.033	2.09

4.3.4   SYNTHESIS, STRUCTURE INVESTIGATIONS, AND SORPTION PROPERTIES OF SAPONITE CLAY IN SITU MODIFIED WITH POLY[4-METHACRYLOYLOXY-(4'-CARBOXY-2-NITRO-5-METHYL) AZOBENZENE] TOWARD CU(II), MN(II), PB(II), AND FE(III) IONS

Preliminarily model studies showed that the composite based on silica gel, *in situ* modified with poly[4-methacryloyloxy-(4'-carboxy-2-nitro-5-methyl) azobenzene], showed the best sorption properties concerning toxic metal ions among the other composites of silica gels *in situ* modified with various polymer azodyes.[19]

Therefore, it was expedient to establish how the saponite properties will improve after the modification of their surface with poly[4-methacryloyloxy-(4'-carboxy-2-nitro-5-methyl)azobenzene] (PMCNM-AzoB). For this purpose, a composite (saponite]–PMCNM-AzoB) based on saponite was synthesized by *in-situ* immobilization of poly[4-methacryloyloxy-(4'-carboxy-2-nitro-5-methyl)-azobenzene]. *In-situ* immobilization was realized during the initiated (AIBN was used as an initiator) radical polymerization of 4-methacryloyloxy-(4'-carboxy-2-nitro-5-methyl)azobenzene (MCNM-AzoB) monomer in the presence of saponite.

The scheme of *in-situ* immobilization process during the polymerization of poly[4-methacryloyloxy-(4'-carboxy-2-nitro-5-methyl)-azobenzene] on the saponite surface is shown in Figure 4.21.

The fact that the MCNM-AzoB polymerization reaction took place on the saponite surface was established by comparative analysis of the IR spectra of the original saponite and the saponite]–PMCNM-AzoB composite. It is shown that in the spectrum of the composite, there are new absorption bands in the region from 1405 to 2962 cm^{-1}. The absorption band at 1405 cm^{-1} can be attributed to the characteristic vibration of the $\upsilon(N{=}N)$ bond of azo-group. The absorption band at 1651 cm^{-1} corresponds to stretching asymmetric vibrations of the nitro group $\upsilon_{ac}(NO_2)$. The absorption bands in the region from 2833 to 2962 cm^{-1} can be attributed to the stretching vibrations of sp^3-hybridized carbon $\upsilon(C{-}H)$ of the main polymer chain.[16] These absorption bands presented in the spectrum of the composite confirm the presence of a polymer PMCNM-AzoB on the saponite surface.

A thermogravimetric analysis of the synthesized composite was carried out in order to determine the concentration of the immobilized polymer. The thermogram of saponite]–PMCNM-AzoB is shown in Figure 4.22. These data obtained under similar conditions were compared with the thermogram of original saponite. As one can see, a thermal destruction of the immobilized polymer occurs in the temperature range from 220 to 480°C. Decomposition with an exothermic effect occurs in the range from 400 to 500°C. At the same time, about $13.5 \pm 0.1\%$ of composite weight is lost. Taking into account the fact that original saponite loses 10.5 ± 0.1 wt% in this temperature range, the weight fraction of the immobilized polymer in the synthesized composite was calculated as 3.0 ± 0.1 wt.%.

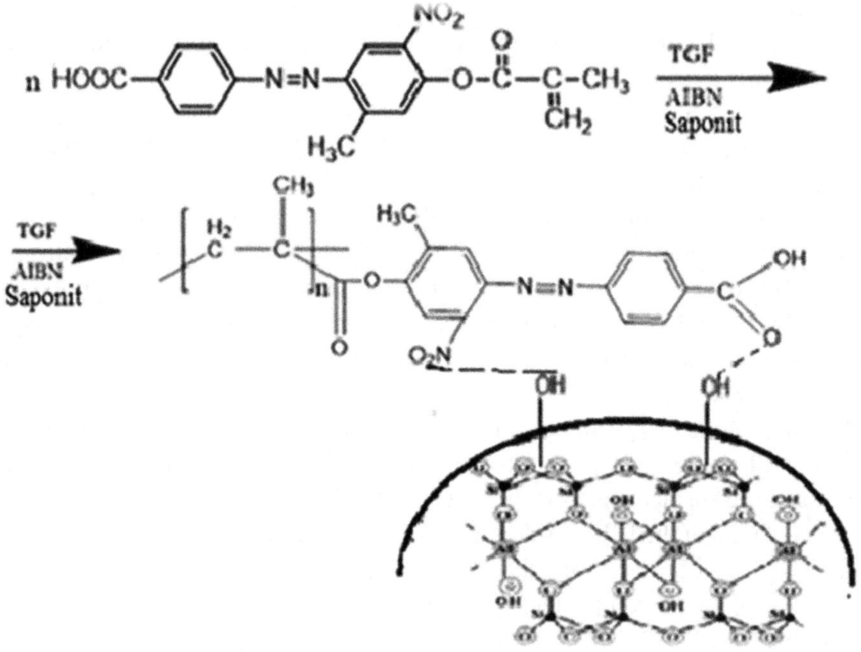

FIGURE 4.21 Scheme of *in-situ* immobilization process of poly[4-methacryloyloxy-(4'-carboxy-2-nitro-5-methyl)-azobenzene] during the polymerization of MCNM-AzoB on the saponite surface.

A carboxyl group content in the polymer PMCNM-AzoB immobilized on the saponite surface was determined by acid–base titration with 0.01 N NaOH. According to the results of the obtained experimental titration data (excluding titration data of original saponite), it was found that the number of carboxyl groups, which is equal to the number of active centers of the immobilized polymer, is 0.26 mmol/g. The method of low-temperature adsorption–desorption of nitrogen was used to determine parameters of the saponite surface after immobilization of the polymer PMCNM-AzoB. Comparison of the obtained isotherms of the original and modified saponite allows us to state that they are similar and belong to type IV isotherms and at the same time exhibit H3-type hysteresis according to the IUPAC classification.[14] The shapes of both isotherms indicate that saponite pores consist of plane-parallel layers. The obtained data indicate that the immobilized polymer practically does not affect the structure of the saponite surface.

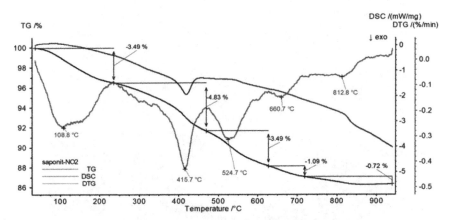

FIGURE 4.22 Thermogram of synthesized composite [saponite]–PMCNM-AzoB.

The obtained nitrogen adsorption–desorption isotherms for original saponite and the synthesized composite were processed using several computer programs. The values of the surface area, average volume, and surface pore size of the synthesized composite calculated by computer processing of the isotherm data are shown in Table 4.4. The data obtained indicate that after poly[4-methacryloyloxy-(4'-carboxy-2-nitro-5-methyl)-azobenzene] immobilization, saponite surface area decreased from 42 to 8 m^2/g. From the data in this table, we can also conclude that polymer modification "smoothes out" the pores of the original mineral, reducing their depth and expanding them. This is evidenced by a simultaneous 2.2-fold increase in the size of pores and a halving of their volume.

Changes in saponite surface morphology after modification with PMCNM-AzoB were monitored using SEM. An SEM photo of the composite surface at a magnification of 10,000 times is shown in Figure 4.23. As we can see, the polymer on the mineral surface has an "island" or "globular" placement.

As was expected, synthesized composite saponite]–PMCNM-AzoB should exhibit sorption capacity for metal ions due to the complexing activity of the immobilized polymer and due to the adsorption of the original complexes in the pores of the mineral. The study of the sorption properties

of this composite to Cu(II), Mn(II), Pb(II), and Fe(III) ions was carried out
in a static mode in an aqueous medium without any buffer addition, similarly
to the experimental conditions used for the composite based on silica with
in-situ immobilized PMCNM-AzoB.[20]

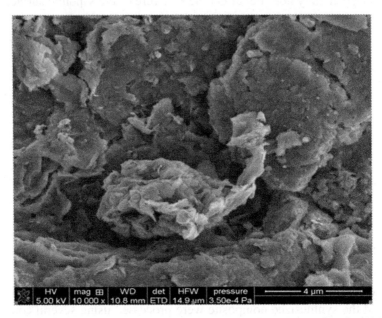

FIGURE 4.23   SEM-photo of composite saponite]–PMCNM-AzoB (magnification 10,000×).

Dependence of ions Cu(II), Pb(II), Fe(III), and Mn(II) sorption degree on
the contact time with the surface of saponite]–PMCNM-AzoB in the static
mode shows that Pb(II) ions are quantitatively sorbed during the first minutes
of contact with the surface of the composite, Cu(II) ions—within an hour,
and Mn(II) and Fe(III) ions—within a day of contact. At the optimal sorption time, sorption isotherms of these ions on the saponite]–PMCNM-AzoB
surface and the original saponite were constructed. The sorption isotherms
of Pb(II) ions on the saponite surface (1) and the saponite]–PMCNM-AzoB
composite (2) are shown in Figure 4.24.

All of the isotherms belong to the 2L type, which is typical for porous
inorganic matrices modified with complexing substances, and allowed to
establish the sorption capacity of saponite]–PMCNM-AzoB for each of
the investigated ions. In Table 4.6, it is shown a comparison of the sorption
capacity values for the original saponite and synthesized composite to Cu(II),
Pb(II), Mn(II), and Fe(III) ions found from the data of sorption isotherms.

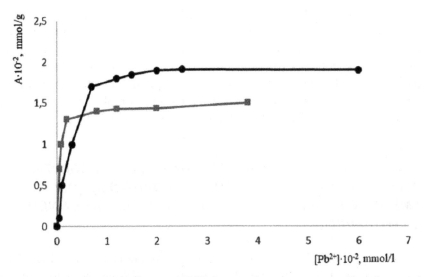

FIGURE 4.24 Sorption isotherms of Pb(II) ions on the parent saponite (1) and composite saponite]–PMCNM-AzoB (2) surface.

TABLE 4.6 Sorption Capacity Comparison for Original Saponite and Saponite]–PMCNM-AzoB toward Cu(II), Mn(II), Pb(II), and Fe(III) Ions.

Cation	Sorption capacity			
	Original saponite		Saponite]–PMCNM-AzoB	
	mmol/g	mg/g	mmol/g	mg/g
Fe(III)	0.014	0.78	0.016	0.89
Pb(II)	0.016	3.31	0.019	3.93
Mn(II)	0.016	0.88	0.017	0.94
Cu(II)	0.017	1.09	0.035	2.24

The data in this table allow us to state that the sorption capacity of saponite after modification with poly[4-methacryloyloxy-(4′-carboxy-2-nitro-5-methyl)-azobenzene] toward Cu(II) ions increases by more than 2 times, toward Pb(II)—by 18% and toward Fe(III) ions—only by 14%. Undoubtedly, this is the result of immobilization on the saponite surface of an azopolymer capable of complex formation. The increase in the sorption capacity of the modified saponite, in comparison with the original one toward Mn(II) ions, is within the experimental error. Therefore, it can be assumed that surface modification of saponite with poly[4-methacryloyloxy-(4′-carboxy-2-nitro-5-methyl)-azobenzene] does not lead to improvement of its sorption capacity for Mn(II) ions.

Comparing the data from Table 4.6 with the concentration of complexation-active centers found according to the data of acid–base titration (0.26 mmol/g) in the composite, it can be stated that $13 \pm 1\%$ of these active centers took part in the complex formation with Cu(II) ions, and only $6 \pm 1\%$ with Fe(III) and Pb(II) ions. Such results can be explained by steric hindrances (low accessibility of active centers for "large" aqua complexes) as a result of the globular arrangement of the polymer on the saponite surface.

4.3.5   SYNTHESIS, STRUCTURE INVESTIGATIONS, AND SORPTION PROPERTIES OF SAPONITE CLAY IN SITU MODIFIED BY POLY (8-METHACROYLOXYQUINOLINE) TOWARD CU(II), PB(II), MN(II), AND FE(III) IONS

A composite saponite]-poly(8-methacroyloxyquinoline) (saponite]–POQ-MA (*in situ*)) was synthesized by radical polymerization of 8-methacroyloxyquinoline using AIBN as an initiator in the presence of saponite from the Tashkiv deposit (Khmelnytskyi region, Ukraine).[21] The reaction scheme of *in-situ* polymerization of 8-oxyquinoline methacrylate on the saponite surface is similar to the one depicted in Figure 4.8. By adsorption on the saponite surface of the preliminarily obtained poly(8-methacroyloxyquinoline), synthesized similarly, but without a solid carrier, a composite saponite]–POQ-MA (*absorb*) was synthesized.

Mobilization of poly(8-methacroyloxyquinoline) on the saponite surface for both composites was confirmed by IR spectroscopy and thermogravimetric analysis combined with mass spectrometry. The obtained thermograms of saponite]–POQ-MA (*in situ*) and saponite]–POQ-MA (*absorb*) composites are shown in Figures 4.25 and 4.26, respectively.

Comparison of thermograms shown in Figures 4.25 and 4.26 shows that the thermal destruction of both composites has a similar nature and occurs in the temperature range from 100 to 700°C, which is caused by the destruction of the immobilized polymer and the loss of structured water in the composition of saponite. But the majority of immobilized poly (8-methacroyloxyquinoline) is destroyed in the temperature range from 350 to 500°C. At the same time, the saponite]–POQ-MA composite (*in situ*) loses $20.6 \pm 0.1\%$ of its weight. Taking into account the fact that the original saponite loses 10.5 ± 0.1 wt% in this temperature range, the weight fraction of the immobilized polymer in the composite saponite]–POQ-MA (*in situ*) is 10.1 ± 0.1 wt%. The total weight loss for the composite saponite]–POQ-MA (*absorb*) according to the data in Figure 4.26 is $33.5 \pm 0.1\%$, and then the weight fraction of adsorbed polymer in the composite is 23.0 ± 0.1 wt%. Therefore, poly(8-methacroyloxyquinoline) is almost completely adsorbed on the saponite surface. In case of *in-situ* immobilization method, the yield of the polymerization reaction is about 70%.

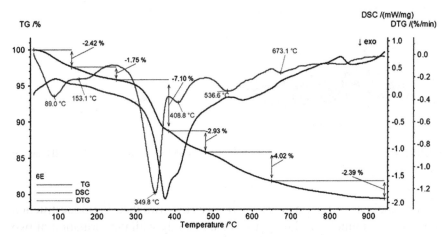

FIGURE 4.25   Thermogram of composite saponite]–POQ-MA (*in situ*).

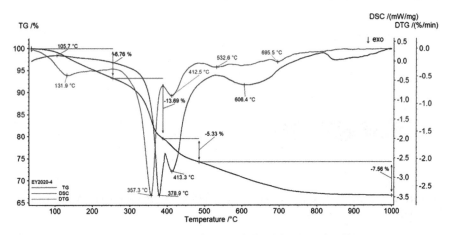

FIGURE 4.26   Thermogram of composite saponite]–POQ-MA (*absorb*).

In order to detail the process of thermodestruction of poly(8-methacroyloxyquinoline) *in-situ* immobilized on the saponite surface as well as adsorbed, thermograms of both composites combined with the mass spectrum in 3D format (Figs. 4.27a and 4.28a) and mass spectra in 2D formats at different temperatures were obtained, examples of which are shown in Figures 4.27b and 4.28b.

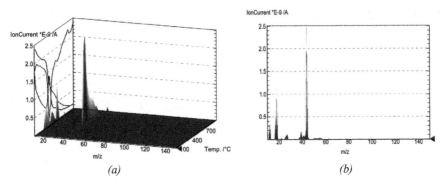

(a)   *(b)*

FIGURE 4.27   (a) TG–MS-3D data of saponite]–POQ-MA *(in situ)* and (b) MS-2D data of saponite]–POQ-MA *(in situ)*.

The analysis of the obtained mass spectra shows that thermodestruction of the *in-situ* immobilized polymer occurs mainly with the formation of two main compounds—carbon dioxide and water, which correspond to the most intense mass peaks at 44 and 18 Da (see Fig. 4.27b).

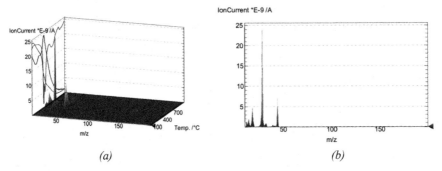

(a)   *(b)*

FIGURE 4.28   (a) TG–MS-3D data of saponite]–POQ-MA *(absorb)* and (b) MS-2D data of saponite]–POQ-MA *(absorb)*.

As can be seen from Figure 4.28b, three intense peaks are present in the mass spectrum of the composite based on saponite with adsorbed poly (8-methacroyloxyquinoline), which indicates that the thermal destruction of the immobilized polymer occurs with the formation of CO and/or nitrogen, carbon dioxide, and water. Moreover, the signal at 28 Da is the most intensive in contrast with the spectrum of *in-situ* immobilized polymer with the most intensive signal at 44 Da.

The analysis of saponite surface parameters after loading of poly (8-methacroyloxyquinoline) in various ways was carried out by methods of low-temperature nitrogen adsorption–desorption and SEM. The values of specific surface area and the average volume calculated by computer processing of low-temperature nitrogen adsorption–desorption isotherm data of composites saponite]–POQ-MA (*in situ*) and saponite]–POQ-MA (*absorb*) and surface pore size calculated from the pore distribution diagrams are given in Table 4.4. It follows from the data of the Table 4.4, the specific surface area and average pore volume significantly decrease, and the average pore size increases sharply both in the case of *in-situ* immobilization and adsorption of poly(8-methacroyloxyquinoline) on the saponite surface. This indicates that the surface porosity of the original mineral after immobilization of poly(8-methacroyloxyquinoline) is sharply reduced, which is a logical result of fixing a large amount of nonporous polymer on its surface.

At the same time, adsorbtion on the saponite surface twice the mass of the polymer in comparison with *in-situ* immobilized leads to half the value of the specific surface area of the modified saponite.

SEM photos of composites saponite]–POQ-MA (*in situ*) and saponite]–POQ-MA (*absorb*) surfaces are presented in Figures 4.29 and 4.30, respectively. Comparison of SEM photos shows that a smaller amount of *in-situ* immobilized poly(8-methacroyloxyquinoline) covers the porous surface of the mineral with a thin layer, only slightly smoothing the pores of the saponite surface, while a large amount of the adsorbed polymer tightly covers the surface of the mineral, significantly reducing its porosity, which is in good agreement with the data presented in Table 4.4. Thus, the obtained results of a low-temperature nitrogen adsorption–desorption and SEM correlate well with each other.

The difference in the morphology of the saponite surface after poly (8-methacroyloxyquinoline) immobilization in various ways definitely affects the sorption properties of the synthesized composites to metal ions.

The sorption capacity study of the synthesized composites toward Pb(II), Cu(II), and Fe(III) ions should be carried out similarly to the study of the sorption properties of composites based on clinoptilolite, namely, in an aqueous medium without adding any buffers to the initial solutions of nitrates of these metals.

All isotherms made it possible to unambiguously establish the sorption capacity of the synthesized composites for each of the studied metal ions. The values of the sorption capacity for Pb(II), Cu(II), and Fe(III) ions, determined from the sorption isotherms, are shown in Table 4.7.

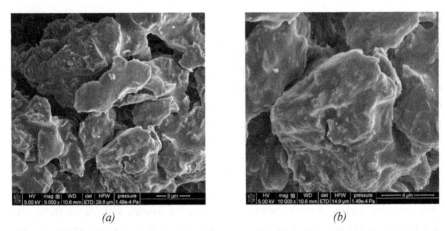

(a)   *(b)*

FIGURE 4.29   SEM photo of the composite saponite]–POQ-MA (*in situ*) surface: a (magnification 5000×) and b (magnification 10,000).

(a)   *(b)*

FIGURE 4.30   SEM-photo of the composite saponite]–POQ-MA *(absorb)* surface: (a) magnification 2500× and (b) magnification 50,000×.

TABLE 4.7   Sorption Capacity toward Cu(II), Mn(II), Pb(II), and Fe(III) Ions Comparison for Original Saponite and Composites Based on Saponite with *In-Situ* Immobilized and Absorbed Poly(8-Methacroyloxyquinoline).

Cation	Sorption capacity (mmol/g)		
	Original saponite	Saponite]–POQ-MA (absorb)	Saponite]–POQ-MA (in situ)
Cu^{2+}	0.017	0.075	0.064
Fe^{3+}	0.014	0.027	0.025
Pb^{2+}	0.016	0.016	0.021

As can be seen from Table 4.7, the sorption capacity of saponite after adsorption of poly(8-methacroyloxyquinoline) on its surface increases toward Cu(II) ions by 4.4 times and toward Fe(III) ions by 1.9 times instead to Pb(II) ions remains unchanged. A comparison of the obtained data with the sorption capacity of saponite modified *in situ* with poly(8-methacroyloxyquinoline) proves that the sorption capacity of saponite]–POQ-MA (*absorb*) for Cu(II) ions is slightly higher (by 1.17 times) than that of saponite modified *in situ* with this polymer. Regarding Fe(III) ions, the sorption capacity of both composites is practically the same. And for Pb(II) ions, the sorption capacity of composite saponite]–POQ-MA (absorb) is lower than that of saponite, *in-situ*-modified poly(8-methacroyloxyquinoline).

The obtained results can be explained as follows: original saponite removes these metal ions due to their physical adsorption into the pores. According to Ref. [8], in the case of *in-situ* modification, the polymer partially covers the surface of the mineral. Therefore, the adsorption of metal ions occurs by two parallel mechanisms—physical adsorption into the pores and due to the formation of multiligand complexes with polymeric molecules. It leads to an improvement in the sorption characteristics of the composite for all studied ions. However, a large amount of synthesized polymer in the composite surface of mineral is almost completely covered by the polymer. In such a case, the first mechanism of adsorption (into the pores of the mineral) is practically negligible and extraction of metal ions occurs only through the formation of various ligand complexes with adsorbed polymer molecules.

4.3.6 SYNTHESIS, STRUCTURE INVESTIGATIONS, AND SORPTION PROPERTIES OF SAPONITE CLAY IN SITU MODIFIED BY COPOLYMER POLY(4-VINYLPYRIDINE-CO-STYRENE) WITH MOLAR RATIO 4-VP:ST = 3:1 TOWARD CU(II), PB(II), MN(II), AND FE(III) IONS

Pyridine and its derivatives are well-known ligands which form strong complexes with most transition metal ions. Processes of complex formation are also typical for pyridine-containing polymers. Therefore, the modification of the surface of porous inorganic materials by pyridine-based polymers is a perspective approach to produce new composite materials that can be used as effective sorbents of metal cations due to their complexation with nitrogen atoms of pyridine function in immobilized polymer.[22–24]

Saponite-based composite was synthesized using original mineral saponite from Tashkiv deposit (Ukraine) by *in-situ* immobilization of 4-vinylpyridine (4-VP) and styrene (St) copolymer. Copolymerization was carried out by radical polymerization of 4-VP and styrene solution in CCl_4 at the starting molar ratios of comonomers St:4-VP = 1:3 in the presence of saponite according to the method described in Ref. [25].

On a basis of comparison of thermograms for the original saponite and organo-mineral composite, it was proved that thermal decomposition predominantly occurs in the temperature range 150–630°C. The weight fraction of copolymer in the synthesized material was calculated as 8.5 ± 0.1 wt%.[25]

It was established that copolymer immobilization on a saponite surface leads to a decrease of the surface area from 41 to 3 m^2/g. At the same time, the average pore volume decreased from 0.085 to 0.012 cm^3/g, and the average pore diameter decreased from 8 to 1.4 nm (see Table 4.4).

Changes in the surface morphology of saponite after its modification with a copolymer poly(4-vinylpyridine-*co*-styrene) were recognized using SEM technic. As one can see from SEM photo presented in Figure 4.31, the copolymer on the saponite surface is located in the form of elongated and aggregated short chains.

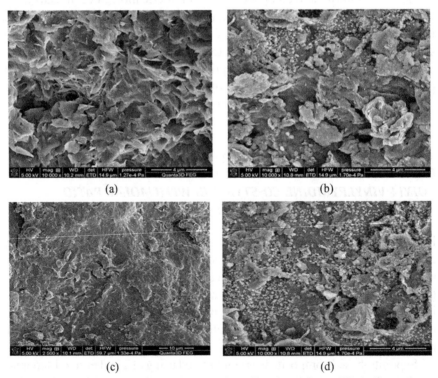

(a) (b)

(c) (d)

FIGURE 4.31 SEM photo of the original saponite (a, b) and saponite modified with copolymer poly(4-vinylpyridine-*co*-styrene) and (c, d) surface (magnification 10,000× and 2500×).

The sorption characteristics of the synthesized composite toward Cu(II), Pb(II), Mn(II), and Fe(III) ions were investigated in a static mode in aqueous solutions without buffer addition. Based on the experimental dependence of sorption degree of Cu(II), Pb(II), Mn(II), and Fe(III) ions on the *in-situ*-modified saponite surface on the contact time in the static mode, it was established that Cu(II) ions are quantitatively sorbed by the composite within 120 min of contact. After that, the sorption equilibrium is established. Fe(III) ions are sorbed by 75% during 120 min of contact, and their quantitative sorption is achieved only within a day. Sorption of Pb(II) and Mn(II) ions increases gradually during the day and reaches its maximum at the level of 70% (see Figure 4.32).

In order to establish the sorption capacity values of the synthesized composite toward mentioned above ions of transition metals, their sorption isotherms were built and the value of the sorption capacity toward each of metal ion was calculated. The values of the sorption capacity of saponite modified *in situ* with a copolymer of 4-VP and styrene for Pb(II), Cu(II), Mn(II), and Fe(III) ions, determined from the sorption isotherms, are shown in Table 4.8.

As one can see from Table 4.8, the sorption capacity of saponite increases by 29%, 21%, and 20% compared to Cu(II), Fe(III), and Pb(II) ions, respectively, after *in-situ* surface modification with 4-VP-containing copolymer.

This growth can be explained by complexation of metal ions with pyridine moiety of an immobilized copolymer. Since the stability of pyridine complexes of Cu(II), and Fe(III) is higher than for Pb(II) ions, it increases the sorption capacity of the composite toward Cu(II) and Fe(III) ions.

As follows from Table 4.8, the growth of sorption capacity of modified saponite in comparison with the original mineral toward Mn(II) ions is within the experimental error. Since the stability of Cu(II) and Fe(III) pyridine complexes is higher than that of Pb(II), it could be an explanation of higher sorption capacity increase toward Cu(II) and Fe(III) ions in comparison with Pb(II). An increase of the sorption capacity for the modified saponite in comparison with the original saponite toward Mn(II) ions is within the experimental error. Therefore, it can be asserted that the modification of saponite surface with a copolymer poly(4-vinylpyridine-*co*-styrene) does not lead to an improvement in sorption capacity toward Mn(II) ions.

Nevertheless, *in-situ* immobilization of poly(4-vinylpyridine-*co*-styrene) on the saponite surface from the Tashkiv deposit was obtained as an inexpensive sorbent suitable for wastewater treatment from excess concentrations of Cu(II), Pb(II), and Fe(III) ions.

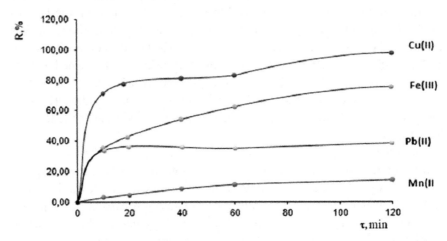

FIGURE 4.32   Dependence of Cu(II), Pb(II), Mn(II), and Fe(III) ions adsorption degree from the contact time with modified saponite in static mode:

Experimental conditions: $m_{sorb} = 0.1$ g, $V_{solution} = 25$ ml, $m^0_M = 100$ µg.

TABLE 4.8   Adsorption Capacity Comparison for Original Mineral Saponite and Saponite with *In-Situ* Modified Surface toward Cu(II), Mn(II), Pb(II), and Fe(III) Ions.

Cation	Adsorption capacity			
	Original saponite		Modified saponite	
	mmol/g	mg/g	mmol/g	mg/g
Fe(III)	0.014	0.78	0.017	0.95
Pb(II)	0.015	3.11	0.018	3.73
Mn(II)	0.016	0.88	0.018	0.97
Cu(II)	0.017	1.09	0.022	1.41

Comparing the data in Tables 4.4–4.8, the following conclusions can be drawn:

- all saponite-based composites have a higher sorption capacity toward such toxic ions as Pb(II), Cu(II), and Fe(III) (the mentioned ions have polluted of surface natural waters of Ukraine in multi-ton quantities and their removal using cheap sorbents still remains actual) compared to the original saponite;

- among all synthesized composites based on saponite, composites with *in-situ* immobilized and adsorbed poly[8-oxyquinoline methacrylate] have the highest sorption capacity for Pb(II), Cu(II), and Fe(III) ions. At the same time, the composite with in-situ immobilized poly[8-oxyquinoline methacrylate] is the most effective for the sorption of Pb(II) ions, and the composite with adsorbed poly[8-oxyquinoline methacrylate] is the most effective for Cu(II) and Fe(III) ions.

The obtained data can be explained in the same way as for organo-mineral composites based on poly(8-methacroyloxyquinoline): adsorption of transition metal ions on the synthesized composite surface could be realized in two ways—by metal ions complexation with immobilized organic polymer and absorption of metal ions into the pores and channels of minerals. When the presynthesized polymer is adsorbed on the surface, a significant degree of coverage of the mineral surface is achieved, while the second mechanism stops working. Therefore, the sorption of metal ions occurs only due to complexation. In the case of *in-situ* immobilization of the polymer, pores of the mineral carrier remain open and the sorption of metal ions occurs in both ways and better sorption characteristics of the organo-mineral sorbents could be achieved.

4.4 CONCLUSION

Vast amounts of industrial, agricultural, and domestic wastewater need to be treated before being discharged into the hydrosphere. Polymer-modified clay composites remain the adsorbents of choice for heavy metal and cationic dye removal. The specificity and selectivity of clay composites for tested pollutants mainly depend on the corresponding adsorption mechanism, which in turn is determined by the modification method employed. Ion exchange, electrostatic attraction, hydrogen bonding, and pore-filling are the predominant mechanisms controlling the adsorption of targeted pollutants. Ion exchange, electrostatic attraction, hydrogen bonding, and pore-filling are the predominant mechanisms controlling the adsorption of targeted pollutants. Modification of natural clays is an alternative to enhancing their properties for application in wastewater treatment. Overall, modified clay composites represent promising candidates for the removal of heavy metals, phenolic compounds, and dyes from various types of industrial wastewater.

KEYWORDS

- adsorption
- *in-situ* immobilization
- polymer
- organo-mineral composite
- wastewater

REFERENCES

1. Petrenko, O. V.; Yanovska, E. S.; Terebilenko K. V.; Stus N. V. *Green Chemistry*; Kyiv University: Kyiv, Ukraine, 2020; 239 pp.
2. Smith, J. A.; Tillman, F. D.; Bartelt-Hunt, S. L.; et al. Evaluation of an Organoclay, an Organoclay-Anthracite Blend, Clinoptilolite, and Hydroxy-Apatite as Sorbents for Heavy Metal Removal from Water. *J. Bull. Environ. Contam. Toxicol.* **2004,** *72*(6), 1134–1141.
3. Valpotic, H.; Gracner, D. Zeolite Clinoptilolite Nanoporous Feed Additive for Animals of Veterinary Importance: Potentials and Limitations. *Period. Biol.* **2017,** *119*, 159–172.
4. Valaskova, M.; Martynkova, G. S. Vermiculite: Structural Properties and Examples of the Use. In Clay Minerals in Nature—Their Characterization, Modification and Application. *Publ. Intech. Open Sci.* **2012,** *11*, 209–238.
5. Sokol, H.; Sprynskyy, M.; Ganzyuk A.; Raks, V.; Buszewski B. Structural, Mineral and Elemental Composition Features of Iron-Rich Saponite Clay from Tashkiv Deposit (Ukraine). *Colloids Interfaces* **2019,** *3* (1), 10.
6. Nikitina M. V.; Popova, F.; Nakvasina, E. N.; Romanov, E. M.; Zhuravleva, E. A. Possibility Determination of Using Saponite in Agriculture. *IOP Conf. Ser.: Earth Environ. Sci.* **2021,** *723*, 022016.
7. Carretero, M. I.; Gomes, C. S. F.; Tateo F. Clays and Human Health. *Handbook of Clay Science*; Elsevier Ltd.: Amsterdam, 2006; pp 717–741.
8. Yanovska, E. S.; Zatovskyi, I. V.; Slobodianyk, M. S. Scientific Basis of Waste-Free Technology of Further Purification of Industrial Wastewater from Mixtures of Heavy Metal Ions. *Environ. Ecol. Life Saf.* **2008,** *5*, 50–54.
9. Tertykh, V. A.; Polishchuk, L. M.; Yanovska, E. S., Dadashev, A. D. Concentration of Anions by Silica Adsorbents with Immobilized Nitrogen-containing Polymers. *Adsorp. Sci. Technol.* **2008,** *26* (1–2), 59–68.
10. Budnyak, T. M.; Yanovska, E. S.; Tortikh, V.; Kichkiruk, O. Y. Adsorption Properties of Natural Minerals with In Situ Immobilized Polyaniline with Respect to the Anionic Forms of Mo(VI), W(VI), Cr(VI), As(V), V(V) and P(V). *Voprosy khimii khim. Tekhnol.* **2010,** *5*, 43–47.
11. Budnyak, T. M.; Yanovska, E. S.; Tertikh, V. A.; Vozniuk, V. I. Adsorption Properties of the Sokirnytsky Clinoptilolite-Polyaniline Composite Relative to the Anions of Elements of Groups V and VI of the Periodic Table Mendeleev. *Doklady NAN Ukrainy* **2011,** *3*, 141–145.

12. Yanovska, E. S. Comparison of Adsorption Properties of Natural Minerals Modified with Polyaniline. *Surface* **2015,** *7* (22), 173–185.

13. Yanovska, E.; Savchenko, I.; Polonska, Y.; Ol'khovik, L. Sternik, D.; Kychkiruk, O. Sorption Properties for Ions of Toxic Metals of Carpathian Clinoptilolite (Ukraine), *In Situ* Modified by Poly[*N*-Carboxyphenyl)methacrylamide. *New Mater., Compd. Appl.* **2017,** *1,* 45–53.

14. Parfitt, G. D.; Rochester, C. H. *Adsorption from Solution at the Solid/Liquid Interface*; Academic Press: London, New York, 1983, 416 p.

15. Savchenko, I.; Yanovska, E.; Sternik, D.; Kychkiruk, O. Synthesis of an Organo-Inorganic Composite Based on Clinoptilolite (Ukraine) *In Situ* Modified by Poly[8-Oxyquinoline Methacrylate] with Respect to Toxic Metal Ions. *Funct. Mater.* **2021,** *28,* 1–7.

16. Socrates, G. *Infrared Characteristic Group Frequencies*; John Wright & Sons Ltd.: Bristol, 1980; 153 p.

17. Babko, A. K;, Pylypenko, A. T. *Photometric Analysis*; Khimiya: Moscow, 1968; 388 p.

18. Polonska, Y.; Yanovska, E.; Savchenko, I.; Sternik, D.; Kychkiruk, O.; Ol'khovik, L. Sorption Properties for Toxic Metal Ions of Saponite In Situ Modified with Poly[*n*-(4-Carboxyphenyl) Methacrylamide]. *Ukr. Khim. Zhurnal.* **2018,** *84* (2), 67–72.

19. Polonska, Y.; Yanovska, E.; Savchenko, I.; Sternik, D.; Kychkiruk, O.; Ol'khovik, L. *In Situ* Immobilization on the Silica Gel Surface and Adsorption Capacity of Polymer-Based Azobenzene on Toxic Metal Ions. *Appl. Nanosci.* **2019,** *9* (5), 657–664.

20. Polonska, Y.; Yanovska, E.; Savchenko, I.; Sternik, D.; Kychkiruk, O.; Ol'khovik, L. In Situ Immobilization on the Silica Gel Surface and Adsorption Capacity of Poly[4-Methacroyloxy-(4'-Carboxy-2'-Nitro)-Azobenzene] on Toxic Metals Ions. *Mol. Cryst. Liq. Cryst.* **2018,** *671* (1), 164–174.

21. Yanovska, E.; Savchenko, I.; Sternik, D.; Kychkiruk, O. Adsorption Properties of Natural Alumosilicate Ukrainian Minerals, *In Situ* Modified by Poly[8-Methacroyloxy-Quinoline] to Pb(II), Mn(II), Cu(II) and Fe(III) Ions. *Mol. Cryst. Liq. Cryst.* **2021,** *717* (1), 1–13.

22. Zhang, M.; Xu, T.; Tian, T.; et al. A Composite Polymer of Polystyrene Coated with Poly(4-Vinylpyridine) as a Sorbent for the Extraction of Synthetic Dyes from Foodstuffs. *Anal. Methods* **2020,** *12,* 3156–3163.

23. Behbahani, M.; Bide, Y.; Bagheri, S.; et al. A pH Responsive Nanogel Composed of Magnetite, Silica and Poly(4-Vinylpyridine) for Extraction of Cd(II), Cu(II), Ni(II) and Pb(II). *Microchim. Acta* **2016,** *183* (1), 111–121.

24. Fang, J.; Gu, Z.; Gang, D.; Liu, C.; Ilton, E. S.; Deng, B. Cr(VI) Removal from Aqueous Solution by Activated Carbon Coated with Quaternized Poly(4-Vinylpyridine). *Environ. Sci. Technol.* **2007,** *41* (13), 4748–4753.

25. Yanovska, E.; Nikolaeva, O.; Kondratenko, O.; Vretik, L. Sorption Properties of Saponite Clay, *In Situ* Modified by Poly(4-Vinylpyridine-*co*-Styrene), Towards Cu(II), Cd(II), Pb(II), Mn(II) and Fe(III) ions. *French-Ukrain. J. Chem.* **2019,** *7* (2), 153–159.

PART II
Porous Functional Materials

CHAPTER 5

Synthesis and Investigation of Cross-Linked Hydrogels Based on Chitosan and Polyacrylamide

O. NADTOKA, PAVLO VIRYCH, and NATALIYA KUTSEVOL

Taras Shevchenko National University of Kyiv, Kyiv, Ukraine

ABSTRACT

The synthesis and physicochemical properties of chemically cross-linked hydrogels based on polyacrylamide and chitosan are considered in the work. Interpenetrating polymer networks of both cross-linked polyacrylamide and polyacrylamide grafted on chitosan by radical polymerization were obtained. It was studied that equilibrium swelling properties depend on the pH value of the solution and the composition of the gels. The chemical structure of the obtained hydrogels was characterized using IR spectroscopy.

5.1 INTRODUCTION

Hydrogels are three-dimensional hydrophilic polymer networks that can hold a large amount of water or biological fluids and have a soft and rubbery consistency, so they resemble living tissues.[1,2] Nowadays, remarkable works published about hydrogels based on interpenetrating polymer networks (IPN hydrogels) by reason of their improved sensitivity and mechanical properties that distinguish them from hydrogels based on mono-networks. An IPN is a mixture of two or more polymers in a network, where at least one of the systems is synthesized in the presence of the other.[3] This conducts to the

Mechanics and Physics of Porous Materials: Novel Processing Technologies and Emerging Applications.
Chin Hua Chia, Tamara Tatrishvili, Ann Rose Abraham, & A. K. Haghi (Eds.)

formation of a physically cross-linked network, where the polymer chains of the other system become entangled or penetrate the network of the first polymer. Each individual network retains its individual properties, so it is possible to observe synergistic improvements in physical and chemical properties.[4] The mechanical properties, swelling capacity, and reusability of hydrogels were further improved by synthesizing gels using both natural and synthetic polymers.

Hydrogels based on interpenetrating polymeric networks can be divided into the following classes: IPN hydrogels based on polysaccharides (alginate, starch, and derivatives, other polysaccharides), IPN hydrogels based on protein, IPN hydrogels based on synthetic polymers.

According to the chemical synthetic methods, hydrogels can be classified into (a) simultaneous IPN, when the reagents of both networks are mixed and two networks are synthesized simultaneously and independently of each other, by irrespective methods;[4,5] (b) sequential IPN, which are usually obtained by swelling a single polymer network in a solution containing a mixture of monomer, initiator, and activator in the presence or absence of a cross-linking agent.[3,6,7] When a linear polymer (synthetic or biopolymer) is entrapped in a matrix, a hydrogel of semi-IPN is formed.

Chitosan (Ch) (Figure 5.1), a linear cationic polysaccharide obtained by alkaline deacetylation of chitin, consisting of units β-(1→4)-2-amino-2-deoxy-D-glucopyranose and β-(1→4)-2-acetamido-2-deoxy-D-glucopyranose randomly distributed along the polymer chain. It is used by scientists due to its specific biological properties, such as biodegradability, biocompatibility, and antibacterial activity.

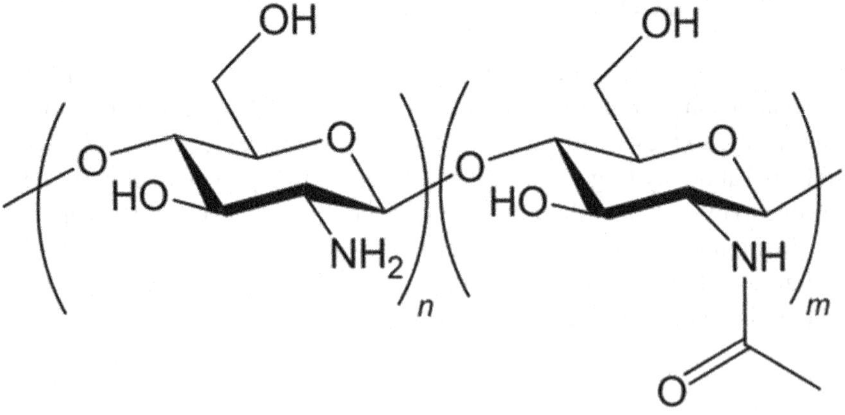

FIGURE 5.1   Chemical structure of chitosan.

The ratio between the two units is considered the deacetylation degree. When the deacetylation degree of Ch reaches approximately 50%, it becomes soluble in an aqueous acidic medium.[8] Amino groups (p*K*a from 6.2 to 7.0) are fully protonated in acids with p*K*a less than 6.2, which makes Ch soluble. Ch is insoluble in water, organic solvents, and alkalis, and soluble in acids such as acetic, nitric, hydrochloric, and phosphoric acids. When Ch is dissolved in an acidic medium, the amino groups in the chain are protonated and the polymer becomes cationic, allowing it to interact with different types of molecules as a single cationic polysaccharide. It is suggested, that this positive charge is responsible for its antimicrobial activity through interaction with the negatively charged cell membranes of microorganisms.[8]

Due to the high content of amino and hydroxyl functional groups, Ch also attracts attention as a biosorbent, which shows a high potential for the adsorption of proteins, dyes, and metal ions.[9,10] Also, during the creation of new and more effective drug release systems, IPN hydrogels based on Ch and other polysaccharides or their derivatives were synthesized. In particular, cellulose[11] or acrylamide-g-dextran[12] was first mixed with Ch, and then Ch was selectively cross-linked with glutaraldehyde.

The production of Ch-based hydrogels is accomplished by mixing Ch solutions with crosslinking agents or charged polymers under specific reaction conditions, such as pH, temperature, and so on, to form a viscous gel. Physical cross-linking of Ch through H-bonding interaction without the use of toxic cross-linking agents is the preferred approach due to safety in clinical practice. However, by controlling the amount of chemical cross-linking agents, such as glutaraldehyde, or using biologically permissible cross-linking agents, such as genipin, biocompatible hydrogels can be made.

Ch hydrogels are also prepared by cross-linking with glutaraldehyde, diglycidyl ether, diisocyanate, diacrylate, etc.,[13] but their use in pharmaceuticals is limited because these hydrogel products must undergo strict purification before administration. In some cases, cross-linking reagents can also interact with the drug, deactivating or limiting its therapeutic level.

Although the cross-linking density, hydrogel porosity and rigidity can be adjusted in IPN hydrogels based on Ch according to the intended application, the encapsulation of sensitive biopharmaceuticals turns out to be complicated. During the preparation of IPN, toxic agents that initiate or catalyze polymerization or cross-linking can lead to reduced biocompatibility. The removal of initiators and catalysts can be improved by washing with solvents, ionic substances, or changing the pH after the reaction. However, their complete removal is a difficult task, which makes their clinical application problematic.

When cross-linking during the polymerization of nonionic monomers in the presence of Ch, such monomers as acrylamide (AA),[14] *N*-isopropylacrylamide,[15] *N,N*-dimethylacrylamide,[16] and 2-hydroxyethyl methacrylate[17] are most often used. Variations in mechanical properties and water content are expected for them. One of the main goals is their use in biomedical applications, such as controlled drug release systems and as scaffolds in tissue engineering.

Our chapter presents the creation and research of polymer hydrogels based on polyacrylamide and Ch as materials that can satisfy the requirements for wound dressings, in particular, to be suitable for the atraumatic treatment of affected skin areas, ensuring their disinfection and treatment.

5.2   EXPERIMENTAL METHODS AND MATERIALS

5.2.1   MATERIALS

Ch with a deacetylation degree >90% (Glentham Life Sciences), AA, *N,N'*-methylene-bis-acrylamide (MBA), acetic acid, nitric acid (HNO_3), and cerium(IV) ammonium nitrate (CAN) (Sigma-Aldrich) were used for the synthesis. Double-distilled water was used to synthesize and study the swelling kinetics of the hydrogels.

5.2.2   SYNTHESIS OF HYDROGELS

A series of chemically cross-linked hydrogels were synthesized by radical polymerization of AA in the presence of the cross-linking agent MBA, which was taken at a concentration of 0.4% of the total mass of AA and Ch (Figure 5.2). The redox system Ce(IV)/HNO_3 was used as an initiator. Ch (0.05, 0.1, or 0.2 g) was placed in 15 ml of distilled water in a 50-ml beaker; then 1 ml of concentrated acetic acid was added dropwise. The mixture was left for 24 h for the complete dissolution of Ch. After that, when heated to 50°C, argon was passed for 20 min, while stirring on a magnetic stirrer. About 10 ml of an aqueous solution of AA (3.5–3.35 g) and MBA was added to the reaction mixture and continued to pass argon for 10 min. CAN was dissolved in 1 ml of 0.125 N HNO_3 and added to the reaction mixture dropwise while stirring and passing argon. The reaction mixture was left overnight and further the formed hydrogels were transferred to distilled water to remove unreacted monomers, ethanoic acid, and initiator residues. Distilled water was changed three times a day for several days.

As a result, three samples of hydrogels based on interpenetrating polymer networks with different concentrations of Ch were obtained. They were designated as PAA-Ch-0.05, PAA-Ch-0.1, PAA-Ch-0.2, according to a contained amount of Ch.

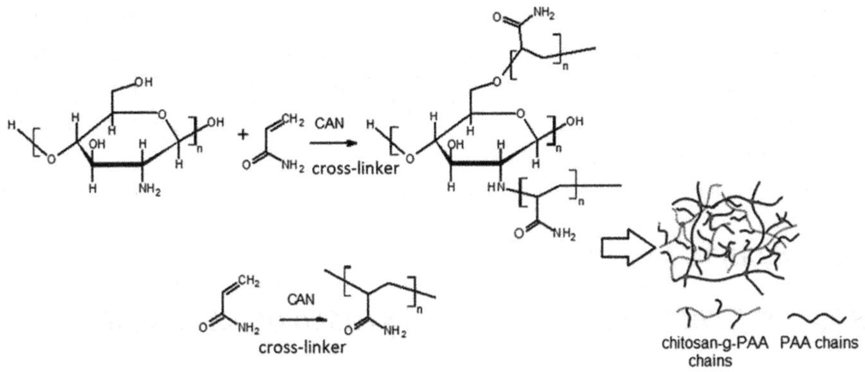

FIGURE 5.2 Scheme of a possible reaction.

For comparison, a hydrogel based on cross-linked PAA polyacrylamide was obtained.

The obtained samples were characterized by FTIR spectroscopy and the swelling behavior in different environments was investigated.

5.2.3 FTIR SPECTRA

FTIR spectra of the obtained hydrogels were recorded in KBr powder using a BrukerVector-22 spectrometer in the range of 400–4000 cm^{-1}.

5.2.4 SWELLING STUDY

Swelling properties are quantitatively characterized by the swelling degree S (%):

$$S = \frac{m - m_0}{m_0} \times 100, \%$$ (5.1)

where m is the mass of the swollen sample, g; m_0 is the mass of the polymer sample before swelling.

The rate of limited swelling is described by the first-order kinetic equation (5.2):

$$\frac{dS}{d\tau} = k(S - S_{max})$$ (5.2)

where $d\alpha/d\tau$ is the rate of limited swelling in time τ; S is the swelling degree in time τ; S_{max} is the maximum swelling degree. On the graph, this equation is represented by a straight line in coordinates $dS/d\tau - S$, where the segment cutting off the ordinate is equal to kS_{max}, and the tangent of the angle of inclination of the line is equal to k. At the same time, the segment that cuts off the straight line is equal to S_{max}.

5.3   RESULTS AND DISCUSSION

5.3.1   SYNTHESIS OF IPN HYDROGELS BASED ON AA AND CH

Ch-based hydrogels are obtained using homo- and copolymerization methods to obtain networks, which possess the functionality of both materials. During synthesis, Ch previously swells in an aqueous solution of monomers, which are then radically polymerized to form a physically entangled polymer network known as an interpenetrating polymer network. The scheme of a possible polymerization reaction is shown in Figure 5.3. It indicates the formation of a free macroradical initially with the AA monomer, which is continuously added to the Ch molecule. Cross-linking during the graft copolymerization of AA on Ch can lead to interpenetrating networks of cross-linked chitosan-g-polyacrylamide and cross-linked polyacrylamide, which are entangled polymers. The synthesis of IPN hydrogels based on AA and Ch took place according to the following possible scheme:

Samples of hydrogels with different concentrations of Ch are marked as PAA-Ch-0.05, PAA-Ch-0.1, and PAA-Ch-0.2. Cross-linked PAA, which represents a mono-network, was obtained as a sample for comparison. During the synthesis, the concentration of the cross-linker remained constant and was 0.4% as optimal for obtaining cross-linked hydrogels based on PAA.[18]

5.3.2   FTIR CHARACTERIZATION OF HYDROGELS

In the IR spectrum of AA (Figure 5.3), valence vibrations of the NH group of the primary amide are observed: asymmetric at 3198 cm^{-1} and symmetric at 3198 cm^{-1}. The overlap of the C=O valence vibration band and the first amide band is observed at 1680 cm^{-1}. The second amide band is observed at 1617 cm^{-1}; the band of N–H deformation out-of-plane vibration corresponds to 700–600 cm^{-1}.

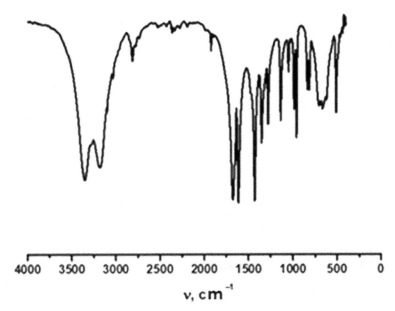

FIGURE 5.3   FTIR spectrum of acrylamide.

In the FTIR spectra of the studied samples (Figures 5.4 and 5.5), valence vibrations of NH groups at 3400 cm^{-1} are observed. The peak at 2900 cm^{-1} indicates the presence of CH$_2$ groups in polymer chains and cross-links. Deformation and valence vibrations of the amide group are observed at 1600–1660 cm^{-1}.

The structure of chitosan-g-polyacrylamide IPN hydrogels was confirmed by comparing their FTIR spectra with the spectrum of Ch (Fig. 5.5). The spectrum of Ch shows characteristic absorption bands at 1655 cm^{-1} (amide) and 1597 cm^{-1} (amine). These two peaks are slightly shifted and are more distinct in the hydrogel spectrum, as confirmed by the peak at 3206 cm^{-1} (amide).

In addition, bending vibrations of C–H and O–H are observed at 1300–1400 cm^{-1} with a sharp peak at 1383 cm^{-1}. However, this sharp peak corresponds to the O–H group, confirmed by two sharp peaks at 1159 and 1082 cm^{-1} (alcohol and ester valence vibrations, respectively) and which are not present in the hydrogel spectrum. This means that the hydroxyl group of Ch is a better site for the reaction with the cross-linking agent and the grafting of AA caused by the lower steric hindrance of the primary hydroxyl group.

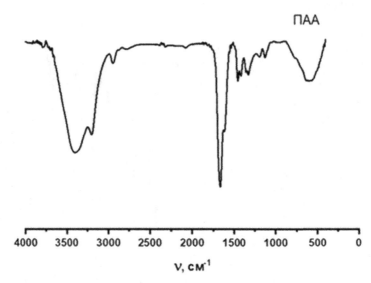

FIGURE 5.4   IR spectrum of cross-linked PAA.

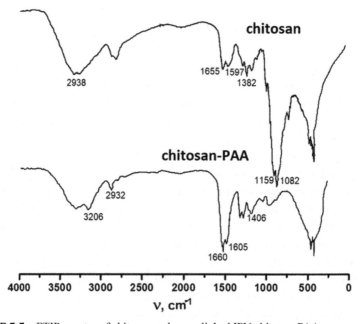

FIGURE 5.5   FTIR spectra of chitosan and cross-linked IPN chitosan-PAA.

5.3.3 SWELLING PROPERTIES OF HYDROGELS

Transparent gel samples were dried and their swelling kinetics in distilled water at a temperature of 25°C was investigated.

As shown in Figure 5.6, the swelling percentage first increased and then reached a plateau. The equilibrium swelling percentage was reached after immersing the IPN hydrogel in water for 3×10^3 min. This behavior can be explained by the high hydrophilicity of polyacrylamide chains and the high capacity of the chitosan-g-polyacrylamide structure to retain water. Thus, the difference in osmotic pressure between the gels and the solvent can be reduced due to the strengthening of the polymer–solvent interaction. As a result, the rate of absorption from water will increase. However, a further increase in time leads to a balance of osmotic pressure, so swelling will remain constant.

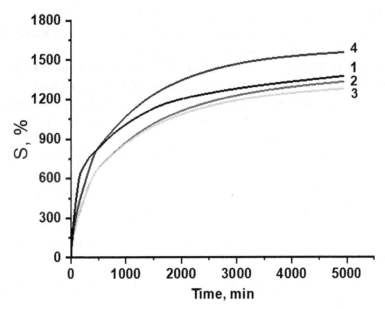

FIGURE 5.6 Swelling curves of 1—PAA; 2—PAA-Ch-0.05; 3—PAA-Ch-0.1; and 4—PAA-Ch-0.2 in distilled water.

As is known, Ch is positively charged at low pH, but uncharged and insoluble at neutral and high pH. As can be seen from Figure 5.6, the slight presence of Ch in PAA network and its gradual increase leads to a decrease in hydrophilicity and, accordingly, a decrease in the swelling degree. This feature is confirmed by the results obtained for pure cross-linked PAA, taken as a sample for comparison. The equilibrium swelling degree S_{eq} and the swelling rate constant k for PAA are higher than for samples containing Ch. As can be seen in Table 5.1 for PAA, PAA-Ch-0.05, and PAA-Ch-0.1, S_{eq} is 1380%, 1335%, and 1280%, respectively.

TABLE 5.1   Hydrogel Swelling Parameters.

| Sample | % of chitosan | S_{eq} (%) | | | | k (min^{-1}) (in dist. water) |
		pH = 4.76	pH =7	pH = 9.25	0.8% NaCl	
PAA	0	1650	1380	2750	1360	0.4232
PAA-Ch-0.05	1.4	1600	1335	2600	1300	0.3976
PAA-Ch-0.1	1.8	1550	1280	2500	1210	0.3624
PAA-Ch-0.2	3.6	1780	1535	3100	1500	0.4555

However, a significant concentration of Ch, on the contrary, led to a significant increase in the equilibrium swelling degree in the hydrogel, as shown for the sample PAA-Ch-0.2, S_{eq} is 1535%.

Obviously, this fact can be explained by the decrease in cross-linking during polymerization. A higher concentration of Ch in an aqueous solution (>3.6%) during polymerization led to the formation of colloidal solutions with high viscosity, which prevents the movement of free radicals and monomer molecules and their effective interaction.

In this regard, it was not possible to obtain high-quality samples with an even higher concentration of Ch.

A similar dependence was observed for the IPN hydrogels based on Ch and polyacrylamide obtained by other methods.[19]

Figure 5.7 shows the swelling of PAA-Ch-0.1 depending on pH. The effect of pH on the swelling degree was studied in buffer solutions.

Ammonium acetate buffer solutions were used:

1. Acetate buffer mixture (CH_3COOH (acid) + CH_3COONa (base)), pH = 4.76;
2. Ammonia buffer mixture ($NH_3 \cdot H_2O$ (base) + NH_4Cl (acid)), pH=9.25.

The highest equilibrium swelling degree S_{eq} = 2500% was achieved in an alkaline medium (pH = 9.25). In acidic conditions, the amide groups of AA are converted into carboxyl groups (COO–). Therefore, the hydrogel consists of Ch as a cationic (NH_3^+) and acrylic acid as an anionic (COO$^-$) polyelectrolyte.[20] However, in acidic conditions at pH = 4.76, the cationic nature of the hydrogel (NH_3^+) prevails. Therefore, $NH_3^+–NH_3^+$ electrostatic repulsion increases the osmotic pressure inside the gel relative to the solution. The gel achieves osmotic pressure equilibrium by swelling. But at pH = 9.25, the percentage of swelling increased, probably due to the predominance of the anionic nature of the hydrogel (COO^-).

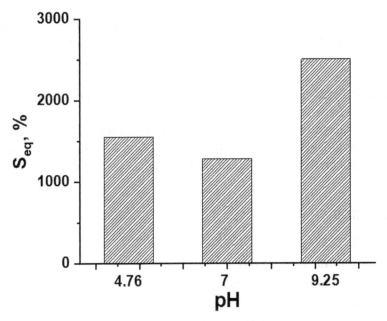

FIGURE 5.7   Swelling degree at different pH for PAA-Ch-0.1.

The swelling of hydrogels was studied in a physiological solution of 0.8% sodium chloride. It was found that the hydrogel swelling degree in physiological solution is less than in distilled water and is 1210% (Table 5.1). This well-known phenomenon can be explained by a decrease in the osmotic pressure between the internal and external solutions of the mesh in the presence of salt, and therefore the percentage of swelling decreases. A similar observation was previously obtained in hydrogels based on PAA and dextran-g-PAA.[21]

The phenomenon of the dependence of swelling of Ch-based hydrogels on various conditions allows changing the content of water or solution in IPN by changing the medium. These features of "smart" hydrogels make it possible to use them as biomaterials to create new and more effective drug release systems.

5.4 CONCLUSION

Chemically cross-linked hydrogels based on interpenetrating polymer networks of Ch and polyacrylamide, which differed in the quantitative composition of polymers, were synthesized by radical polymerization using a redox system as initiator. In the swelling study of hydrogels, it was found that a slight increase in the concentration of Ch during synthesis naturally reduces the swelling degree of the samples. This regularity is contravened when significant amounts of Ch are used during synthesis. It was shown that Ch-containing hydrogels are pH-sensitive and swell best in acidic and alkaline mediums, which indicates their behavior as polyelectrolytes due to a change in the degree of their ionization (as polyampholytes).

The obtained results indicate a high potential ability of the synthesized polymer hydrogels based on polyacrylamide and Ch to meet the requirements of materials for biomedical applications.

KEYWORDS

- **cross-linked hydrogels**
- **interpenetrating polymer network**
- **chitosan**
- **swelling**
- **polyacrylamide**

REFERENCES

1. Hoffman, A. S. Hydrogels for Biomedical Applications. *Adv. Drug Deliv. Rev.* **2002,** *43,* 3–12.
2. Peppas, N. A.; Bures, P.; Leobandung, W.; Ichikawa, H. Hydrogels in Pharmaceutical Formulations. *Eur. J. Pharm. Biopharm.* **2000,** *50,* 27–46.
3. Myung, D.; Waters, D.; Wiseman, M.; Duhamel, P. E.; Noolandi, J.; Ta, C. N.; Frank, C. W. Progress in the Development of Interpenetrating Polymer Network Hydrogels. *Polym. Adv. Technol.* **2008,** *19,* 647–657.
4. Ignat, L.; Stanciu, A. Advanced Polymers: Interpenetrating Polymer Networks. In *Handbook of Polymer Blends and Composites*; Kulshreshtha, A. K., Vasile, C., Eds.; Rapra Technology: Shrewsbury, 2003; pp 275–280.

5. Wang, J. J.; Liu, F. Enhanced Adsorption of Heavy Metal Ions onto Simultaneous Interpenetrating Polymer Network Hydrogels Synthesized by UV Irradiation. *Polym. Bull.* **2013**, *70*, 1415–1430.

6. Chivukula, P.; Dušek, K.; Wang, D.; Duškova-Smrčkova, M.; Kopečkova, P.; Kopeček, J. Synthesis and Characterization of Novel Aromatic Azo Bond-Containing pH-Sensitive and Hydrolytically Cleavable IPN Hydrogels. *Biomaterials* **2006**, *27*, 1140–1151.

7. Hoare, T. R.; Kohane, D. S. Hydrogels in Drug Delivery: Progress and Challenges. *Polymer* **2008**, *49*, 1993–2007.

8. Pillai, C. K. S.; Paul, W.; Sharma, C. P. Chitin and Chitosan Polymers: Chemistry, Solubility and Fiber Formation. *Prog. Polym. Sci.* **2009**, *34* (7), 641–678.

9. Crini, G.; Badot, P.-M. Application of Chitosan, a Natural Aminopolysaccharide, for Dye Removal from Aqueous Solutions by Adsorption Processes Using Batch Studies: A Review of Recent Literature. *Prog. Polym. Sci.* **2008**, *33*, 399–447.

10. WanNgah, W. S.; Teong, L. C.; Hanafiah, M. A. K. M. Adsorption of Dyes and Heavy Metals by Chitosan Composites: A Review. *Carbohydr. Polym.* **2011**, *83*, 1446–1456.

11. Cai, Z.; Kim, J. Cellulose–Chitosan Interpenetrating Polymer Network for Electro-active Paper Actuator. *J. Appl. Polym. Sci.* **2009**, *114*, 288–297.

12. Rokhade, A. P.; Patil, S. A.; Aminabhavi, T. M. Synthesis and Characterization of Semi-Interpenetrating Polymer Network Microspheres of Acrylamide Grafted Dextranand Chitosan for Controlled Release of Acyclovir. *Carbohydr. Polym.* **2007**, *67*, 605–613.

13. Hoare, T.; Kohane, D. Hydrogels in Drug Delivery: Progress and Challenges. *Polymer* **2008**, *49* (8), 1993–2007.

14. Bonina, P.; Petrova, T. S.; Manolova, N. pH-Sensitive Hydrogels Composed of Chitosan and Polyacrylamide—Preparation and Properties. *J. Bioact. Compat. Polym.* **2004**, *19*, 101–116.

15. Alvarez-Lorenzo, C.; Concheiro, A.; Dubovik, A. S.; Grinberg, N. V.; Burova, T. V.; Grinberg, V. Y. Temperature-Sensitive Chitosan–Poly(*N*-Isopropylacrylamide) Interpen-etrated Networks with Enhanced Loading Capacity and Controlled Release Properties. *J. Control. Release* **2005**, *102*, 629–641.

16. Ramesh Babu, V.; Hosamani, K. M.; Aminabhavi, T. M. Preparation and *In-Vitro* Release of Chlorothiazide Novel pH-Sensitive Chitosan-*N,N*-Dimethylacrylamide Semi-interpenetrating Network Microspheres. *Carbohydr. Polym.* **2008**, *71*, 208–217.

17. Ha, Y. A.; Lee, E. M.; Ji, B. C. Mechanical Properties of Semi-Interpenetrating Polymer Network Hydrogels Based on Poly(2-Hydroxyethyl Methacrylate) Copolymer and Chitosan. *Fibers Polym.* **2008**, *9*, 393–399.

18. Nadtoka, O.; Virych, P.; Kutsevol, N. Synthesis and Absorption Properties of Hybrid Polyacrylamide Hydrogels. *Mol. Cryst. Liq. Cryst.* **2021**, *719* (1), 84–93.

19. Saber-Samandari, S.; Gazi, M.; Yilmaz, E. UV-Induced Synthesis of Chitosan-*g*-Polyacrylamide Semi-IPN Superabsorbent Hydrogels. *Polym. Bull.* **2012**, *68*, 1623–1639.

20. Yazdani-Pedram, M.; Retuert, J.; Quijada, R. Hydrogels Based on Modified Chitosan, Synthesis and Swelling Behavior of Poly(Acrylic Acid) Grafted Chitosan. *Macromol. Chem. Phys.* **2000**, *201*, 923–930.

21. Nadtoka, O.; Virych, P.; Kutsevol, N. Investigation of Swelling Behavior of PAA and D-PAA Hydrogels. *Springer Proc. Phys.* **2020**, *247*, 47–60.

CHAPTER 6

Study of Dehydration Behavior of Silica Xerogel and Aerogel by Diffuse Reflectance Fourier Transform Spectroscopy

SENEM YETGIN[1] and DEVRIM BALKOSE[2]

[1]*Department of Food Engineering, Kastamonu University, Kastamonu, Turkey*

[2]*Department of Chemical Engineering, İzmir Institute of Technology, Urla, İzmir, Turkey*

ABSTRACT

The pore size, surface area, and density of silica gel were controlled by drying silica alcogel in ambient conditions and using supercritical ethanol extraction to obtain silica xerogel and silica aerogel, respectively, in the present study. The samples were characterized by N_2 gas adsorption at $-196°C$, scanning electron microscopy and Fourier transform infrared (FTIR) spectroscopy and diffuse reflectance infrared Fourier transform (DRIFT) spectroscopy. The effect of time on surface characteristics of silica aerogel up to three-month storage time was investigated by N_2 gas adsorption. The functional groups of the xerogel and silica aerogel and a commercial mesoporous silica gel were examined by FTIR transmission spectroscopy. The effects of pressure, relative humidity, and temperature on hydration–dehydration behavior of silica aerogel and silica xerogel were investigated using a temperature and pressure-controlled chamber in diffused reflectance cell of FTIR spectrophotometer. It was observed that free water was eliminated from the samples

Mechanics and Physics of Porous Materials: Novel Processing Technologies and Emerging Applications.
Chin Hua Chia, Tamara Tatrishvili, Ann Rose Abraham, & A. K. Haghi (Eds.)
© 2024 Apple Academic Press, Inc. Co-published with CRC Press (Taylor & Francis)

under vacuum at ambient temperature. Hydrogen-bonded OH groups and isolated OH groups were eliminated partially by heating the samples up to 500°C. The samples heated up to 500°C and cooled to ambient conditions had still had the capability of adsorbing free water.

6.1 INTRODUCTION

Aerogels applied in different fields, such as[1,2]

- supercapacitors,
- -refrigerators, hydrophobic adsorbents for nonpolar compounds,
- Cerenkov radiation detector media in high energy physics, or inertial confinement fusion targets for thermonuclear fusion reactions,
- thermal insulation in solar window systems,
- catalytic supports,
- acoustic barriers.

For silica aerogels and xerogels, the drying process was studied for the first time by Ru et al.[3] In such a drying approach, the conventional and CO_2 along with ethanol supercritical drying method or (SCD methods) was used.

It should be noted that silica is often used as inorganic simple oxides. Nowadays, silica aerogels have been studied progressively due to their unique properties. For this reason, they have the potential to be used in different technological sectors. It is worth mentioning that our nanostructured material in the form of silica aerogels has the following characteristics[4]:

- a low density,
- a high porosity,
- a high specific surface area,
- an insignificant dielectric constant, and
- appropriate heat insulation properties.

As reported by previous studies that aerogel has

- elevated thermal insulation value 0.005 W/m K,
- high porosity 80–99.8%, low density ~0.003 g/cm³,
- high specific surface area between 500 and 1200 m²/g,
- ultralow dielectric constant $k = 1.0$–2.0, and
- modest index of refraction practically equal to 1.05.

An aerogel is the outcome of the removal of the liquid part of an alcogel without harming the spectroscopy solid material (frequently by supercritical removal). In the course of SCD, mainly under elevated pressure and elevated *T*, not merely drying even so the chemical reaction of alcogel component and supercritical extraction takes place. This drying process is related to the removal of the solvent from wet gel.[5] The texture of the dried material will be conserved in this process. Practically speaking, it strongly lower the pore collapse. SCD comprises heating a gel in a device such as autoclave, prior to the pressure and temperature outreached the critical temperature T_c and critical pressure P_c of the liquid entrapped in the gel pores. It should be noted that the aerogel shows an elevated surface area along with remarkable nanoporosity. This is observed by the lack of capillary forces. The surface forces captivating the solid as well as the liquid phases were not evident. This is, in general, due to gas phase and the solid existed over the critical (*T*). During drying, the shrinkage of the gels is driven by the capillary pressure, *P*. This is shown by Equation (6.1). Capillary force can also be evaluated by contact angle (please refer to Eq. (6.2)):

$$P = \frac{\gamma_{LV}}{(r_P - \delta)}$$

$$P = \frac{2\gamma x Cos\theta}{r_p} \tag{6.1}$$

where γ_{lv} is the surface tension of the pore liquid, r_p is the pore radius (this can be represented by Eq. (2)), and δ is the thickness of a surface adsorbed layer.[6]

$$r_P = \frac{2V_p}{S_p} \tag{6.2}$$

where V_p is the pore volume and S_p is the surface area.

Iswar et al.[7] investigated the effect of aging silica alcogels before drying on pore volume of silica gels. The specific surface area decreased with aging time due to Oswald ripening phenomena in conventional dried gels. However, it was not affected in supercritical carbon dioxide–dried gels. The capability of the gel particle network to hold out against irreversible pore collapse during ambient pressure drying is increased by aging of alcogels.

DNA adsorption by silica aerogel was investigated by Yetgin and Balkose.[8,9] Silica aerogel adsorbed more DNA than mesoporous DNA due to its high pore volume.[8,9] Dehydration behavior of zeolites and zinc borates[11] was also investigated by diffuse reflectance infrared Fourier transform (DRIFT) spectroscopy.

He et al.[12] investigated the effect of heat treatment in hydrophobic silica gels. Silica hydrogels from water glass exchanged solvent and surface was modified with trimethylchlorosilane, and dried in ambient conditions. The obtained hydrophobic aerogels were heated in a temperature-controlled oven. When the heat treatment temperature was lower than 350°C, the properties of the silica aerogels showed little change. With a further increase in the heat treatment temperature, secondary particles consolidation with each other and mesopores were demolished and surface area decreased.[12]

Alcaniz-Monge et al.[13] studied the impact of pore size scattering on water adsorption for a series of relatively dehydroxylated silica gels with various pore structures. It was observed that unspecified silica gels present adsorption of water mainly in the first range of correlative pressures (correlative pressure: 0.0–0.6), while other silica gels also represent adsorption of water at elevated relative pressures. Comparison of the N_2 and water adsorption isotherms indicates that only silica gels contain mesoporosity; meanwhile, it shows water adsorption at relative pressures higher than 0.6. Nevertheless, an adequate relation among the primary part of water adsorption isotherms (relative pressure: 0.0–0.3) and CO_2 adsorption isotherms has been determined (the greater the CO_2 adsorption capacity, the greater the water adsorption capacity at low pressure).[13] DRIFT spectroscopic characterization of silica gels was carried out in a spectrometer Mattson infinities (MI60). Powdered samples were located in a controlled environment chamber available in this equipment. Measurements were carried out under N_2 flow (60 mL/min) and at a temperature of 623 K. The signal related to OH groups appears in the region between 3800 and 3000 cm^{-1}. Each sample shows a sharp peak around 3750 cm^{-1} attributed to free OH groups, and a broad band between 3700 and 3000 cm^{-1}, related to hydrogen-bonded OH groups.[13] Garbasi et al.[14] investigated surface modification of silica gel with methyl trichlorosilane (MTCS), dimethyl dichloro silane (DCDMS), trimethyl chlorosilane (TMCS), and tetrachlorosilane (TCS). DRIFT spectroscopy was used to examine the modified silica gels. Isolated surface hydroxyls are responsible for the narrow band at 3746 cm^{-1}, internal hydroxyls give rise to the band at 3650–3670 cm^{-1} and vicinal hydroxyls correspond to the absorption at 3540–3550 cm^{-1}. DRIFT spectra of silica samples treated with TMCS, DCDMS, MTCS, or TCS showed two main effects: the disappearance or decrease of the band assigned to isolated hydroxyls and the occurrence of a series of absorptions in the 2900–3100 cm^{-1} region. The strongest peak, occurring at 2962 cm^{-1}, was attributed to asymmetric stretching of methyl, while its companion near 2920 cm^{-1} is due to CH_3 symmetric stretching. A third small peak at 2858 cm^{-1} was probably due to CH$_2$ stretching, originated by some residual *n*-heptane used to wash silica after treatments. Such bands provide direct evidence for methyl silyl formation at the silica surface.[14]

While Fourier transform infrared (FTIR) gives information about the vibrations of the bulk material, DRIFT spectroscopy gives information about the surface of the material. DRIFT spectroscopy was used for quantification of maximal chemisorption of 3-aminopropylsilyl groups on silica by Bukleski et al.[15] and for identification of chemically modified silica by Fursgerber et al.[16] and high-temperature properties of silica by Chunxiao et al.[17] Catalyst evaluation for hydrocarbon synthesis was made by DRIFT spectroscopy.[18–20] Lewis and Brønsted acid groups on silica/zirconia catalyst were investigated using DRIFT spectroscopy by Ma et al.[21] Lewis acid cites were converted by hydrogen treatment to Brønsted acid cites of tungsten oxide supported on mesoporous spherical silica nanoparticles as shown by *in-situ* ammonia adsorption in DRIFT cell of infrared spectrophotometer.[22] Amine-functionalized silica was synthesized and shown by DRIFT spectroscopy; they had C–H stretching band at a range of 2850–3000 cm^{-1} and C–N stretching mode at ~1250 cm^{-1}.[23]

Wang et al.[24] classified the silanol groups of silica as isolated, geminal, chained, and terminal. The engineering properties of silica surface alongside with its physical and chemical properties are highly affected by the surface silanol groups. Therefore, the surface of silica under ambient condition was covered by a layer of adsorbed water molecules. The absorbed water was combined with hydrogen-bonded (H-bonded) silanol groups emerging in a broad band in infrared spectrum (3700–3000 cm^{-1}). All samples annealed using the temperature range of 480°C to remove the water effect. Throughout, the thermal treatment, the surface of silica experienced many transitions that were

- removal of weakly H-bonded hydroxyls,
- isolated hydroxyls in vicinity,
- the elimination of water, and
- removal of strongly H-bonded hydroxyls

In the present study, silica aerogel and silica xerogel were obtained by supercritical ethanol drying of the silica alcogel from tetraethoxysilane and by conventional drying. The gel samples were characterized by nitrogen gas adsorption, and scanning electron microscopy (SEM), transmission infrared spectroscopy to observe the drying method on their properties. The effect of aging time on aerogels surface properties was also aimed to be investigated. The dehydrating behavior of silica gel samples by heating under high vacuum were investigated *in situ* by DRIFT spectroscopy.

6.2   EXPERIMENTAL METHODS AND MATERIALS

6.2.1   *MATERIALS*

Silica (Sigma Aldrich-Silica gel, Grade 7744 pore size 60 Å, 70–230 mesh) was used as commercial mesoporous silica gel.

Silica aerogel was prepared by controlled hydrolysis and condensation reactions of tetraethyl orthosilicate (TEOS). In Table 6.1, properties of used materials for silica aerogel and xerogel preparation are given.

TABLE 6.1   Properties of Used Materials for Silica Aerogel and Xerogel Preparation.

Materials	Abbreviation	Specifications
TEOS	$Si(OC_2H_5)_4$	98%, $M = 208.3$, $d = 0.934$ g/cm³ (Aldrich)
EtOH	C_2H_5OH	99.8%, $M = 46.07$, $d = 0.79$ g/cm³ (Riedel)
Ammonium hydroxide	NH_4OH	28–30% NH3, $M = 35.05$, $d = 0.9$ g/cm³ (Aldrich)
Hydrochloric acid	HCl	37% $M = 36.46$, $d = 1.9$ g/cm³ (Merck)

TEOS = Tetraethyl orthosilicate, EtOH = ethyl alcohol.

6.2.2   *METHODS*

6.2.2.1   *SILICA AEROGEL*

Silica aerogel was obtained using the following two steps of sol–gel procedure:

At first, hydrochloric acid was applied as an acid catalyst of the hydrolysis.

Then, the TEOS, ethanol (EtOH), distilled water, and hydrochloric acid were all mixed in a 250-cm³ glass bottle at a molar ratio of $1:6:4:10^{-3}$ (to be stirred for 30 min, until a hydrolysis solution was reached).

For the second step:

- ammonia (as a base catalyst) was added drop-wise into the solution (resulting from the first step, at the molar ratio of TEOS:NH_4OH equals to $1:10^{-2}$).
- A white precipitate was seen right away.
- The mixture was more stirred for 2 min (and then kept for gelation).

The set gel is known as the alcogel and contained some amount of water.

- The alcogel was washed five times by holding in contact with 40 cm³ fresh ethanol for 5, 7, 9, 12, and 39 days for the removal of water.

- Right after the washing approach, the alcogel, having 84.3% ethanol, 1.1% water, and 14.6% silica, was dried by two techniques: ethanol SCD technique and normal-condition drying technique.
- The samples were considered as aerogel and xerogel (as respectively based on drying technique).

Ethanol SCD system is shown in Figure 6.1.

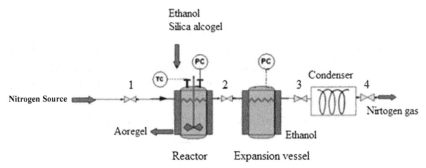

FIGURE 6.1 Ethanol supercritical drying system.

The supercritical ethanol drying set-up is shown in Figure 6.1. The system has a temperature-controlled 300-cm³ PARR 4561 reactor and a 300-cm³ PARR expansion vessel. In a typical experiment, approximately 5.0 g of silica alcogel and 100 cm³ ethanol was placed in the reactor. First nitrogen gas was passed from the system by opening valves 1, 2, 3, and 4 to remove oxygen from the system. Then all the valves were closed and the reactor is heated to 250°C and 7.2 MPa pressure. The supercritical temperature and pressure of ethanol are 241°C and 6.36 MPa, respectively. Thus ethanol was brought to a supercritical state. The ethanol and silica alcogel mixture was stirred at 600 rpm during heating. The valve 2 is opened at this temperature. The gauge pressure in the system was decreased to zero in this process. The ethanol in supercritical state is expanded into expansion vessel and condensed there as a liquid. Heating is stopped and the system was opened when it was cooled to room temperature. The silica aerogel particles were collected from the reactor and the ethanol phase was taken from the expansion vessel.[25]

Silica aerogel that was in the cooled first vessel was transferred to a desiccator for storage. The surface characteristics of stored gels up to three months were measured to see the effect of aging time.

6.2.2 SILICA XEROGEL

The washed alcogel was dried at ambient conditions to obtain the xerogel.

6.2.3 CHARACTERIZATION

6.2.3.1 NITROGEN GAS ADSORPTION

Specific surface area, pore diameter, and pore volume of the commercial silica gel, silica aerogel, and silica xerogel were obtained by using the physical adsorption of nitrogen technique at 77 K, using Micromeritics ASAP 2010. Pore-size analysis of mesoporous materials from adsorption isotherms is a specific technique based on capillary condensation, evaporation, and associated hysteresis phenomena of the adsorbed gases such as nitrogen (N_2).

6.2.3.2 TAPPED DENSITY

The tap densities of commercial silica gel and silica aerogel were measured by filling a known volume of a glass cylinder and by weighing the samples.

6.2.3.3 TRANSMITTANCE FTIR SPECTROSCOPY

Transmittance infrared spectra of commercial silica gel, silica aerogel, and silica xerogel were obtained by the standard potassium bromide (KBr) pellet method in the region of 400–4000 cm^{-1} after 40 scans at 4 cm^{-1} resolution at room temperature on a spectrometer equipped with a deuterated triglycine sulfate detector (FTS 3000 MX, Digilab Excalibur Series). Prior to the analyses, 1 mg of dehydrated silica gel samples was ground with 150 mg of KBr and the powder mixtures were pressed into pellets of 13 mm in diameter under 4 t/cm^2 by a hydraulic press.

6.2.3.4 TEMPERATURE-PROGRAMMED DIFFUSE REFLECTANCE FTIR SPECTROSCOPY

The dehydration behavior of the samples was determined by an in situ temperature-programmed DRIFT infrared spectroscopy method. The samples were heated in situ from room temperature up to 500°C at a heating rate of 2°C/min under vacuum (10^{-2} mbar) in a praying mantis diffuse reflection attachment (Harrick Scientific Products Inc., Ossing, NY) equipped with a high-temperature, low-pressure reaction chamber (HVC-DRP, Harrick Scientific Products Inc.). The reaction chamber has two CaF_2 windows and one glass window for observation. The temperature of the sample in the sample cup of the reaction chamber was measured by a K-type thermocouple and controlled by a low-voltage heating cartridge using a single-loop proportional–integral–derivative (PID) temperature controller (Series 989, Watlow, Winona, MN). The pressure in the chamber was measured by a Pirani vacuum gauge (measurement range: 5×10^{-4}–1,000 mbar, Thermovac TR 216 S, Oerlikon Leybold, Cologne, Germany) and Penning vacuum gauge (measurement range: 10^{-9}–10^{-3} mbar, PTR 225, Oerlikon Leybold). The DRIFT spectrum of the sample was obtained at different temperatures during its dynamic heating. The ambient conditions in the laboratory were 20–25°C temperature and 60% relative humidity air at 1 atm pressure. A 0.1-g sample of silica aerogel or xerogel was placed on the sample holder and after its spectrum at room temperature and 1 atm pressure was taken, the sample chamber was evacuated down to the 0.1 Pa pressure and heated up to 500°C at 2°C/min^{-1} rate at 0.1 Pa pressure. The spectra were recorded at 50 scans at a resolution of 8 cm^{-1} in the 400–4000-cm^{-1} range. The background spectra were recorded using KBr as sample. The spectra of the samples were also taken after cooling to room temperature under vacuum and opened to the atmosphere after the dehydration experiment.

6.2.3.5 SCANNING ELECTRON MICROSCOPY

SEM images were taken by Philips SFEG 30S and FEI Qanta 250 FEG instruments.

6.3 RESULTS AND DISCUSSION

6.3.1 NITROGEN ADSORPTION

The adsorption isotherm of N_2 adsorption at 77 K is shown for mesoporous silica in Figure 6.2. The isotherm is Type IV of the UIPAC classification. Due to condensation along with evaporation of liquid nitrogen from mesopores, hysteresis loop is formed by adsorption/desorption isotherms and depends directly to the shape and size of pores.[26] Hysteresis cycle was, in general, determined for materials with interrelated pore networks and different sizes and shapes. The isotherms of silica, alumina, and hydroxyapatite have hysteresis loops, which are agreeable with the presence of mesopores. Relative pressure of $P/P°$ if higher than 0.6 is related to capillary condensation in mesopores, which is characteristic of type IV isotherms.[3]

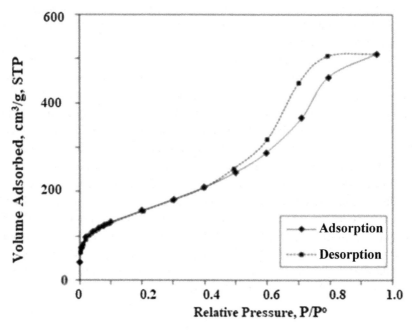

FIGURE 6.2   Nitrogen adsorption and desorption isotherm of mesoporous silica gel.

Hysteresis loop formed by adsorption/desorption isotherms due to condensation and evaporation of liquid nitrogen from mesopores depends upon the shape and size of pores. Hysteresis cycle was commonly observed for materials with interrelated pore networks with different size and shape. According to these data, single point, BET and Langmuir surface areas, average pore diameter, total pore volume, and micropore volume of the adsorbents have been listed in Table 6.2.

Silica aerogel was tested many times for surface area:

- The first test performed after SCD was completed.
- The second test was done after one-month aging.
- The third test was performed after two months aging in a desiccator.

TABLE 6.2   Properties of Silica Gels Determined by Nitrogen Adsorption.

Properties	Sigma-Aldrich silica gel	Aerogel (1)	Aerogel (2)	Aerogel (3)	Xerogel (4)
Single-point surface area (m^2/g)	556	1055	740	691	922
BET surface area (m^2/g)	571	1107	756	708	945
Langmuir surface area (m^2/g)	787	1549	1012	981	1390
Average pore diameter (nm) ($4V/A$ by BET)	5.5	4.2	4.3	4.4	2.6
Single-point total pore volume (cm^3/g)	0.79	1.19	0.81	0.79	0.62
Max micropore volume (cm^3/g)	0.20	0.36	0.28	0.26	0.35

Sample codes 1, 2, and 3 in Figure 6.3 and Table 6.2 referred to super-critical dried silica aerogel. (1), (2), and (3) were for as prepared aerogel, two months aged and three months aged aerogel, respectively. From curve (4), it is observed that the silica which was dried conventionally (xerogel). The total pore volume of prepared aerogel was equal to 1.19 cm^3/g, and it reduced to 0.81 and 0.79 cm^3/g on one and two months aging time. BET surface area of the aerogel was also reduced for aging period of one month and two months from 1549 to 1012 m^2/g and 981 m^2/g, respectively. It was observed that the mesoporous structure was collapsed with time. Neverthe-less, it is clear that SCD is a preference method to gain high-surface area than conventional dried silica (xerogel) structure.

Total pore volume of silica aerogel and commercial silica gel were calculated from their tapped densities. The tapped bulk densities of the commercial silica gel and silica aerogel were determined as 0.56 and 0.09 g/cm^3, respectively. Considering random packing particles void volume fraction as 0.33 and density of silica as 2.2 g/cm^3, the total pore volumes were calculated as 0.73 and 7 cm^3/g, respectively, for commercial silica gel and silica aerogel, respectively. While the total pore volumes determined by nitrogen adsorption and from bulk density measurements were close to each other for silica gel (0.76 and 0.73 cm^3/g, respectively), silica aerogel had pore volumes of 7 cm^3/g from bulk density measurement and 1.19 g/cm^3 from N_2 adsorption.

6.3.2   *MORPHOLOGY OF SILICA AEROGEL AND XEROGEL*

Produced silica aerogel and xerogel morphology were also investigated by SEM technique. According to obtained micrographs in Figure 6.4, silica aerogel has larger agglomerates of particles than conventional dried xerogel. Conventional drying particles were attracted to each other by the surface tension of liquid filling the empty spaces between particles. The surface tension of liquid ethanol is 22 mN/m at 25°C.[26] In case of ethanol vapor, there is no attraction between the particles since no liquid surface tension is present. Thus, the particle agglomerate keeps its original size during the removal of supercritical ethanol.

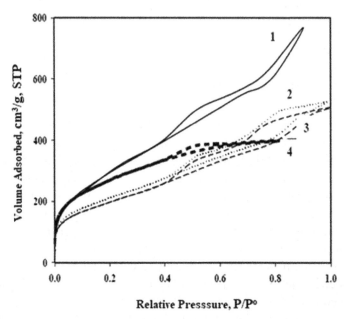

FIGURE 6.3 Nitrogen adsorption and desorption isotherms of (1) as prepared aerogel, (2) aerogel after two months, (3) aerogel after three months, and (4) as prepared xerogel.

(a)   (b)

FIGURE 6.4 SEM micrograph of aerogel (a) and xerogel (b).

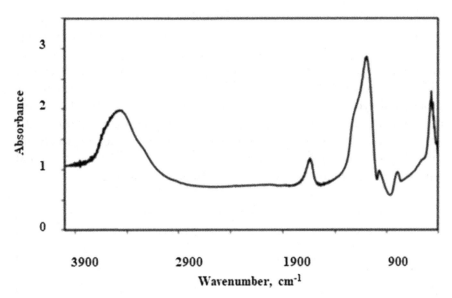

FIGURE 6.5 FTIR spectrum of commercial silica gel.

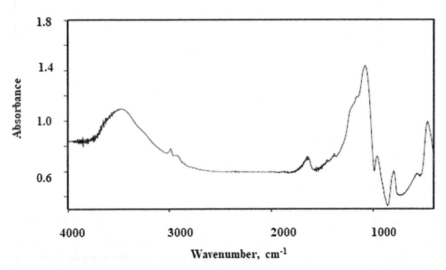

FIGURE 6.6 FTIR spectrum of silica xerogel.

FTIR spectra of commercial silica, silica xerogel, and silica aerogel are seen in Figures 6.5–6.7. All three different types of silica gel had the same functional groups in the bulk material. The presence of water can be observed from the band at 1640 cm^{-1}, assigned to the bending vibration of H_2O.[27] The peak at 3400 cm^{-1} is related to the stretching of hydrogen-bonded residual silanol (Si–OH) groups and hydrogen-bonded OH groups of adsorbed water.[27] In other words, this wavenumber is relevant to the –OH vibrations of molecular water that is physically adsorbed in the network. This band mostly reflects stretching vibration of hydrogen-bonded Si–OH groups and hydrogen-bonded OH groups of adsorbed water, while the peak at 1630 cm^{-1} related to bending water vibration.[2] The weak band near 970 cm^{-1} was assigned to the oscillating oxygen atoms in the silica network, including silanol groups and broken Si–O–Si bridges.[2] The peak at 1080 cm^{-1} wavenumber has been related Si–O–Si bond stretching.

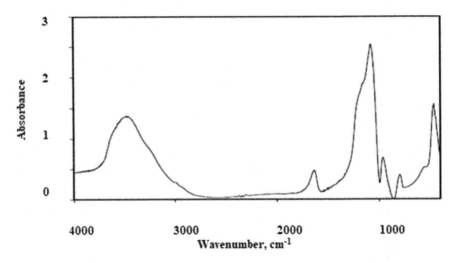

FIGURE 6.7   FTIR spectrum of silica aerogel.

6.3.3   DRIFT SPECTRA OF SILICA GELS

Information about the phenomena occurring on the surface of silica particles on vacuum application, heating, cooling, and pressurizing was obtained by DRIFT spectroscopy. The samples were placed to the sample holder of Harrick high-temperature cell at ambient temperature and pressure, and the pressure in the sample cell was lowered using the vacuum system. The temperature was increased at 2°C/min rate up to 500°C. Then, the system under vacuum was cooled to room temperature and opened to ambient atmosphere. The DRIFT spectra of the sample at ambient conditions, at room temperature, 100, 200, 300, 400, 500°C, cooled to room temperature at 10^{-2} mbar and at ambient conditions (60% relative humidity, 20 or 25°C) were taken to study the dehydration and hydration behaviors of silica aerogel and silica xerogel.

The DRIFT spectrum of aerogel when placed in the DRIFT cell at 1 atm pressure and 20°C is shown in Figure 6.8. The broad peak at 3400 cm^{-1} belonged to hydrogen-bonded OH groups, the small peak at 1640 cm^{-1} belonged to the bending vibration of adsorbed H$_2$O molecules.[14,15] The small peak at 1420 cm^{-1} could be due to the presence of CH$_2$ groups in aerogel. When the pressure of the system was lowered to 10^{-2} mbar, the peak at 1640 cm^{-1} lowered in intensity indicating desorption of H$_2$O molecules from the surface.

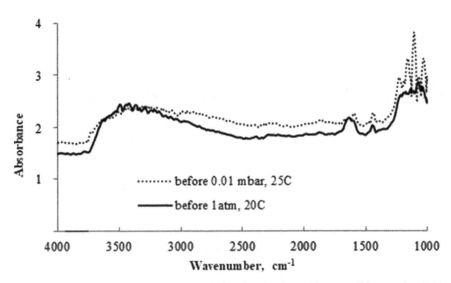

FIGURE 6.8 DRIFT spectra of aerogel before heating in ambient conditions and at 0.01 mbar and 20°C.

When the aerogel sample is heated under vacuum, the intensity of the hydrogen-bonded OH groups at 3400 cm^{-1} is lowered and a sharp peak belonging to isolated OH groups at 3700 cm^{-1} is observed as shown in Figure 6.9.

When the heated aerogel is up to 500°C cooled to room temperature under vacuum, its DRIFT spectrum does not change as seen in Figure 6.10. However, when it is open to ambient conditions at 1 atm and room temperature, it adsorbs moisture from the air and hydrogen-bonded OH group's intensity and intensity of H_2O bending in DRIFT spectrum increase.

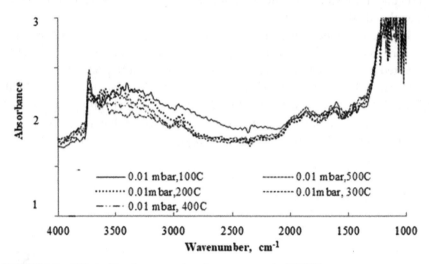

FIGURE 6.9   Effect of heating aerogel under vacuum on its DRIFT spectrum.

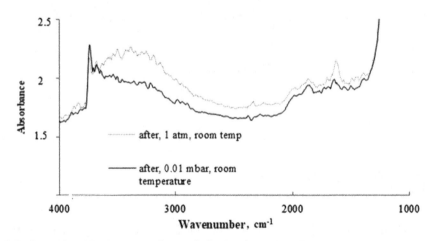

FIGURE 6.10   DRIFT spectra of aerogel after heating and cooling to room temperature.

The xerogel also has a broad hydrogen-bonded OH group peak at 3400 cm^{-1} and H_2O bending peak at 1640 cm^{-1} as seen in Figure 6.11. When vacuum is applied, the intensity of hydrogen-bonded OH groups peak is lowered and the H_2O bending peak at 1640 cm^{-1} disappears indicating desorption of adsorbed water under vacuum. Two small peaks at 2984 and 2916 cm^{-1} are seen under vacuum. These peaks are due to asymmetric and symmetric stretching vibrations of CH$_3$, the ethyl group of entrapped ethyl alcohol or unconverted Si(OCH$_2$CH$_3$)$_4$ to SiO$_2$ in aerogel.

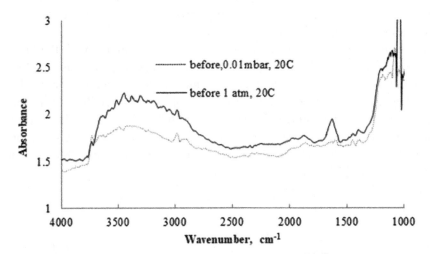

FIGURE 6.11   DRIFT spectra of xerogel before heating at 1 atm and 0.01 mbar at 20°C.

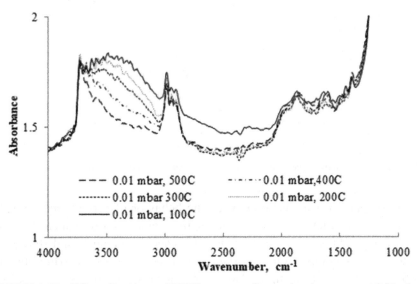

FIGURE 6.12   Effect of heating on DRIFT spectrum of xerogel under vacuum at 0.01 mbar.

When the xerogel is heated under vacuum, the broad hydrogen-bonded OH group peak at 3400 cm^{-1} and isolated OH group peak at 3700 cm^{-1} peak becomes observable as seen in Figure 6.12. The peaks at 2976, 2908, and 2887 cm^{-1} indicated the presence of CH_3 and CH_2 groups in heated xerogel.[28–30] The intensities of these peaks do not change with temperature indicating the thermal stability of these groups. These groups remained in xerogel sample even at high temperatures.

The isolated OH group peak at 3700 cm^{-1} is combined with the broad hydrogen-bonded OH groups peak when the xerogel is heated up to 500°C and then cooled to room temperature and opened to ambient air. Also, bending vibration of H_2O at 1640 cm^{-1} is observed at 1640 cm^{-1} as seen in Figure 6.13. The ambient air had around 60 % relative humidity and the dehydrated xerogel sample has the ability to adsorb moisture. The vibrations of CH_3 and CH_2 groups are also observed at 2974, 2929, and 2882 cm^{-1} in moisture-adsorbed samples.

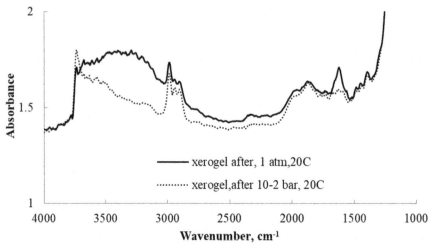

FIGURE 6.13   DRIFT spectrum of xerogel after heating and cooling to room temperature.

A comparison was made between the behavior of aerogel and xerogel for heating them under vacuum. The relative values of absorbance at 3400/ absorbance 1100 cm^{-1} were calculated for each temperature and shown in Figure 6.14 for comparison. Absorbance at 1100 cm^{-1} is related to the amount of Si–O groups in the samples. However, since CaF_2 windows were used in high-pressure and high-temperature chamber, its transparency range should also be considered. Since CaF_2 is transparent in mid-IR range down to 1050 cm⁻¹ (https://kinecat.pl/wp-content/uploads/2012/11/crystal_ref.pdf), taking 1100 cm⁻¹ peak as reference would be in the safe limit. The hydrogen-bonded OH group concentration in xerogel decreased continuously for xerogel, and for aerogel, it first increased from 20 to 100°C remained nearly constant up to 400°C, and it decreased to its initial value at 500°C.

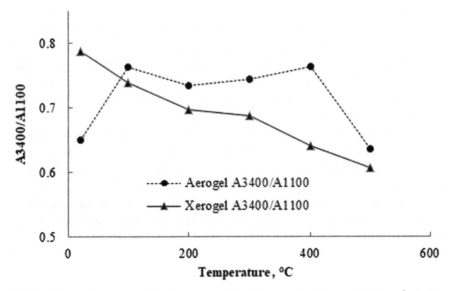

FIGURE 6.14   The change of hydrogen-bonded OH groups absorbance at 3400 cm⁻¹ relative to absorbance at 1100 cm⁻¹ versus temperature.

The isolated OH groups concentration change with temperature was also compared for aerogel and xerogel in Figure 6.15. The isolated OH group concentration in xerogel increased with temperature when it is heated up to 100°C and then its concentration is increased at a slower rate up to 400°C and at 500°C, it decreased indicating elimination of isolated OH groups from silica hydrogel. He et al. observed pore collapse of silica aerogel above 350°C. However, pore collapse elimination of isolated OH groups was observed according to the present study.[12]

6.4   CONCLUSION

The pore size, surface area, and density of silica gel were controlled by drying silica alcogel in ambient conditions and using supercritical ethanol extraction to obtain silica xerogel and silica aerogel, respectively, in this study. The samples were characterized by N_2 gas adsorption at $-196°C$, SEM, and FTIR spectroscopy. The effect of time on surface characteristics of silica aerogel up to three-month storage time was investigated by N_2 gas adsorption. The pore volume and surface area of silica aerogel decreased with storage time as indicated by nitrogen gas adsorption. The functional groups of the xerogel and silica aerogel and a commercial mesoporous silica gel were examined by FTIR transmission spectroscopy. The effects of pressure, relative humidity, and temperature on hydration–dehydration behavior of silica aerogel and silica xerogel were investigated using a temperature and pressure-controlled chamber in diffuse reflectance (DRIFT) cell of FTIR spectrophotometer. The presence of isolated OH groups at 3700 cm^{-1}, hydrogen-bonded OH groups at 3400 cm^{-1}, and CH_3 and CH_2 groups were observed in heated samples. It was observed that free water was eliminated from the samples under vacuum at ambient temperature as indicated by the disappearance of 1640 cm^{-1} H_2O vibration peak. Hydrogen-bonded OH groups and isolated OH groups were eliminated partially by heating the samples up to 500°C. The samples heated up to 500°C and cooled to ambient conditions and opened to ambient atmosphere at 60% relative humidity still had the capability of adsorbing free water.

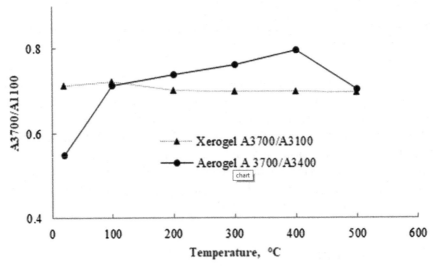

FIGURE 6.15 The change of isolated OH groups absorbance at 3700 cm^{-1} relative to absorbance at 1100 cm^{-1} for aerogel and xerogel.

KEYWORDS

- **silica aerogel**
- **silica xerogel**
- **DRIFT**
- **bound water**

REFERENCES

1. Pierre, A. C.; Pajonk, G. M. Chemistry of Aerogels and Their Applications. *Chem. Rev.* **2002,** *102* (11), 4243–4265. DOI: 10.1021/cr0101306.

2. Estella, J.; Echeverría, J. C.; Laguna, M. L.; Garrido, J. J. Effect of Supercritical Drying Conditions in Ethanol on the Structural and Textural Properties of Silica Aerogels. *J. Porous Mater.* **2008,** *15* (6), 705–713. DOI: 10.1007/s10934-007-9156-9.

3. Ru, Y.; Guoqiang, L.; Min, L. Analysis of the Effect of Drying Conditions on the Structural and Surface Heterogeneity of Silica Aerogels and Xerogel by Using Cryogenic Nitrogen Adsorption Characterization. *Micropor. Mesopor. Mater.* **2010,** *129* (1–2), 1–10. DOI: 10.1016/j.micromeso.2009.08.027.

4. Soleimani Dorcheh, A. S.; Abbasi, M. H. Silica Aerogel; Synthesis, Properties and Characterization. *J. Mater. Process. Technol.* **2008,** *199* (1–3), 10–26. DOI: 10.1016/j.jmatprotec.2007.10.060.

5. Land, V. D.; Harris, T. M.; Teeters, D. C. Processing of Low Density Silica Gel by Critical Point Drying or Ambient Pressure Drying. *J. Non-Crystal. Solids* **2001,** *283* (1–3), 11–17. DOI: 10.1016/S0022-3093(01)00485-9.

6. Brinker, C. J.; Scherer, G. W. *Sol–Gel Science: The Physics and Chemistry of Sol–Gel Processing*; Academic Press: Cambridge, MA, 1990.

7. Iswar, S.; Malfait, W. J.; Balog, S.; Winnefeld, F.; Lattuada, M.; Koebel, M. M. Effect of Aging on Silica Aerogel Properties. *Micropor. Mesopor. Mater.* **2017,** *241*, 293–302. DOI: 10.1016/j.micromeso.2016.11.037.

8. Yetgin, S.; Balkose, D. DNA Adsorption on Silica Aerogel. In *International Porous and Powder Materials Symposium Proceedings*, Polat, M., Tanoğlu, M., Eds.; İzmir, Turkey, 2013; pp 624–627.

9. Yetgin, S.; Balkose, D. Calf Thymus DNA Characterization and Its Adsorption on Different Silica Surfaces. *RSC Adv.* **2015,** *5* (71), 57950–57959. DOI: 10.1039/C5RA01810B.

10. Narin, G.; Balköse, D.; Ülkü, S. Characterization and Dehydration Behavior of a Natural, Ammonium Hydroxide, and Thermally Treated Zeolitic Tuff. *Drying Technol.* **2011,** *29* (5), 553–565. DOI: 10.1080/07373937.2010.507913.

11. Alp, B.; Gönen, M.; Savrık, S. A.; Balköse, D.; Ülkü, S. Dehydration, Water Vapor Adsorption and Desorption Behavior of $Zn[B_3O_3(OH)_5]\cdot H_2O$ and $Zn[B_3O_4(OH)_3]$. *Drying Technol.* **2012,** *30* (14), 1610–1620. DOI: 10.1080/07373937.2012.701258.

12. He, S.; Huang, Y.; Chen, G.; Feng, M.; Dai, H.; Yuan, B.; Chen, X. Effect of Heat Treatment on Hydrophobic Silica Aerogel. *J. Hazard. Mater.* **2019,** *362,* 294–302. DOI: 10.1016/j.jhazmat.2018.08.087.

13. Alcañiz-Monge, J.; Pérez-Cadenas, M.; Lozano-Castelló, D. Influence of Pore Size Distribution on Water Adsorption on Silica Gels. *J. Porous Mater.* **2010,** *17* (4), 409–416. DOI: 10.1007/s10934-009-9317-0.

14. Garbassi, F.; Balducci, L.; Chiurlo, P.; Deiana, L. A Study of Surface Modification of Silica Using XPS, DRIFT and NMR. *Appl. Surf. Sci.* **1995,** *84* (2), 145–151. DOI: 10.1016/0169-4332(94)00469-2.

15. Bukleski, M.; Ivanovski, V.; Hey-Hawkins, E. A Direct Method of Quantification of Maximal Chemisorption of 3-Aminopropylsilyl Groups on Silica Gel Using DRIFT Spectroscopy. *Spectrochim. Acta A: Mol. Biomol. Spectrosc.* **2015,** *149,* 69–74. DOI: 10.1016/j.saa.2015.04.026.

16. Fuchsgruber, A.; Lindner, W.; Dietl, R. Characterization of Chemically Modified Silica-Gels by DRIFT Spectroscopy. *Mikrochim. Acta* **1988,** *95* (1–6), 123–126. DOI: 10.1007/BF01349735.

17. Chunxiao, S.; Shouchun, Z.; Yonggang, J.; Junzong, F.; Jian, F. High Temperature Properties of Silica Aerogel. *Rare Met. Mater. Eng.* **2016,** *45,* 210–213.

18. Huang, J.; Qian, W.; Ma, H.; Zhang, H.; Ying, W. Highly Selective Production of Heavy Hydrocarbons over Cobalt–Graphene–Silica Nanocomposite Catalysts. *RSC Adv.* **2017,** *7* (53), 33441–33449. DOI: 10.1039/C7RA05887J.

19. Shee, D.; Mitra, B.; Chary, K. V. R.; Deo, G. Characterization and Reactivity of Vanadium Oxide Supported on TiO_2–SiO_2, Mixed Oxide Support. *Mol. Cat.* **2018,** *451,* 228–237. DOI: 10.1016/j.mcat.2018.01.020.

20. Shao, Z. D.; Cheng, X.; Zheng, Y. M. Facile Co-precursor Sol–Gel Synthesis of a Novel Amine-Modified Silica Aerogel for High Efficiency Carbon Dioxide Capture. *J. Colloid Interface Sci.* **2018,** *530,* 412–423. DOI: 10.1016/j.jcis.2018.06.094.

21. Yetgin, S. DNA Adsorption on Silica, Alumina and Hydroxyapatite and Imaging of DNA by Atomic Force Microscopy. *Ph.D. Thesis*; Izmir Institute of Technology, İzmir, 2013.

22. Ma, Y.; Wang, Y.; Wu, W.; Zhang, J.; Cao, Y.; Huang, K.; Jiang, L. Slurry-Phase Hydrocracking of a Decalin–Phenanthrene Mixture by MoS_2/SiO_2–ZrO_2 Bifunctional Catalysts. *Ind. Eng. Chem. Res.* **2021,** *60* (1), 230–242. DOI: 10.1021/acs.iecr.0c04999.

23. Boonpai, S.; Wannakao, S.; Panpranot, J.; Jongsomjit, B.; Praserthdam, P. Active Site Formation in WO_x Supported on Spherical Silica Catalysts for Lewis Acid Transformation to Brønsted Acid Activity. *J. Phys. Chem. C* **2020,** *124* (29), 15935–15943. DOI: 10.1021/acs.jpcc.0c03657.

24. Elimbinzi, E.; Nyandoro, S. S.; Mubofu, E. B.; Osatiashtiani, A.; Manayil, J. C.; Isaacs, M. A.; Lee, A. F.; Wilson, K. Synthesis of Amine Functionalized Mesoporous Silicas Templated by Castor Oil for Transesterification. *MRS Adv.* **2018,** *3* (38), 2261–2269. DOI: 10.1557/adv.2018.347.

25. Wang, C.; Yang, Q.; Wang, J.; Zhao, J.; Wan, X.; Guo, Z.; Yang, Y. Application of Support Vector Machine on Controlling the Silanol Groups of Silica Xerogel with the Aid of Segmented Continuous Flow Reactor. *Chem. Eng. Sci.* **2019,** *199,* 486–495. DOI: 10.1016/j.ces.2019.01.032.

26. Greeg, S. J.; Sing, K. S. W. *Adsorption, Surface Area and Porosity*; Academic Press: Cambridge, MA, 1982.

27. Gonçalves, F. A. M. M.; Trindade, A. R.; Costa, C. S. M. F.; Bernardo, J. C. S.; Johnson, I.; Fonseca, I. M. A.; Ferreira, A. G. M. PVT, Viscosity, and Surface Tension of Ethanol:

New Measurements and Literature Data Evaluation. *J. Chem. Thermodyn.* **2010,** *42* (8), 1039–1049. DOI: 10.1016/j.jct.2010.03.022.

28. Fidalgo, A.; Farinha, J. P. S.; Martinho, J. M. G.; Ros, M. E.; Ilharc, L. M. The Influence of the Wet Gels Processing on the Structure and Properties of Silica Xerogels. *Micropor. Mesopor. Mater.* **2005,** *84* (1–3), 229–235. DOI: 10.1016/j.micromeso.2005.04.021.

29. Kristiansen, T.; Mathisen, K.; Einarsrud, M. A.; Bjørgen, M.; David, G.; Nicholson, D. G. Single-Site Copper by Incorporation in Ambient Pressure Dried Silica Aerogel and Xerogel Systems: An X-ray Absorption Spectroscopy Study. *J. Phys. Chem. C* **2014,** *118,* 2439–2453.

30. Kristiansen, T.; Mathisen, K. On the Promoting Effect of Water During NO$_x$ Removal over Single-Site Copper in Hydrophobic Silica APD-Aerogels. *J. Phys. Chem. C* **2014,** *118* (5), 2439–2453. DOI: 10.1021/jp406610v.

CHAPTER 7

The Effect of Small Additives of Carbon Nanotubes with the Original and Oxidized Surface on the Mechanical Properties and Durability of the Polyepoxy Composite as Porous Media

D. STAROKADOMSKY[1,2], V. DIAMANT[3], and M. RESHETNYK[4]

[1]*Chuiko Institute of Surface Chemistry, National Academy of Sciences (NAS), Kiev, Ukraine*

[2]*Institute of Geochemistry & Mineralogy, NAS, Kiev, Ukraine*

[3]*Vernadsky Institute of General & Inorganic Chemistry NAS, Kiev, Ukraine*

[4]*National Nature—Historical Museum, NAS, Kiev, Ukraine*

ABSTRACT

In this chapter the effect of small additives of carbon nanotubes is studied. This experimental investigation is directed to the formation of black glossy compositions with an aggregative distribution of filler. From the experimental studies we observed that application of carbon nanotubes enhanced resistance in chemically aggressive environment along with some remarkable improvement of fire-resistant composites.

It is shown that the introduction of 0.1–2 wt% multiwall carbon nanotubes (CNTs) leads to the formation of black glossy compositions with an aggregative distribution of filler. Starting from 1 wt% inhomogeneous distribution leads to a sharp increase of viscosity and deterioration of operational properties. It is established that the introduction of 0.1–2 wt% CNT with

Mechanics and Physics of Porous Materials: Novel Processing Technologies and Emerging Applications.
Chin Hua Chia, Tamara Tatrishvili, Ann Rose Abraham, & A. K. Haghi (Eds.)
© 2024 Apple Academic Press, Inc. Co-published with CRC Press (Taylor & Francis)

the original and the oxidized surface as a rule does not allow to increase the studied mechanical properties (compressive and flexural strength/modulus, microhardness, wear). At the same time, the introduction of CNT allows you to increase resistance in chemically aggressive environment (concentrated nitric acid) and fire-resistant composites (growth with increasing concentration of CNT). These effects are better observed for filling by CNT with oxidized surface.

CNT is an allotropic form of carbon that forms tubular nanosized particles. The experimental discovery of CNTs is officially attributed to Ijima in 1991.[1] However,[2] back in 1952, in the USSR and the USA electronic images of carbon substances formed on the surface of stainless steel were obtained as a result of thermal decomposition of CO. These images were clearly visible tubular carbon formations,[2] which in modern terminology should be called nanotubes (Fe—a typical catalyst for CNT growth by chemical vapor deposition).

In 2002, CNTs were first synthesized at our Chuyko Institute of Surface Chemistry,[3] and subsequently, these methods of obtaining them were improved.[4–7]

In CNT, each carbon atom sp2 is hybridized and bound to three adjacent hexagonal lattice atoms. Graphene layers form cylindrical structures[8,20] with different numbers of concentric cylinders: one, two, or several—as shown in Figures 7.1 and 7.2.

a—Single-layer CNT   b—Two- and multilayer CNT

FIGURE 7.1 Different types of nanotubes: one (a), two (b), and (c) multiwall.

(a)

(b)

FIGURE 7.2 SEM images of multilayer CNT powder: (a) Nanocyl NC7000 and (b) Baytubes C150P.

Thus, CNTs have been in the spotlight for 30 years. *A priori*, CNT can be called as "advanced carbon fibers," and composites filled with them can be observed as a *carboplastics* with improved properties. There are many valuable research studies available on investigation of epoxy-CNT compositions.[9–17]

However, the issues of their influence on the properties of polymer composites have been studied fragmentarily so far. Theoretically, the introduction of nanotubes into a thermosetting polymer (e.g., polyepoxide or polyester) should dramatically improve a number of properties—due to the construction of new structures and phases. The most significant effects take place for the electrical and thermal properties of composites.

Regarding CNT–polyepoxide composites, there are ambiguous experimental data in the literature. Thus, according to Ref. [12], the introduction of CNTs in epoxy resin led to a *deterioration* of physical and mechanical properties.

Chakraborty et al.[17] studied the effect of dispersion and distribution of CNT in epoxy. Figure 7.3 clearly shows the *agglomerative and chain* distribution of CNT in the composition, which is characteristic of many nanoparticles. It follows from Figures 7.11 and 7.12 that compression and flexural as strength as modulus change insignificantly.[17]

At the same time, thermal and electrical conductivity are significantly *improved*. In Ref. [9], measurements of the electrical properties of CNT epoxy composites containing low carbon nanotube loads (less than 1%) were shown. It is well observed that the resistivity reduction is up to 40%, while the application of simulated sunlight with the concomitant surge in temperature reached to a maximum decrease of 58%. In our work, it was shown that the introduction of CNTs can change the properties of polyepoxides, giving them higher resistance to aggressive environments.[12]

Our team has significant experience in the study of polymer composites with carbon materials.[12,18,19] The potential of CNT as a structure-strengthening filler is considered significant, so the experimental search continues. In our case, it was of interest to trace the effect of CNTs with different surface natures on the mechanical properties of composites and their behavior in aggressive environments.

7.1 METHODS AND REACTIVES

7.1.1 POLYMERIC MATRIX

The investigated CNTs were incorporated into an epoxy resin produced in Poland (epoxy rubber) and cured under normal conditions, with polyethylene–polyamine (Chech production) in a ratio of 6:1. The compositions were mixed with filler and, after brief evacuation, poured into molds. At the end of the curing process at room temperature (after 7 days), they were heat treated at 75°C (6 h), and then some samples—at another 170°C (1 h).

7.1.2 CARBON NANOTUBES

The CNTs produced in our laboratories (D = 20–40 nm, L = 10–30 mcm, 10–15 layers, purity up to 95%, surface area 110 m^2/g) were taken for study in the original and oxidized form. Oxidation was carried out by the action of concentrated HNO_3; as a result, the content of –COOH groups in CNT was ≈1.2%.[12] TUBALL (Luxembourg, Russian production) and OCSiAl (China production) CNTs were also taken for comparison.

Multilayer carbon nanotubes (BSWTs) for polymer filling were synthesized by CVD method,[5] to obtain a product with homogeneous reproducible properties used a reactor that rotates constantly during synthesis. Using previously developed methods (Melezhik O., Sementsov Y.),[3–7] on the basis of the research and production enterprise "TM Spetsmash" in Kyiv, conducted a synthesis of CNT batches on catalysts with different ratios of metals.

7.1.3 MECHANIC AND RESISTANCE TESTS

- *Bending test performed* (GOST 56810-2015). Size of plates used: $6 \times 1 \times 0.2$ cm^3.
- Bending fracture was performed based on $L = 3$ cm of a DI-1 bending testing machine.
- The elastic modulus and the value of strength were determined (based on the test results).
- Compression test performed (GOST 4651-2014, ISO 604:2002), using Louis Shopper press-machine until complete destruction of samples achieved.
- Based on the results observed from the compression test, the strength was determined using: $f = P/s$

Here P is the load in kgf, s is the area equal to 0.332 cm^2, and E is the modulus: $E = f/e$ (where e is the value of relative elongation observed).

- The value of the Brinell microhardness (based on the standards of GOST 9012-59, ISO 6506-1: 2005) was observed as resistance load while a steel hemisphere (with the diameter of $d = 3$ mm) was immersed in a sample-plate (with considering the $b = 1.5$ mm) to 10–60 μm.
- Specimens were tested in aggressive environments using the standard swelling method (ISO 62: 2008, GOST 4650-2014).
- Tablets $1 \times 1 \times 0.2$ cm^3 were located in an aggressive environment.
- The change in their weight was periodically recorded, which was converted into % of swelling ($q = (m - m_0)/m_0 \times 100\%$).

7.2 MICROSCOPY AND VISUAL ANALYSIS

A typical multilayer CNTs[12,20] were observed as shown in Figures 7.3–7.6.

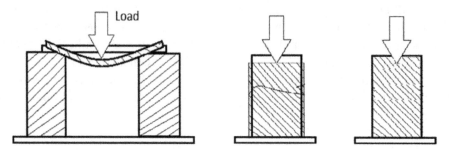

FIGURE 7.3 Visualization of mechanical tests for compression (left), bending and micro-hardness (right).

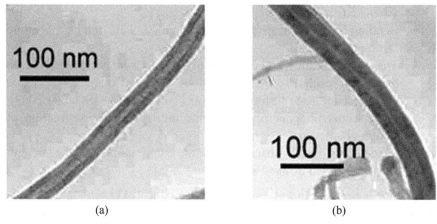

(a)   (b)

FIGURE 7.4 CNT: a—initial; b—after oxidation.

At the macro level, the introduction of CNT is similar to the action of graphite: the composite turns black, and at concentrations above 1 wt% becomes viscous.

From Figure 7.6, we can see that at 0.1–1 wt%, CNT are distributed relatively homogeneous, with aggregates 50–500 mcm and bubbles ≤3 mm. A boundary CNT layer forms around air inclusions. But after 1 wt%, the macro-distribution of CNTs is nonuniform. Indeed, at such concentrations, the composite becomes viscous and difficult to form; it appears a lot of large bubbles and inhomogeneities.

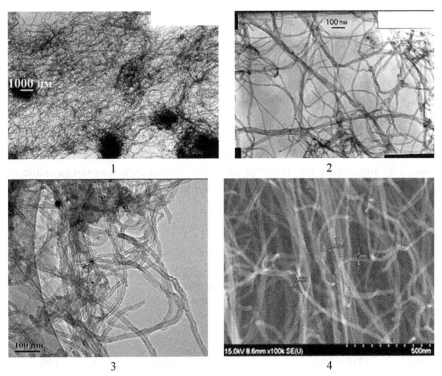

FIGURE 7.5 TEM images of CNT samples obtained on the Al_xFeMo_y—catalysts (1–3) and Chinese CNT (4).

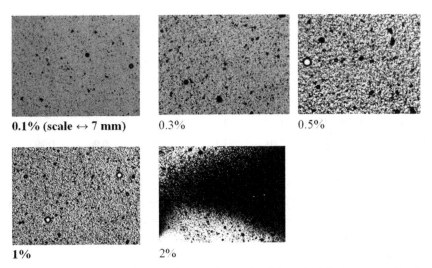

0.1% (scale ↔ 7 mm)   0.3%   0.5%

1%   2%

FIGURE 7.6 Photographic images of composites with 0.1–2 wt% CNTs. 100× magnification (with an actual screen length of 7 cm).

7.3 STRENGTH OF COMPOSITES

The tests show that the introduction of 0.1–2% CNTs does not lead to significant changes in the compressive strength of nonthermally tested epoxy polymer (EP) samples. There are "weakening" concentrations (0.3%). As for "enhancer concentration" (0.1%), they should obviously be sought in the field of small additives (up to 0.1%).

After heat treatment, the strength increases markedly for the unfilled polymer, and even more so for almost all filled samples (Table 7.1). This is especially noticeable for oxidized samples. However, it was not possible to increase the strength above 14% even after heat treatment.

TABLE 7.1 Values of the Destructive Load C of Compressive Strength Tests (Cubes of 1 cm). "Oh"—oxidized; "Ch"—Chinese CNT.

	C, 75°C 6 h			C_t, 75° 6 h + 170° 3 h		
	C_{aver} and C_{max} (kgf/cm²)	C_{aver} (rel. un.)	≈Modulus (rel. un.)	C_{aver} (kgf/cm²)	$C_{aver, rel. un.}$	C/C_t (rel. un.)
H	1250–1370	1	1	1320	1	1.06
0.1%	1380–1450	1.11	1.2	1290	0.98	0.94
0.3%	1230–1240	1	1	1340	1.02	1.09
0.5%	1180–1190	0.95	0.9	1330	1	1.13
1%	1290–1310	1.03	1.1			
2%	1270–1290	1.01	1	1260	0.96	0.99
0.1% ox	1240–1260	1	1	1420	1.1	1.15
0.3% ox	1150–1180	0.9	0.9	1330	1	1.17
0.5% ox	1240–1260	1	1	1380	1.05	1.11
0.5% ch. ox	1310–1340	1.05	1.1	1390	1.05	1.06
1% ox	1250–1250	1	1			
2% ox	1270–1300	1.01	1.1	1260	0.96	0.99

Filling by CNT increases the bending strength rarely (up to 10–15% for "0.1% ox," Table 7.2). But the modulus when bending by filling can be significantly increased—by 15–20%, and in some cases by 50% (for 0.3%). It is also sometimes possible to slightly increase the microhardness and abrasion resistance (Table 7.2).

TABLE 7.2 Mechanical Parameters of Composites.

75°C, 6 h	Bending strength (kgf/mm^2)	Bending modulus (kgf/cm^2 ×10^5)	Microhardness (XF)*	Wear resistance (rel. un.)
H	11.8	0.34	65	1
0.1%	11.5	0.28	65	0.9
0.3%	11.3	0.52	70	1.1
0.5%	11.2	0.37	66	0.9
0.5% ch.	11.7	0.43	68	1
1%	9.6	0.24	60	0.9
2%	8.3	0.3	60	0.9
0.1% ox	12.2	0.38		
0.3% ox		0.41		
0.5% ox	10.3	0.36		
1% ox	9.1	0.25		
2% ox	10.2	0.34		
0.5% ch. ox	11.4	0.31		

**First estimation.*

7.4 CHEMICAL RESISTANCE OF COMPOSITES

7.4.1 STABILITY IN ACETONE

Acetone is known to be one of the most aggressive media for polyepoxides. Despite the obvious absence of chemical interactions with EP, acetone can completely "break" the polymer structure in 1–2 days due to physical sorption. The CNT filling gives a chance to strengthen the structure of the polymer to the action of aggressive media.

However, it has been experimentally found that nanotubes do not contribute to the strengthening of epoxy rubber in acetone. Moreover, the destruction of samples is faster—only a sample of 0.3 ok and "highly filled" 2 and 2 ox (Table 7.3) withstand more than 26 days, on a par with an unfilled analog. In a number of cases, the destruction of a sample (by which we mean its division into several parts) occurs much faster than without filling—for example, for samples 0.1, 0.5, 0.5 Koc and 1. The dynamics of the behavior of composites with author's nanotubes in acetone is generally similar to the behavior of composites with industrial nanotubes (0.5 K and 0.5 Kox).

The change in the degree of swelling practically does not change after the introduction of nanotubes (Table 7.3).

It can be seen that oxidized nanotubes give composites more resistance to acetone than nonoxidized ones (Table 7.3). This agrees with the assumption about the composite strengthening effect of –COOH groups on the surface of oxidized SNTs.

TABLE 7.3   Swelling of Samples of Epoxy Rubber in Acetone.

Days	0%	0.1	0.1 ox	0.3	0.3 ox	0.5	0.5 ox	0.5 ch.	0.5 ch. Ox	1	1 ox	2	2 ox
0	0.0	0.0	0.0	0.0	0.0	0.0	0.0	0.0	0.0	0.0	0.0	0.0	0.0
0.02	0.0	0.6	0.4	0.3	−0.5	0.0	0.0	0.0	0.0	0.0	0.0	0.0	0.0
0.17	2.5	4.0	2.5	4.3	1.4	2.0	2.0	3.2	4.7	1.9	1.4	0.4	0.3
1	10.5	**18.3**	11.8	11.1	6.4	11.1	6.0	4.6	17.8	7.9	7.1	1.2	3.1
2	13.7		15.1	8.1	8.6		7.8	6.8			8.3	3.3	4.2
10	20.0				20.5		25.0	20.3			24.1	18.2	10.8
16	20.0				20.5			21.4			24.1	21.9	15.0
26	19.6				19.4							21.1	16.8

However, from Figure 7.7, it is noticeable that most of the filled samples decompose faster than the unfilled polyepoxide.

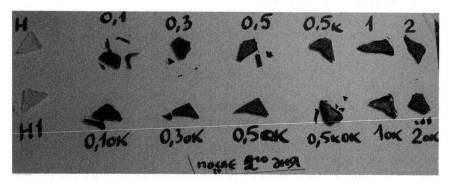

FIGURE 7.7   View of the samples after two days of endurance in acetone.

7.5   CONCLUSIONS

1. The introduction of CNTs leads to a significant change in both the consistency/viscosity of composition and the morphology of polymer

composite. Agglomerates, bubbles (bordered by nanotubes), and inhomogeneities appear in the composite. After 2 wt%, the form-ability of the curable composition is significantly more difficult, and the resulting composites are heterogeneous due to caverns and nonvacuumated air bubbles.

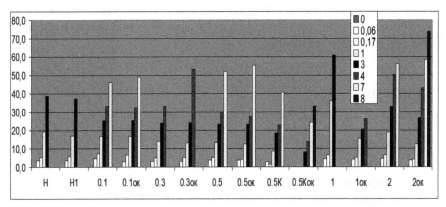

FIGURE 7.8   Swelling of samples in concentrated nitric acid.

TABLE 7.4   Sample lifetime in conc. HNO_3.

	H	H1	0.1	0.1 ox	0.3	0.3 ox	0.5	0.5 ch.	0.5 ch. ox	1	1 ox	2	2 ox
Days	3	3	7	7	4	4	7	7	7	3	4	7	8
Destr.	Oo	Oo	÷	÷	§	§	÷	÷	÷	÷	§	§	§

Designations of destruction: Oo—foams, §—smears and spreads, ÷—spalls from the main mass of the sample are observed.

2. Experimentally, no large-scale changes (±5–10%) in the strength properties (compression, bending, Young's modules, microhardness, wear) of the composites after CNT filling were revealed. It is shown that after hard heat treatment, the compressive strength increases for the unfilled polymer, and even stronger for almost all filled samples.

3. Chemical resistance tests have shown that initial CNT do not improve the resistance of the composite in acetone solutions. At the same time, oxidized nanotubes give composites more resistant to acetone than unoxidized ones. At the same time, the introduction of nanotubes can significantly increase the resistance to acid swelling and liquid-phase oxidation (e.g., exposure to concentrated nitric acid). As in the

case of acetone, oxidized nanotubes usually give a more resistant composite to concentrated nitric acid than nonoxidized ones.

ACKNOWLEDGMENTS

We thank researcher Gavrilyuk Natalia (Ph.D., surface chemistry of CNT-filled polyolefins) for providing images of nanotubes and methods for their fabrication for open publication.

KEYWORDS

- **carbon nanotubes**
- **fillers**
- **composites**
- **fire resistance**
- **environment**

REFERENCES

1. Iijima, S. Helical Microtubules of Graphitic Carbon. *Nature* **1991,** *354* (6348), 56–58. DOI: 10.1038/354056a0.
2. Rakov, E. G. *Nanotubes and Fullerenes*; University Book, Logos: M, 2006; pp 1–376.
3. Brichka, S. Y.; Prikhod'ko, G. P.; Brichka, A. V.; Ogenko, V. M.; Chuiko, A. A. Matrix Synthesis of N-Containing Carbon Nanotubes. *Theor. Exp. Chem.* **2002,** *38* (2), 114–117. DOI: 10.1023/A:1016092101771.
4. Melezhik, O. V.; Kovalenko, O. O.; Sementsov, Y. I.; Yanchenko, B. B. Method of Obtaining Catalyst for Chemical Deposition of Carbon Nanotubes from Gas Phase. *Patent of Ukraine No. 83532.* Registered in the State Register of Patents of Ukraine for Inventions, July 25, 2008.
5. Gunko, G. S.; Sementsov, Y. I.; Melezhik, O. V.; Prikhod'ko, G. P.; Pyatkovskiy, M. L.; Gavrylyuk, N. A.; Cartel, M. T. CVD-Method and Equipment for MWCNT Obtaining. In *International Meeting "Clusters and Nanostructured Materials" (CNM-2)*; Uzhgorod, Ukraine, 2009; p 158.
6. Melezhik, A. V.; Smykov, M. A. Influence of Parameters of Technological Regimes on Cultivation of Carbon Nanotubes by Catalytic Pyrolysis of hydrocarbons. *Bull. TSTU* **2010,** *16* (4), 918–923.
7. Pat. Melezhik, A. V.; Tkachev, A. G. Method for Production of Carbon Nanotubes and Reactor for Their Production (RU 2493097), IPC 8 CBB. Patent EE, *31/02* (B82B),

3/00, B82Y 40/00 of NanoTechCenter LLC. No. 2010124350/05, stated 15.06.2010, publ. 09/20/2013.

8. Dresselhaus, M. S.; Lin, Y. M.; Rabin, O.; Jorio, A.; Souza Filho, A. G.; Pimenta, M. A.; Saito, R.; Samsonidze, G.; Dresselhaus, G. Nanowires and Nanotubes. *Mater. Sci. Eng. C* **2003**, *23* (1–2), 129–140.

9. Earp, B.; Hubbard, J.; Tracy, A.; Sakoda, D.; Luhrs, C. Electrical Behavior of CNT Epoxy Composites Under In-Situ Simulated Space Environments. *Compos. B: Eng.* **2021**, *219*, 108874.

10. Sun, X. M.; Sun, H.; Li, H. P.; Peng, H. Developing Polymer Composite Materials: Carbon Nanotubes or Graphene? *Adv. Mater.* **2013**, *25* (37), 5153–5176.

11. Hsu, S.-H.; Wu, M.-C.; Chen, S.; Chuang, C.-M.; Lin, S.-H.; Su, W.-F. Synthesis, Morphology and Physical Properties of Multi-walled Carbon Nanotube/Biphenyl Liquid Crystalline Epoxy Composites. *Carbon* **2012**, *50* (3), 896–905.

12. Starokadomsky, D.; Zhuravsky, S.; Tkachenko, A. The Effect of Small Additions of Carbon Nanotubes with Original and Oxidized Surfaces on the Mechanical Properties of an Epoxy Composite. *Plast. Massy* **2017**, *9–10*, 24–27.

13. Roy, S.; Petrova, R. S.; Mitra, S. Effect of Carbon Nanotube (CNT) Functionalization in Epoxy-CNT Composites. *Nanotechnol. Rev.* **2018**, *7* (6), 475–485.

14. Naik, N. K.; Pandya, K. S.; Kavala, V. R.; Zhang, W.; Koratkar, N. A. High-Strain Rate Compressive Behavior of Multi-walled Carbon Nanotube Dispersed Thermoset Epoxy Resin. *J. Compos. Mater.* **2015**, *49* (8), 903–910.

15. Vahedi, F.; Shahverdi, H. R.; Shokrieh, M. M.; Esmkhani, M. Effects of Carbon Nanotube Content on the Mechanical and Electrical Properties of Epoxy-Based Composites. *New Carbon Mater.* **2014**, *29* (6), 419–425.

16. Siddiqui, N. A.; Sham, M.-L.; Tang, B. Z.; Munir, A.; Kim, J.-K. Tensile Strength of Glass Fibres with Carbon Nanotube–Epoxy Nanocomposite Coating. *Compos. A* **2009**, *40* (10), 1606–1614.

17. Chakraborty, A.; Plyhm, T.; Barbezat, M.; Necola, A.; Terrasi, G. Carbon Nanotube (CNT)-Epoxy Nanocomposites: A Systematic Investigation of CNT Dispersion. *J. Nanopart. Res.* **2011**, *13* (12), 6493–6506. DOI: 10.1007/s11051-011-0552-3.

18. Gorelov, B. M.; Gorb, A. M.; Polovina, O. I.; Nadtochiy, A. B.; Starokadomskiy, D. L.; Shulga, S. V.; Ogenko, V. M. Impact of Few-Layered Graphene Plates on Structure and Properties of an Epoxy Resin. *Nanostruct. Nanotech. Nanotechnol.* **2016**, *14* (4), 527–537.

19. Gorelov, B.; Gorb, A.; Nadtochiy, A.; Starokadomsky, D.; Kuryliuk, V.; Sigareva, N.; Shulga, S.; Ogenko, V.; Korotchenkov, O.; Polovina, O. Epoxy filled with Bare and Oxidized Multi-layered Graphene Nanoplatelets: A Comparative Study of filler Loading Impact on Thermal Properties. *J. Mater. Sci., 3.* DOI: 10.1007/s10853-019-03523-72.

20. Gavrilyuk, N. A. Physicochemical Properties of Polytetrafluoroethylene and Polypro-pylene Filled with Multilayer Carbon Nanotubes. Qualifying Scientific Work on the Rights of the Manuscript. *PhD. Dissertation on a Specialty 01.04.18: "Physics and Chemistry of a Surfaces"*; Institute of Surface Chemistry, O. O. Chuyka of the National Academy of Sciences of Ukraine, 2019.

CHAPTER 8

Investigation of Coating Based on Melamine Formaldehyde Lacquer Modified by Cu–C Porous Mesocomposite and Ammonium Polyphosphate

I. N. SHABANOVA[1], T. M. MAKHNEVA[1], V. I. KODOLOV[2], R. A. MAKHNEV[2], N. S. TEREBOVA[1], and S. G. BYSTROV[1]

[1]*The Udmurt Federal Research Centre of the Ural Branch of the Russian Academy of Sciences, Izhevsk, Russia*

[2]*Kalashnikov Izhevsk State Technical University, Izhevsk, Russia*

ABSTRACT

The present chapter is dedicated to the investigation of the carbonization of melamine formaldehyde lacquer modified by ammonium polyphosphate and copper–carbon mesocomposite for creating anticorrosion intumescent fireproofing coating. Thorough investigations of the obtained materials modified by mesocomposite are of scientific and practical interest. The investigations were carried out with the use of X-ray photoelectron spectroscopy, X-ray diffraction, atomic force microscopy, optical microscopy, and, also, thermal and fire tests according to the State Standard. The investigation and test results indicate a significant improvement in thermal stability and fire-protecting properties of coating modified by ammonium polyphosphate and copper–carbon mesocomposite due to the improvement of the quality of formed coked foam.

Mechanics and Physics of Porous Materials: Novel Processing Technologies and Emerging Applications.
Chin Hua Chia, Tamara Tatrishvili, Ann Rose Abraham, & A. K. Haghi (Eds.)
© 2024 Apple Academic Press, Inc. Co-published with CRC Press (Taylor & Francis)

8.1   INTRODUCTION

Paint- and lacquer coatings are widely used in machine-building industry, radio electronics, aviation, shipbuilding, industrial and dwelling construction, etc. In construction, light aluminum structures are frequently used. However, such structures are prone to corrosion and are not fire-resistant.

Melamine formaldehyde lacquer (MFL) is used to protect metals against corrosion. The intensive development of the national economy makes it necessary to increase the volume of the production of coating compositions with improved quality indices.

Investigations in the sphere of the creation of anticorrosive and fire-proofing coatings by modifying coating compositions with the use of nanoscale additives are very important. For protecting aluminum structures against fire swellable, fire-proofing coatings are frequently used; under the action of fire, such coatings transform into coked foam with low thermal conductivity. Such coked foam protects the metal at high temperatures.[1,2]

In the present chapter, to improve the MFL thermal stability and fire-proofing properties, additives were used, such as ammonium polyphosphate (APP) and copper–carbon mesocomposite (Cu–C MC).

APP is widely used as a fire-retardant agent in the production of special-purpose paintwork materials: fire-retardant paints, dopes, lacquers, and mastic.[3] APP initiates the formation of coke at the thermal decomposition of a lacquer film. Cu–C MC is added for increasing the strength and adhesive strength of the formed layer of coked foam.

The purpose of the chapter is the comparative investigation of MFL coating modified by a fire-retardant system containing APP and Cu–C MC.

8.2   EXPERIMENT

The subjects of the present investigation were three samples of MFL coating deposited onto the surface of aluminum.

The following samples were studied:

1.   MFL layer on aluminum.
2.   MFL layer + 0.01% of Cu–C MC.
3.   MFL layer + 20% of APP $((NH_4PO_3)_n$ + 0.005% of Cu–C MC.

The X-ray photoelectron spectroscopy (XPS) studies were performed on an X-ray electron magnetic spectrometer with resolution of 10^{-4} and luminosity of 0.085% at the excitation by Al K_α-line 1486.5 eV.[4]

The X-ray phase analysis was conducted on a diffractometer D2 PHASER with Bragg–Brentano geometry and a linear counter LYNXEYE. The X-ray phase analysis of the samples was carried out in Cu K_α radiation; the obtained diffractograms were analyzed using the database PDF-2/2020 of the International Center for Diffraction Data.

The samples were also studied by atomic force microscopy (AFM) on a scanning probe microscope SOLVER PRO in the contact mode. The images of the surface topography were obtained in the height mode. The image dimensions were 5, 10, and 30 μm^2. Images in the height mode permits to obtain information about the three-dimension structure of the studied surface.

The adhesion force (F_{adh}) between the probe needle and studied samples was measured (Table 8.1). The measurements were carried out at 15 points of the surface. The adhesion force was evaluated by the height of the "step" on the curves of the probe detachment from a sample. The optical images of the surface of the studied samples were obtained on an optical microscope OLYMPUS SZ-STB2.

8.3 RESULTS AND DISCUSSION

The peculiar features of the structure of surface protective layers have been studied; the layers have been obtained by depositing the MFL coating and MFL compositions modified by 0.01% of Cu–C MC and also by 20% of APP $(NH_4PO_3)_n$ + 0.005% of Cu–C MC on aluminum.

The XPS spectra of the C 1s, O 1s, N 1s, and Al 2p core levels are obtained for the studied samples. The spectra are obtained without heating of the samples and at step heating with the step of 100°C in the spectrometer chamber in vacuum.

The XPS study of the aluminum surface (Figure 8.1a–c) coated with the layer of nonmodified MFL shows that without heating the C 1s spectrum maximum corresponds to the bond C–H (285 eV) in hydrocarbons (Figure 8.1a, spectrum 1).

In C 1s spectrum, heating at 100°C, a "bump" appears in the region of 283.8 eV (Figure 8.1a, spectrum 2), which corresponds to C–C bond with sp hybridization and can indicate the beginning of carbonization (Figure 8.1c). In the O 1s spectrum, without heating (Figure 8.1b, spectrum (1), there are two components with binding energies 532.4 eV characteristic of C–O bond[5] and 534.0 characteristic of $O_{adsorbed}$ (C–O–H).[6] Heating at 200°C, maximum appears at 530 eV characteristic of metal oxides. Without heating, in the N 1s spectrum, there is an intensive maximum with binding energy 400 eV (N–H–O) (Figure 8.1c, spectrum 1). Heating at 100°C, maximum with binding energy 399.0 eV characteristic of N–H bond appears (Figure 8.1c, spectrum 2). The Al 2p spectrum is not revealed.

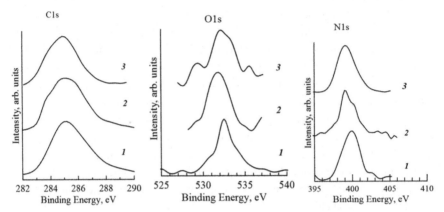

FIGURE 8.1   The XPS spectra of the sample surface No. 1: (a) C 1s: **1**—without heating; **2**—heating at 100°C; **3**—heating at 200°C; (b) O 1s: **1**—without heating; **2**—heating at 100°C; **3**—heating at 200°C; and (c) N 1s: **1**—without heating; **2**—heating at 100°C; **3**—heating at 200°C.

The study of sample 2 of MFL coating modified by Cu–C MC (Figure 8.2a–c) shows that without heating, the C 1s spectrum maximum (Figure 8.2a) corresponds to C–H bond (285 eV) in hydrocarbons; a "bump" in the region 283 eV can indicate the appearance of C–C component with sp hybridization of valence electrons at room temperature. In the region 288 eV, maximum characteristic of C–O in C(O)OH appears. Heating at 100°C, the maximum C–O is retained in addition to the main maximum. Heating at 200°C in the spectrum of carbon, there is one component which can be attributed to C–H bond (285 eV) in hydrocarbons.

Without heating, in the O 1s spectrum (Fig. 8.2b), there are two components with binding energies 532.2 and 534.5 eV characteristic of $O_{adsorbrd}$ in the bond (C–O–H). Heating at 100°C, the maximum at 532.2 eV (C–O) decreases and the contribution of the adsorbed oxygen in (C–O–H) increases. This can indicate the separation of water from the coating. Heating at 200°C, the adsorbed oxygen contribution decreases.

Without heating, in the N 1s spectrum, there are two intensive maxima with binding energies 398.5 eV (Cu–N) and 400.5 eV (N–H–O) (Figure 8.2c, spectrum 1). Heating at 100°C, the maximum with the binding energy 400.5 eV characteristic of N–H–O bond is retained. The Al 2p spectrum is not revealed.

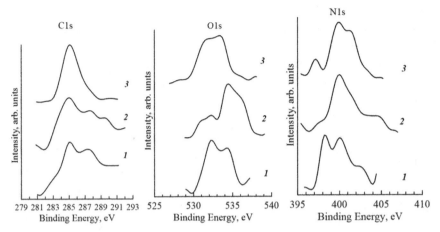

FIGURE 8.2   The XPS spectra of sample surface No. 2: (a) C 1s: **1**—without heating; **2**—heating at 100°C; **3**—heating at 200°C; (b) O 1s: **1**—without heating; **2**—heating at 100°C; **3**—heating at 200°C; (c) N 1s: **1**—without heating; **2**—heating at 100°C; and **3**—heating at 200°C.

The XPS results for sample no. 3 of MFL modified by 20% of APP($NH_4PO_3)_n$ and 0.005% of Cu–C MC are presented in Figures 8.3 and 8.4.

Without heating, the C 1s spectrum maximum (Figure 8.3a, spectrum 1) corresponds to C–H bond (285 eV) in hydrocarbons. In the region 288 eV, there is a maximum characteristic of C(O)OH bond. Heating at 100°C, one component C–H is retained (285 eV) in the spectrum of carbon (Figure 8.3a, spectrum 2).

In the O 1s spectrum (Figure 8.3b, spectrum 1), without heating, there are three components with energies of binding 527.5 eV (Al–N–O), 530.2 eV (Al–O) and 534.5 eV ($O_{adsorbed}$ in C–O–H). The Al 2p spectrum for the sample surface is not revealed despite that there are Me–O components in the oxygen spectrum. Heating at 100°C, the contribution of the component from metal oxide grows and the Al 2p spectrum appears, in which there are components with binding energies 73 eV (Al–N) and 75–77 eV (Al–O)[7] (Fig. 8.4b, spectrum 2). Heating at 200°C, in the spectrum of oxygen, the contribution of metal oxide in the region 527.5 eV decreases; at the same time, the maximum characteristic of Al–N disappears in the Al 2p spectrum.

In the N 1s spectrum, there are three maxima with binding energies 397.5 eV (Al–N), 400 eV (NH–O), and 405 eV (N–O) without heating (Figure 8.4a, spectrum 1). Heating at 100°C, two maxima are retained characteristics of Al–N and NH–O. Heating to 200°C leads to the disappearance of the component with binding energy 397.5 eV (Al–N) in the N 1s spectrum (Fig. 8.4a, spectrum 3) and the appearance of a maximum with binding energy 399 eV characteristic of the N–H bond, and the maximum with binding energy 400 eV (NH–O) is retained (Fig. 8.4a, spectrum 3).

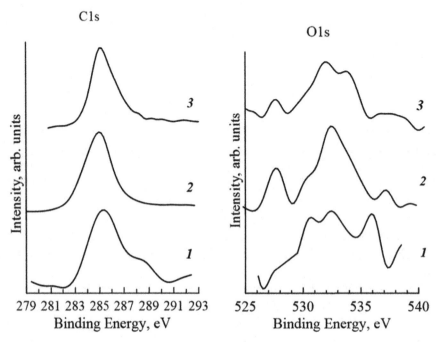

FIGURE 8.3   The XPS spectra for the surface of sample no. 3: (a) C 1s: **1**—without heating; **2**—heating to 100°C; **3**—heating to 200°C; (b) O 1s: **1**—without heating; **2**—heating to 100°C.

No phosphorus is revealed on the surface of the sample. This can be explained by that without heating phosphorus-containing functional groups form coordinate bonds with aluminum and layer-by-layer arrangement is retained; at heating, carbonization and pore formation take place, i.e. phosphorus is in "volume" and above it, there are layers of carbon formed at stimulation by phosphorus-containing products of carbonization.

The phase composition of the surface layer is qualitatively determined starting from a depth of 40–50 μm. The diffractograms obtained for the samples indicate the presence of phases in the composition of the surface layer; the chemical compositions of the phases are shown in Table 8.1.

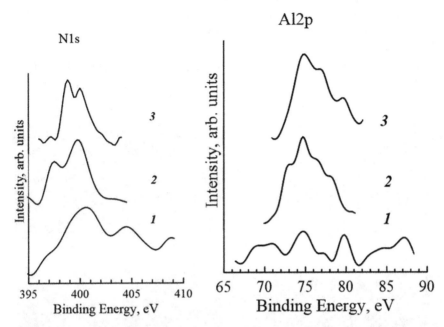

FIGURE 8.4 The XPS spectra for the surface of sample no. 3: (a) N 1s: **1**—without heating; **2**—heating to 100°C; **3**—heating to 200°C; (b) Al 2p: **1**—without heating; **2**—heating to 100°C; **3**—heating to 200°C.

TABLE 8.1 The Phase Composition of the Sample (Al + resin ML92 + NH_4PO_3 + Cu/C) and the Compositions of the Phases.

Composition	P (%)	Cu (%)	O (%)	H (%)	N (%)	C (%)	Al (%)
Al							100
Cu		100					
P		100					
CuP_4O_{11}	34.09	17.48	48,43				
CuC_8		39.81				60.19	
$C_{19}H_{16}$				6.6		93.4	
Cu_2P_{20}	82.98	17.2					
$NH_4H_2(NO_3)_3$				69.88	2.93	27.19	
Cu_3P	13.98	86.02					
$Cu_2(NH4)_2P_6O_{18}H_2O$							

Additionally, diffractograms of the heated samples, phases containing copper, nitrogen, and phosphorus are revealed; the Cu–C MC also contains the said elements, which indicates the structuring influence of Cu–C MC during the formation of the coked foam with corresponding thermophysical properties.

The optical images obtained using the microscope OLYMPUS SZ-STB2O for the studied samples have the dimensions 740 × 580 μm (Figure 8.5).

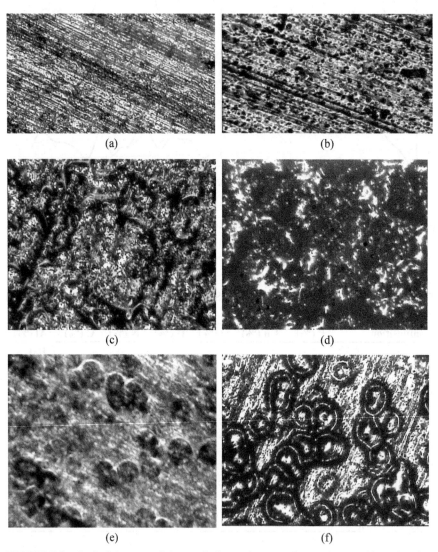

FIGURE 8.5 Optical images of the studied samples; sample no. 1: a—without heating; b—heating to 200°C; sample no. 2: c—without heating, d—heating to 200°C; sample no. 3: e—without heating; f—heating to 200°C.

The samples have been also studied by AFM on the scanning probe microscope SOLVER PRO in the contact mode. The images of the surface topography are obtained in the mode height. The dimensions of the images are 10 μm^2. The adhesion force ($F_{adhesion}$) between the probe needle and the studied samples is measured (Table 8.1). The measurements are performed at 15 points of the surface. The adhesion force is evaluated by the height of the "step" on the curves of the probe detachment from the sample. The surface energies of the studied samples can be compared based of the force of adhesion (Table 8.2).

TABLE 8.2 The Force of Adhesion ($F_{adhesion}$) between the Probe Needle and the Studied Samples and the Error of Measurement (Mean-Square Deviation).

Designation of the samples	Sample composition and the method of processing	$F_{adhesion}$ (nN)	Mean square deviation
1	MFL	2.1	0.1
2	MFL + APP + 0.006%Cu/C	1.8	0.2
5	MFL + 0.01% Cu/C	11.5	0.6
1-2	MFL after heating	4.9	1.3
2-2	MFL + APP + 0.006% Cu/C after heating	4.5	1.8
5-2	MFL + 0.01% Cu/C after heating	2.5	0.2

Sample no. 1 is uniformly coated with an MFL film.

When mesocomposite 0.01% Cu–C is introduced into lacquer (sample no. 2), the coating becomes more uniform.

When APP + 0.006% Cu–C is introduced into lacquer (sample no. 3), the film is less uniform, there are many knobs and wrinkles.

After heating, the coating on sample no. 2 covers the surface of the substrate more uniformly, due to which the coating of sample no. 2 has good protective properties.

8.4 CONCLUSION

Thus, when aluminum sample no. 1 coated with MFL to 100 and 200°C is heated, carbonization of the lacquer film takes place; however, most likely, the formed coke has a uniform fine-pored structure since the Al 2p spectrum from the aluminum substrate coated with lacquer is not revealed.

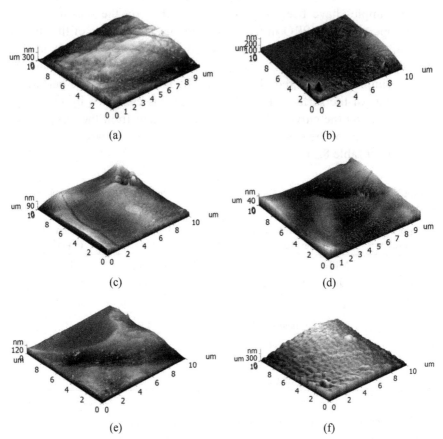

FIGURE 8.6   The AFM images of the studied samples: sample no. 1: a—without heating; b—heating to 200°C; sample no. 2: c—without heating, d—heating to 200°C; sample no. 3: e—without heating; f—heating to 200°C.

In the case when lacquer was modified with the additive Cu–C MC (sample no. 2), the formed coke also has a fine-dispersed structure, since no Al 2p spectrum is revealed from the aluminum substrate coated with modified lacquer.

When aluminum sample no. 3 is heated to 100 and 200°C coated with MFL containing 20% of APP and 0.005% of Cu–C MC in its composition, the Al 2p spectrum appears already at 100°C, which can indicate the formation of pores. Thus, the presence of APP leads to a change in the structure of the coked foam compared with the coke formed on the nonmodified film of lacquer.

At the heating of the MFL film modified by APP and mesocomposite, the formation of foamed coke surface takes place; the surface has lower thermal conductivity,[8,9] which decreases the material's combustibility.

The XPS, X-ray phase analysis, and optical results permit us to conclude that heating at 200°C, the quality of the coked foam and the value of intumescence increase when Cu–C MC is introduced together with 20% of APP into the MFL composition.

Thus, the MFL coating with such fire-proofing composition can function more effectively.

KEYWORDS

- **melamine formaldehyde lacquer**
- **swellable fire-proofing coatings**
- **ammonium polyphosphate**
- **copper–carbon mesocomposite**

REFERENCES

1. Bourbigot, S.; Ducuesne, S. Intumescences and Nanocomposites: A Novel Route for Flame-Retarding Polymeric Materials. In *Flame Retardant Polymer Nanocomposites*; Morgan, A. B., Wilkie, C. A., Eds.; 2007; pp 131–162.
2. Camino, G.; Lomakin, S. In *Fire Regardant Materials*; Horrocks, A., Price, D., Eds.; Cambridge^ CRC Press and Wood head Publishing Ltd., 2001; pp 318–336.
3. Aseyeva, R. M.; Zaikov, G. Y. Combustion of Polymer Materials. *Nauka* **1981**, *M*, 280.
4. Trapeznikov, V. A.; Shabanova, I. N.; Kholzakov, A. V., Ponomaryov, A. G. J. *J. Electron Spectrosc. Relat. Phenom.* **2004**, *383*, 137–140.
5. Aduru, S.; Contarini, S.; Rabalais, J. W. Electron-, X-ray-, and Ion-Stimulated Decomposition of Nitrate Salts. *J. Phys. Chem.* **1986**, *90* (8), 1683–1688. DOI:10.1021/j100399a045.
6. Russat, J. Characterization of Polyamic Acid/Polyimide Films in the Nanometric Thickness Range from Spin-Deposited Polyamic Acid. *Surf. Interface Anal.* **1988**, *11* (8), 414–420. DOI:10.1002/sia.740110803.
7. Taylor, G. A.; Rabalais J. W. *J. Chem. Fiz.* **1981**, *75*, 1735.
8. Kodolov, V. I.; Kodolova-Chukhontseva, V. V. *Chemical Mesoscopy Fundamentals*; Izd. IzhGTU: Izhevsk, 2019; 218 p.
9. Kodolov, V. I.; Semakina, N. V.; Trineeva, V. V. *Introduction in Nanosized Materials Science*. Izd. IzhGTU: Izhevsk, 2018; 474 p.

CHAPTER 9

Natural Fiber-Reinforced Composites: Engineering Properties with Applications in Porous Media

SONIA KHANNA and NARENDER SINGH

Advanced Polymeric Materials Research Laboratory, Department of Chemistry and Biochemistry, Sharda University, Greater Noida, Uttar Pradesh, India

ABSTRACT

Demands for reducing energy consumption and environmental impact have continuously expanded demand for eco-friendly, sustainable, and biodegradable natural fiber-reinforced composites. Natural fibers have attracted researchers, as an alternative to synthetic filler in composites. Due to their low cost, excellent mechanical properties, high specific strength, eco-friendly, and biodegradable characteristics, they are used as a good replacement for conventional fibers. This chapter gives a view of fibers and their application as reinforcement in composites. The mechanical properties and applications of natural fiber-reinforced composites are also discussed.

9.1 INTRODUCTION

Increased environmental awareness, sustainability, and new environmental regulations have provoked people to consider the use of environmentally friendly materials. It has led scientists to explore novel environmentally compassionate and sustainable materials to replace existing synthetic fibers and lessen the dependence on petroleum-based products.[1]

Mechanics and Physics of Porous Materials: Novel Processing Technologies and Emerging Applications.
Chin Hua Chia, Tamara Tatrishvili, Ann Rose Abraham, & A. K. Haghi (Eds.)
© 2024 Apple Academic Press, Inc. Co-published with CRC Press (Taylor & Francis)

In this direction, natural fibers have evolved as an alternative reinforcement in polymer composites. It has captivated the attention of researchers due to its advantages over conventional glass and carbon fibers.[2] They have found use in a variety of engineering applications with an extremely wide variety of properties. Natural fibers are used as a reinforcement or filler material, thereby reducing dependency on nonrenewable resources and energy consumption in production, together with changes in environmental ecological legislation and greater ecological awareness.[3-5] Natural fibers are classified as plant-based and animal-based plant fibers include leaf fibers (pineapple, sisal, and abaca), core fibers (hemp, jute, and kenaf), grass and reed fibers (wheat, corn, and rice), seed fibers (cotton, kapok, and coir), bast fibers (flax, jute, hemp, ramie, and kenaf), and all other types (wood and roots).[6] In this chapter, natural fiber and its application as reinforcement in composites will be discussed.

9.2 NATURAL FIBERS

Natural fibers have evolved as an alternative reinforcement for fiber-reinforced polymer (FRP) composites, due to their[7]:

- nonabrasive properties,
- eco-friendly condition,
- remarkable specific strength,
- remarkable mechanical properties,
- biodegradability characteristic, and
- low cost.

Natural fibers are not synthetic or manmade and can be sourced from plants or animals.[8] Natural fibers originate from the natural source such as animal, plant, and by other geological processes. Natural fibers can be used from both renewable and nonrenewable resources such as sisal, flax, oil palm, and jute to make composite materials, and orientation/arrangement of fibers is important which impact the changes in properties of the material in various terms.

Natural fibers are classified into two types: plant-based fibers and animal-based fibers (Figure 9.1). Natural fibers are those fibers which originate from the natural source such as animal, plant, and by other geological processes. Natural fibers can be sourced from both renewable and nonrenewable resources such as sisal, flax, oil palm, and jute to make composite materials, and one of the most important factors is the orientation/arrangement of fibers which impact the changes in the properties of a material in various terms.

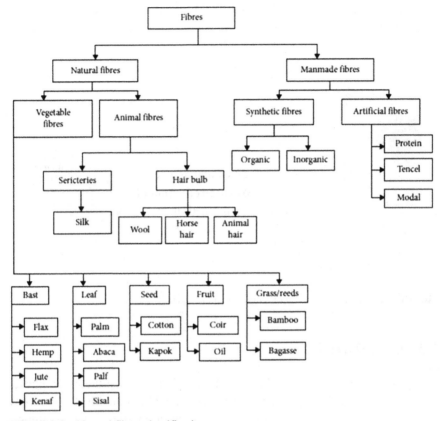

FIGURE 9.1   Natural fibers classification.

Natural fibers have attracted researchers, engineers, and scientists as an alternative reinforcement for FRP composites and have been exploited as a replacement for the conventional fiber, such as glass, aramid, and so on.

Plant fiber containing cellulose is further classified into five parts mainly:

1. Seed fiber—cotton[9,10]
2. Leaf fiber—pineapple and banana[11,12]
3. Bast fiber—ramie, jute, rattan, flax, kenaf, and hemp[13]
4. Fruit fiber—coir
5. Stalk fiber—rice, corn, and wheat[8]

Animal fibers are those which are obtained from animals. Animal fibers are made up of different kinds of proteins such as keratin, fibroin, collagen, and so on. The most widely found examples of animal fibers are silk and wool.

Common and commercially natural fibers in the world and world production have been shown in Figure 9.2.

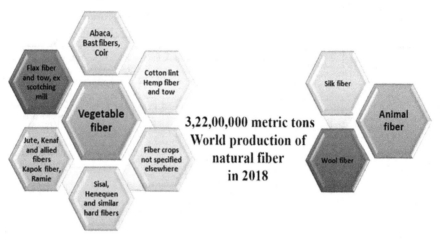

FIGURE 9.2   World production of natural fibers in 2018.

9.3   COMPOSITES

Composites are a mixture of two or more distinct constituents or phases present in reasonable proportions to serve the function. Most commonly composite have ceramic, metallic, or a polymeric matrix. The most important part of a composite is reinforcement that provides strength, stiffness, and the ability to carry a load. It can vary from metals, particulates, fibers, nanoparticles, and so on. The reinforcement is much stronger than the matrix (Figure 9.3).

Fiber-reinforced composites (FRCs) consist of axial particulates embedded in a matrix material resulting in a material with high specific strength and high specific modulus. The strength is obtained by transmitting the applied load from the matrix to the fibers. Hence, interfacial bonding is very important in the case of FRCs. Classic examples of FRCs include fiberglass and wood. FRC is a composite building material that consists of three components:

1.   Fibers as the discontinuous or dispersed phase
2.   Matrix as the continuous phase
3.   Fine interphase region, also known as the interface

Wood is a natural fiber composite comprising cellulose fibers in a matrix of lignin with varying spiral angle.[15] In manufacturing, fiberglass consists of glass fiber reinforcement of unsaturated polyester matrix.[16]

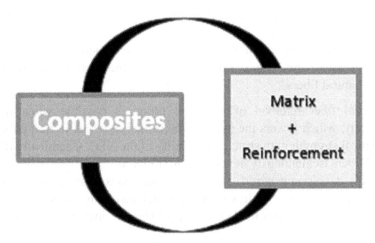

FIGURE 9.3 Components of composites.

Polymer matrix reinforced by natural fibers has various beneficial advantages over the polymer matrix made up of synthetic fibers. They have low production cost, relatively low weight, minimum damage to the functioning equipment, and bearing good mechanical strength, such as flexural strength, tensile strength, and present in sufficient amounts in nature with reasonable prices and renewable resources. They are biodegradable and spread minimum risk to health. Natural FRCs can be produced by incorporating thermosetting or thermoplastic polymer reinforced by natural fiber.[17]

Major issue with the natural fibers when reinforcing to composites is the compatibility factor between the polymer matrix and the natural fiber which is very weak because of the hydrophilic nature of the natural fibers as they consist of cellulose, lignin, pectin, and so on. These substances open the entry for the hydroxyl group which makes the binding weak between the polymer matrix and fiber. To overcome this problem, chemical modification of the fiber is required to improve the binding between natural fiber and the polymer matrix.[9,18]

Some of the treatments are discussed below:

1. *Alkali treatment*—This technique is also known as mercerization. One of the widely used chemical treatments is to reinforce thermoplastics

and thermosets for natural fibers. This method eliminates the lignin, wax, and oils from the outer surface of the fiber cell wall.

2. *Acetylation*—This method modifies the surface of the natural fiber and providing them more hydrophobic nature by incorporating the acetyl group, hence reducing the hydrophilic nature of the natural fiber.

3. *Silane treatment*—In this process, silanes react with the hydroxyl group of the natural fibers, improving the surface quality of the natural fibers.

Natural fiber removal of hydroxyl group is possible when treated chemically, which lowers the hydrophilic behavior, which in turn enhances mechanical strength with structural stability of the NFC. Chemically treated natural fibers tremendously develop natural FRCs.

In this developing world, day-by-day numerous engineering fields are gaining attention toward lightweight material; considering the laws of Earth's environment, composites made of natural fibers are increasingly coming in demand.[10]

9.3.1   NATURAL FRCS

Natural FRCs have gained importance in recent years. In 1930, polymer-based synthetic fibers started in many applications in light of their high stiffness and strength.[19-21] However, the use of synthetic fibers in applications is fading day by day due to their excessive processing cost, nonbiodegradability, and nonskin friendliness behavior.[22-25] The market for reinforced composite materials is estimated to reach 3,738.76 kt by 2022 and is expected compound annual growth rate is 9% over the next seven years. The applications for natural FRCs are rising rapidly because of their many advantageous properties, such as economical cost, biodegradability, and environmentally friendly nature with good mechanical properties.

Natural fibers originate from the natural source such as animal, plant, and by other geological processes. Natural fibers can be used from both renewable and nonrenewable resources such as sisal, flax, oil palm, and jute to make composite materials, and one of the most important factors is the orientation/arrangement of fibers which impacts the changes in properties of the material in various terms. Decomposable polymers and natural fibers begin their life cycle as renewable resources, generally in the form of starch or cellulose.

Fiberglass is probably one of the most familiar reinforcing composite materials that were introduced in 1940, consisting of glass fiber reinforcement of unsaturated polyester matrix.[16,26] In light of ongoing awareness for environmental protection as well as guidelines of the government, synthetic fibers are replaced by natural fibers in composites.

Natural fibers have a cell wall structure divided into three major structural sections. The property of fiber is decided by its microfibril angle and arrangement inside the cell wall. The adhesive property of a fiber is attributed to cellulose and hemicellulose network.[27,28] Natural fibers have certain disadvantages of incompatibility with polymer matrix when used in raw state such as high-water absorption, dead cell, wax, and oil. It can be overcome by modifying the surface of fibers by chemical treatment, enzymatic treatment, or addition of coupling agents to impart better strength in the composites system.[29]

Epoxy resins are a type of thermosetting resins that have a wide range of applications in the aerospace, automotive, and marine industries.[30,31] Brittleness, low impact resistance, and fracture toughness limit its application in high-performance fields and need to be modified to be used in applications. Natural fiber-reinforced epoxy composites show superior mechanical thermal and electrical properties.

Kumar and Srivastava reported Jute fiber-epoxy composite by hand-lay method. The resulting composite exhibited better tensile and compressive strength. The bundle strength of fibers decreased with an increase in the number of fibers and bending strength increased with an increase in the percentage of jute fiber in a bundle. Compressive strength also increases with an increase in the percentage of jute fiber.[32,33]

Yeng-Fong and Chien-Chung prepared polylactic acid (PLA)/banana fiber composites by melt-blending method. Tensile and flexural strength of composites increased to 78.6 and 65.4 MPa on reinforcement with 40 phr fiber.[34,35]

Suresh and coworkers investigated the thermal properties of palmyra fiber reinforced with (treated and untreated) tamarind seed powder epoxy composite. Samples with fiber treatment showed a significant level of strength increase as compared against the untreated (33–67%—hike) specimens. The tamarind seed powder increased the strength of the composite.[11]

Alomayri and Low reported cotton fiber-reinforced geopolymer composites. Optimum enhancements in hardness, compressive strength, and impact resistance were found in composites containing up to 0.5 wt% cotton fibers. The increase in cotton fiber content beyond 0.5 wt% decreased mechanical properties due to fiber agglomeration and poor dispersion of fibers within the matrix.[36,37]

9.3.2 PROCESSING METHODS OF FRCS

Processing of FRC involves the formation of fiber preforms and then reinforcing them in matrix materials done by different techniques as discussed below.

1. **Hand-lay method:** In this method, fiber preforms are placed in a mold coated with a thin film of anti-adhesive material. The matrix material is poured or applied with a brush on the reinforced material. The material is pressed using a roller to ensure increased interaction between slayers of reinforcement and matrix material.
2. **Spray-up method:** In this method, the handgun spray is used to spray resin and chopped fibers on a mold. Finally, a roller is used to fuse fibers into the matrix material.
3. **Resin transfer molding process:** The fiber reinforcement mat or woven roving is arranged at the half bottom of a mold and preheated matrix resin is pumped under pressure through an injector.
4. **Electrospinning process:** In this method, an electrostatic fiber fabrication technique called as electrospinning generates continuous fibers of two nanometers to several micrometers. Polymer solution is then ejected from the spinneret forming a continuous fiber which is collected at the collector.

9.3.3 MECHANICAL PROPERTIES OF NATURAL FIBER-REINFORCED POLYMER COMPOSITES

Alkaline process improves the water resistance property of the pineapple/sisal hybrid fibers. NaOH enhances mechanical strength and thermal stability with 35 wt% of hybrid fiber and 5% NaOH. Flax fibers hybridized with the carbon hence improve the mechanical strength of the epoxy-based composite. Filler incorporation also improves and stabilized the properties of the polymer composites.[18,38]

Green fiber has excellent heat insulation and audile insulation due to the presence of voids and having lignocellulosic structural arrangement in their nature.[39,40] Mechanical property of green fibers is relatively lower than the synthetic fibers, so it can be improved by modification of the surface of the fibers in a proper manner. Mostly, industries are attracted toward green fiber because of its high specific modulus, density at a lower rate, and minimum cost.[41–43] Young's modulus of natural fibers is nearly the same as compared to glass fiber. But, the tensile strength of glass fiber is higher than the natural fibers and opposite for the specific modulus because of these characteristics' natural fiber fits for the natural FRC application (Table 9.1).[44,45]

TABLE 9.1 Mechanical Properties of Various Green Fibers.

Fiber	Specific modulus	Density (g/cm3)	Tensile strength (MPa)	Young's modulus (GPa)	Elongation at break (%)
E-glass	–	2.6	1800–2700	73	2.5
Carbon	–	1.8	3500–5000	260	1.4–1.8
Cotton	–	1.5–1.6	287–800	5.5–12.6	7–8
Pineapple	40	1.53	170	1.44	14.5
Flax	50	1.5	345–1100	27.6	2.7–3.2
Abaca	–	1.5	857	41	1.10
Sisal	22	1.45	468–640	9.4–22	3–7
Hemp	50	1.4	550–900	70	1.6
Kevlar	–	1.4	2758	62	2.5-3.7
Jute	38	1.3-1.45	393–773	13–26.5	1.16–1.5
Coir	–	1.15	131–175	4–6	15–40
Ramie	–	1.0	400–938	61.4–128	1.2–3.8

PLA is produced from lactic acid and is a thermoplastic polymer, and it has the ability to substitute petroleum-based plastic material applications considering environment safety as they have high rigidity and strength, good mechanical properties, and polymer structure and arrangement.[46,47]

Shih et al.[48] produced a new natural fiber-reinforced green composite. They used plastic chopsticks and chemically treated them with coupling agents and the modified fiber was incorporated into the PLA to form the green composite. The green composite preparation was done by the melt-mixing process and characterized by various tests. The weight loss is different at decomposition temperatures and smaller than its original PLA condition because of the plant's nature. Plant fibers started the degradation process of cellulosic substances at approximately 140–370°C. The addition of modified fiber to PLA increases the char residue[48,49] of the material which decreases the thermal conductivity of the burning substances.[50] MRDCF green composites have good tensile strength as compared to original PLA. PLA40 and PLA60 both have a tensile strength greater than the original PLA, TGA, and DSC analysis which shows the high content of natural fiber improves the mechanical properties of the green composites.

Lu et al.[51] studied the polycaprolactone-hydroxyapatite (PCL-HA) porous scaffolds built of 80/20 (PCL-HA) composite by using the method called porogen-based solid freeform method. The ultimate compressive strength was found to be 3.7 ± 0.2 MPa, with a compression modulus of 61.4 ± 3.4 MPa and which is similar to the compressive strength of the trabecular bone. When the HA ratio increased, then the PCL-HA composites' compressive strength increased this reveals that the PCL-HA composites can be used for the regeneration of bone. The tensile strength of the PCL-HA composites increases with the increase in the HA ratio.[52–55]

To extract the cellulose microfiber, various types of natural fibers like jute fiber, coir fiber, red banana, and thespesia lampas plants were used and produce the natural fiFRCs with good mechanical strength.[35,56–58] The natural FRCs have been prepared by treating chemically with different concentrations of the sodium hydroxide and different silane coupling agents, hence improving the mechanical property of the sisal–oil palm hybrid FRCs.[59,60]

9.3.4 FLAME-RETARDANT PROPERTIES OF NATURAL FIBER-REINFORCED POLYMER COMPOSITES

Bachtiar et al.[61] investigated the flame-retardant nature of the natural fiber-reinforced components.

They used two investigated fire retardants, that is, aluminum hydroxide (ALH) and ammonium polyphosphates.[62] The TGA curves of flax FRP composite moved to a higher temperature when treated with the APP and ALH with the increased residue of the material; the composite that contains 30% APP mass content of epoxy gives the best thermal stabilities. The composite with the increased ALH of 40% mass lowers the temperature by releasing water vapor hence increasing the heating capacity[63,64] of the substance.

Wool and Casein both come under the category of the protein-based fiber materials and have shown effective intumescent nature of the substance during the combustion, producing the formation of the char.[65,66] Kim et al.[67] identified that the fiber of wool and intumescent flame retardant combined together to rigid char formation to reduce the heat capacity by 70% and production of smoke of PP-based composites. In determining the mechanical and physical properties of wool, the microfibril-matrix structure plays an important role.[68] In addition to that, inherent low flammability is a needed characteristic of wool as compared to other natural fibers. Moderately high contents of sulfur and nitrogen (3–4 wt%), ignition temperature is high (500–600°C), combustion of heat is low (4.9 kcal/g), and value of limiting oxygen index is high (25.2%), all these boosting the natural fire resistance for the wool.[69,70]

Natural FRP composites were prepared by using the banana fiber[71] extracted from the banana peels and chemically treated with the NaOH 5% and reinforced for epoxy resin Epikote 240 with the various mass ratio (Figure 9.7). Limiting oxygen index and UL-94HB test, it was found that the no holes were observed in the compact structure and air-filled holes would cause the fire resistance. Banana FRP composites showed no reduction in the flame-retardant properties[44–48] of the material and the degree of the resistance is the most stable.

9.4 APPLICATIONS OF NATURAL FRCS

9.4.1 IN AUTOMOBILE

Natural FRP composites have a wide variety in terms of usage in different industries. A bunch of ideas are going around in the market; industries are modifying, building, and producing natural FRP composite in an effective manner. Automobile industry[72,73] is a very well-established example of it as they are using natural fibers to build vehicles for ongoing daily purpose like cars, buses, vans, motorbikes, trucks, and so on. Many inside parts and outside parts are made up of natural fibers because of their high strength, light weighted, minimum cost, and less damage to the earth's environment. Germany-based automobile group Daimler AG, the company that owns the Mercedes-Benz, is using natural fiber for automobile parts. Figure 9.4 shows various parts made of natural fiber in Mercedes-Benz cars as they are using coconut fiber rubber latex composites for the seats of Mercedes-Benz A class model and flax–sisal fiber mat reinforced epoxy doors panels for the Mercedes-Benz E-class model, and many more.[54,74–77] Audi group is also using natural fibers such as flax–sisal-mat-reinforced PU composites to make door trim panels for their models.[54,56,76,77]

Bayerische Motoren Werke AG, popularly known as BMW, is a German-based multinational company that manufactures luxury and sports segment cars and motorbikes. BMW Group used nearly about 10,000 t of natural fibers in the year 2004.[78] Their BMW 7 series each car weighs almost 24 kg of renewable raw materials.

FIGURE 9.4   Automobile parts used by Mercedes-Benz Company
Source: Adapted from Ref.[86]

Parts of car are built by natural FRCs. Doors, hood, front splitter, rear wings, and trunk are all made by flax fiber (Figure 9.5).

Recently BMW i ventures[24] invested big amounts to use natural FRP composites for the upcoming BMW M4 GT4 series. Bcomp is a Swiss clean-tech company that manufactures high-performance reinforced composites for the world. In 2019, BMW collaborated with the Bcomp and used first time cutting-edge reinforcement solutions in BMW M Motorsport in Formula E. The cooling shaft for the BMW iFE.20 was made by flax fiber. Now BMW again setting up their new and thrilling sports car BMW M4 GT4 car (https://www.press.bmwgroup.com/global/article/detail/T0304127EN/the-bmw-ife-20-as-a-tech-lab:-first-race-car-with-parts-made-out-of-renewable-textilefibres?language=en#:   ~:text=cars%20 during%20races.-,The%20BMW%20iFE.,BMW%20M%20Motorsport [accessed on June 22 2022]; https://www.press.bmwgroup.com/global/article/detail/T0377293EN/bmw-i-ventures-invests-in-high-performance-composites-made-from-natural-fibres [accessed on June 22, 2022]), in which most parts of the car are built by the natural FRCs. Bcomp's ampliTex™ and powerRibs™ flax fiber solution (https://www.bmwiventures.com/news/

natural-fibres-for-sustainable-lightweighting-our-investment-in-bcomp [accessed June 23, 2022]) are used for the interior dashboard and on the center console of the car. Doors, hood, front splitter, rear wings, and trunk are all made by flax fiber as shown in Figures 9.5 and 9.6.

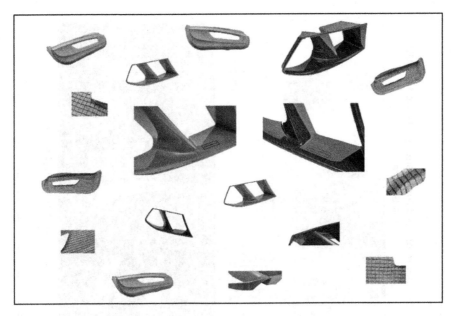

FIGURE 9.5 BMW M4 GT4 parts made from natural fiber (created from https://www. bmwiventures.com/news/natural-fibres-for-sustainable-lightweighting-our-investment-in-bcomp [accessed June 23, 2022]).

9.4.2 SPORTS AND MARINE APPLICATIONS

An important and developing sector is the sports industry; as technologies are growing in the world, the sports industry is also moving very fast toward natural FRP composites considering the properties of natural fiber, such as good durability, high stiff material, high flexural strength, low cost of production, and less hazard to the environment. Many of the sports items are made from natural fiber, for example, snowboards, skies, bicycle frame, badminton rackets, and tennis rackets (https://www.bio-sourced.com/flax-materials-in-sports-and-recreation-a-perfect-fit/ [accessed June 25, 2022]), and so on.

Germany-based company Adidas introduced their new "Greenpadel" racket (https://www.bcomp.ch/news/bmw-unveils-the-new-m4-gt4/ [accessed June

25, 2022]) which is made from the flax FRP composites providing higher strength same as carbon fiber composites. Greenpadel racket have a good flexural and bending strength which gives the player the chance to utilize full energy on the game. Figure 9.7 shows the actual racket photo.

FIGURE 9.6 Left side showing the back view, side view, and font view of the car manufacturing stage and right side after final modification (on track real stage).

Source: Adapted from Ref. [86]

Schwinn bicycles are also made from a blend of natural fiber (https:// www.bio-sourced.com/flax-materials-in-sports-and-recreation-a-perfect-fit/ [accessed June 25, 2022] with carbon fiber as a hybrid fiber. Flax fiber 90% and carbon fiber 10% are used to produce hybrid composites which meet the necessity and strength of the material (https://www.bcomp.ch/ news/greenpadel-from-adidas-with-amplitex-flax-fibres/ [accessed June 25, 2022]).[79] Scotland-based company, Lonely Mountain, uses the same blend of flax fiber and carbon fiber to produce the handmade freeride skies as shown in Figure 9.8.

FIGURE 9.7 Greenpadel racket made from flax fiber natural reinforced polymer composites (created from https://www.bcomp.ch/news/bmw-unveils-the-new-m4-gt4/ [accessed June 25, 2022]).

Image courtesy: Ref. [87]

FIGURE 9.8 Freeride skies made with flax fiber and carbon fiber blend (created from https://www.bcomp.ch/news/greenpadel-from-adidas-with-amplitex-flax-fibres/ [accessed June 25, 2022]).

Fiore et al.[80] investigated that the flax fiber and jute fiber-reinforced epoxy composites were treated chemically by using sodium bicarbonate of 10 wt%, improving the aging resistance against the salt–fog environment conditions.

Bcomp company used flax fiber to build marine parts and boats. Their natural fiber solutions ampliTex™ and powerRibs™ offer benefit factors for the environment and durability of the material; it increases the vibration damping which improves the comfort and reduces the noise (https://www.bcomp.ch/solutions/marine/ [accessed June 25, 2022]).

9.4.3   BUILDING AND CONSTRUCTION

Natural FRP composites are growing rapidly in today's world. Natural fibers contain lumen/voids that help these composites to provide better acoustic and thermal insulation properties especially when used in the construction sector. These composites are relatively eco-friendlier environment (https://www.bcomp.ch/solutions/marine/ [accessed June 25, 2022]) compared to glass fiber or carbon fiber composites. Natural FRP composites have good biodegradability, renewable sources, have high flexural and compression strength, low cost of production, and create fewer problems in the eco-system.

Various natural fibers like coir fiber, cotton, raffia, and sisal fiber are used as a filler to analyze the mechanical properties of the geopolymers.[81] The effect of 0.5 wt% cotton fiber on the mechanical properties of the geopolymer observed by the Alomayri and Low[36] gives good and effective mechanical properties as compared to the untreated one. Kriven et al.[82,83] reported that when bamboo fiber and jut are weaved together, both alkali treated and incorporated in the metakaolin-based geopolymer, it gives enhanced flexural and tensile strength as compared to the untreated composites.

9.4.4   PACKAGING

As natural FRP composites have impressive characteristics such as a biodegradability factor is high, renewable sources, good mechanical properties, vibrational damping, recyclability and low cost because of it, almost all industries are moving toward them. Packaging industry provides protection and coverage to food items using petroleum-based plastics, which are harmful to the environment and have a bad effect on the eco-system, creating problems in the water supply area, sewer systems under the cities, polluting the river and streamlines.[84] Besides this, they are nonrenewable and have high cost of production and finally impending the reduction of petroleum resources. According to Ngo,[85] the reinforcement of the coconut fiber with the natural latex, in place of the synthetic fiber materials hence reducing the usage of the nonrenewable and petroleum-based resources.

9.5 CONCLUSION

Natural fibers are low-cost, recyclable, and low-density eco-friendly material. They show good tensile strength and can be used in place of conventional fibers, such as carbon, glass, and so on. FRCs have gained much importance nowadays in different industries like automobile, aerospace, construction, and sports due to their low weight and superior mechanical properties. Classic fiber-reinforced polymers often pose considerable environmental problems due to difficulty in recycling or nonbiodegradability at the end of their usable lifetime, mainly because of the compounding of miscellaneous and usually very stable fibers and matrices. In contrast, natural fiber and biocomposite materials present considerable environmental benefits due to their biodegradability and, owing to their low cost and high specific mechanical properties making them highly desirable for applications in different fields.

KEYWORDS

- **natural fibers**
- **composites**
- **tensile strength**
- **natural fiber-reinforced composites**

REFERENCES

1. May-Pat, A.; Valadez-González, A.; Herrera-Franco, P. J. Effect of Fibre Surface Treatments on the Essential Work of Fracture of HDPE-Continuous Henequen Fibre-Reinforced Composites. *Polym. Test.* **2013**, *32* (6), 1114–1122. DOI: 10.1016/j.polymertesting.2013.06.006.

2. Saheb, D. N.; Jog, J. P. Natural Fibre Polymer Composites: A Review. *Adv. Polym. Technol.* **1999,** *18* (4), 351–363. DOI: 10.1002/(SICI)1098-2329(199924)18:4<351:: AID-ADV6>3.0.CO;2-X.

3. Ramamoorthy, S. K.; Skrifvars, M.; Persson, A. A Review of Natural Fibres Used in Biocomposites: Plant, Animal and Regenerated Cellulose Fibres. *Polym. Rev.* **2015,** *55* (1), 107–162. DOI: 10.1080/15583724.2014.971124.

4. Sanjay, M. R.; Arpitha, G. R.; Yogesha, B. Study on Mechanical Properties of Natural— Glass Fibre Reinforced Polymer Hybrid Composites: A Review. *Mater. Today Proc.* **2015,** *2* (4–5), 2959–2967. DOI: 10.1016/j.matpr.2015.07.264.

5. Thomason, J. L.; Rudeiros-Fernandez, J. L. A Review of the Impact Performance of Natural Fibre Thermoplastic Composites. *Front. Mater.* **2018,** *5*, 1–18.

6. Li, X.; Tabil, L. G.; Panigrahi, S.; Crerar, W. J. The Influence of Fibre Content on Properties of Injection Molded Flax Fibre-HDPE Biocomposites. *Can. Biosyst. Eng.* **2009,** *8*, 1–10.

7. Groover, M. P. *Fundamental of Modern Manufacturing*, 2nd, ed.; John Wiley & Sons, Inc.: Hoboken, NJ, 2004.

8. Ticoalu, A.; Aravinthan, T.; Cardona, F. A Review of Current Development in Natural Fibre Composites for Structural and Infrastructure Applications. In *Proceedings of the Southern Region Engineering Conference (SREC '10)*, 2010; pp 113–117.

9. Rohit, K.; Dixit, S. A Review—Future Aspect of Natural Fibre Reinforced Composite. *Polym. Renew. Resour.* **2016,** *7* (2), 43–59. DOI: 10.1177/204124791600700202.

10. Mahir, F. I.; Keya, K. N.; Sarker, B.; Nahiun, K. M.; Khan, R. A. A Brief Review on Natural Fibre Used as a Replacement of Synthetic Fibre in Polymer Composites. *Mater. Eng. Research* **2019,** *1*, 86–97.

11. Srinivasan, T.; Kumar, S. B.; Suresh, G.; Ravi, R.; Srinath, S. L.; Paul, A. I.; Vishwesh-waran, M. Experimental Investigation and Fabrication of Palmyra Palm Natural Fibre with Tamarind Seed Powder Reinforced Composite. *IOP Conf. S. Mater. Sci. Eng.* **2020,** *988*, 012022.

12. Bhoopathi, R.; Ramesh, M.; Deepa, C. Fabrication and Property Evaluation of Banana-Hemp-Glass Fibre Reinforced Composites. *Proc. Eng.* **2014,** *97*, 2032–2041. DOI: 10.1016/j.proeng.2014.12.446.

13. Paridah, M. T.; Ahmed, A. B.; SaifulAzry, S. O. A.; Ahmed, Z. Retting Process of Some Bast Plant Fibres and Its Effect on Fibre Quality: A Review. *Bioresources* **2011,** *6* (4), 5260–5281. DOI: 10.15376/biores.6.4.5260-5281.

14. Townsend, T. *World Natural Fibre Production and Employment, Handbook of Natural Fibres*; Woodhead Publishing: Sawston, UK, 2020; pp 15–36.

15. Matthews, F. L.; Rawlings, R. D. Composite Materials: Engineering and Science; Woodhead Publishing: Sawston, UK, 1999.

16. Tsai, S. W.; Hahn, H. T. *Introduction to Composite Materials*; Routledge: Milton Park, 2018.

17. Godara, S. S. Effect of Chemical Modification of Fibre Surface on Natural Fibre Composites: A Review. *Mater. Today Proc.* **2019,** *18*, 3428–3434.

18. Todkar, S. S.; Patil, S. A. Review on Mechanical Properties Evaluation of Pineapple Leaf Fibre (PALF) Reinforced Polymer Composites. *Compos. B Eng.* **2019,** *174*, 106927. DOI: 10.1016/j.compositesb.2019.106927.

19. Begum, K.; Islam, M. Natural Fibre as a Substitute to Synthetic Fibre in Polymer Composites: A Review. *Res. J. Eng. Sci.* **2013,** *2*, 46–53.

20. Keya, K. N.; Kona, N. A.; Koly, F. A.; Maraz, K. M.; Islam, M. N.; Khan, R. A. Natural Fibre Reinforced Polymer Composites: History, Types, Advantages and Applications. *Mater. Eng. Res.* **2019,** *1* (2), 69–87. DOI: 10.25082/MER.2019.02.006.

21. Jagadeesh, D.; Kanny, K.; Prashantha, K. A Review on Research and Development of Green Composites from Plant Protein-Based Polymers. *Polym. Compos.* **2017,** *38* (8), 1504–1518. DOI: 10.1002/pc.23718.

22. Sahu, P.; Gupta, M. K. A Review on the Properties of Natural Fibres and Its Bio-composites: Effect of Alkali Treatment. *Proc. Inst. Mech. Eng. L* **2020,** *234* (1), 198–217.DOI: 10.1177/1464420719875163.

23. Sahu, P.; Gupta, M. K. Lowering in Water Absorption Capacity and Mechanical Degradation of Sisal/Epoxy Composite by Sodium Bicarbonate Treatment and PLA Coating. *Polym. Compos.* **2020,** *41* (2), 668–681. DOI: 10.1002/pc.25397.

24. Mohammed, L.; Ansari, M. N. M.; Pua, G.; Jawaid, M.; Islam, M. S. A Review on Natural Fibre Reinforced Polymer Composite and Its Applications. *Int. J. Polym. Sci.* **2015,** *2015,* 1–15. DOI: 10.1155/2015/243947.

25. Thyavihalli Girijappa, Y. G.; Mavinkere Rangappa, S.; Parameswaranpillai, J.; Siengchin, S. Natural Fibres as Sustainable and Renewable Resource for Development of Eco-friendly Composites: A Comprehensive Review. *Front. Mater.* **2019,** *6,* 226. DOI: 10.3389/fmats.2019.00226.

26. Nicolais, L. Mechanics of Composites. *Polym. Eng. Sci.* **1975,** *15* (3), 137–149. DOI: 10.1002/pen.760150305.

27. Lyczakowski, J. J.; Bourdon, M.; Terrett, O. M.; Helariutta, Y.; Wightman, R.; Dupree, P. Structural Imaging of Native Cryo-Preserved Secondary Cell Walls Reveals the Presence of Macrofibrils and Their Formation Requires Normal Cellulose, Lignin and Xylan Biosynthesis. *Front. Plant Sci.* **2019,** *10,* 1398. DOI: 10.3389/fpls.2019.01398.

28. Asim, M.; Abdan, K.; Jawaid, M.; Nasir, M.; Dashtizadeh, Z.; Ishak, M. R.; Hoque, M. E.; Deng, Y. A Review on Pineapple Leaves Fibre and Its Composites. *Int. J. Polym. Sci.* **2015,** 2015, 1–16. DOI: 10.1155/2015/950567.

29. George, M.; Mussone, P. G.; Alemaskin, K.; Chae, M.; Wolodko, J.; Bressler, D. C. Enzymatically Treated Natural Fibres as Reinforcing Agents for Biocomposite Material: Mechanical, Thermal, and Moisture Absorption Characterization. *J. Mater. Sci.* **2016,** *51* (5), 2677–2686. DOI: 10.1007/s10853-015-9582-z.

30. Shalwan, A.; Yousif, B. F. In State of Art: Mechanical and Tribiological Behaviour of Polymeric Composites Based on Natural Fibres. *Mater. Des.* **2013,** *48,* 14–24. DOI: 10.1016/j.matdes.2012.07.014.

31. Shinoj, S.; Visvanathan, R.; Panigrahi, S.; Kochubabu, M. Oil Palm Fibre (PPF) and Its Composites: A Review. *Ind. Crops Prod.* **2011,** *33* (1), 7–22. DOI: 10.1016/j.indcrop. 2010.09.009.

32. Kumar, A.; Srivastva, A. Preparation and Mechanical Properties of Jute Fibre Reinforced Epoxy Composites. *Ind. Eng. Manage.* **2017,** *6,* 4.

33. Sarikaya, E.; Çallioğlu, H.; Demirel, H. Production of Epoxy Composites Reinforced by Different Natural Fibres and Their Mechanical Properties. *Compos. B* **2019,** *167,* 461–466. DOI: 10.1016/j.compositesb.2019.03.020.

34. Yeong-Fong, S.; Chien-Chung, H. Polylactic Acid (PLA)/Banan Fibre (BF) Biodegradable Green Composites. *J. Polym. Res.* **2011,** *18,* 2335–2340.

35. Ashok, B.; Feng, H.; Rajulu, V. Preparation and Properties of Cellulose/Thespesia Lampas Microfibre Composite Films. *Int. J. Biol. Macromol.* **2019,** *127,* 153–158. DOI: 10.1016/j.ijbiomac.2019.01.041.

36. Alomayri, T.; Low, I. M. Synthesis and Characterization of Mechanical Properties in Cotton Fibre-Reinforced Geopolymer Composites. *J. Asian Ceram. Soc.* **2013,** *1* (1), 30–34. DOI: 10.1016/j.jascer.2013.01.002.

37. Lomelí-Ramírez, M. G.; Valdez-Fausto, E. M.; Rentería-Urquiza, M.; Jiménez-Amezcua, R. M.; Anzaldo Hernández, J.; Torres-Rendon, J. G.; García Enriquez, S. Study of Green Nanocomposites Based on Corn Starch and Cellulose Nanofibrils from Agave Tequilana Weber. *Carbohydr. Polym.* **2018**, *201*, 9–19. DOI: 10.1016/j.carbpol.2018.08.045.

38. Flynn, J.; Amiri, A.; Ulven, C. Hybridized Carbon and Flax Fibre Composites for Tailored Performance. *Mater. Des.* **2016**, *102*, 21–29. DOI: 10.1016/j.matdes.2016.03.164.

39. Nurazzi, N. M.; Asyraf, M. R. M.; Athiyah, S. F.; Shazleen, S. S.; Rafiqah, S. A.; Harussani, M. M.; Kamarudin, S. H.; Razman, M. R.; Rahmah, M.; Zainudin, E. S.; Ilyas, R. A.; Aisyah, H. A.; Norrrahim, M. N. F.; Abdullah, N.; Sapuan, S. M.; Khalina, A. A Review on Mechanical Performance of Hybrid Natural Fiber Polymer Composites for Structural Applications. *Polymers* **2021**, *13* (13), 2170. DOI: 10.3390/polym13132170.

40. Ilyas, R. A.; Sapuan, S. M.; Harussani, M. M.; Hakimi, M. Y. A. Y.; Haziq, M. Z. M.; Atikah, M. S. N.; Asyraf, M. R. M.; Ishak, M. R.; Razman, M. R.; Nurazzi, N. M.; Norrrahim, M. N. F.; Abral, H.; Asrofi, M. Polylactic Acid (PLA) Biocomposite: Processing, Additive Manufacturing and Advanced Applications. *Polymers* **2021**, *13* (8), 1326. DOI: 10.3390/polym13081326.

41. Ilyas, R. A.; Zuhri, M. Y. M.; Aisyah, H. A.; Asyraf, M. R. M.; Hassan, S. A.; Zainudin, E. S.; Sapuan, S. M.; Sharma, S.; Bangar, S. P.; Jumaidin, R.; Nawab, Y.; Faudzi, A. A. M.; Abral, H.; Asrofi, M.; Syafri, E.; Sari, N. H. Natural Fibre-Reinforced Polylactic Acid, Polylactic Acid Blends and Their Composites for Advanced Applications. *Polymers* **2022**, *14* (1), 202. DOI: 10.3390/polym14010202.

42. Malkapuram, R.; Kumar, V.; Negi, Y. S. Recent Development in Natural Fibre Reinforced Polypropylene Composites. *J. Reinf. Plast. Compos.* **2009**, *10*, 1169–1189.

43. Pandita, S. D.; Yuan, X.; Manan, M. A.; Lau, C. H.; Subramanian, A. S.; Wei, J. Evaluation of Jute/Glass Hybrid Composite Sandwich: Water Resistance, Impact Properties and Life Cycle Assessment. *J. Reinf. Plast. Compos.* **2014**, *33* (1), 14–25. DOI: 10.1177/0731684413505349.

44. Asyraf, M. R. M.; Ishak, M. R.; Sapuan, S. M.; Yidris, N.; Ilyas, R. A.; Rafidah, M.; Razman, M. R. Potential Application of Green Composites for Cross Arm Component in Transmission Tower: A Brief Review. *Int. J. Polym. Sci.* **2020**, *2020*, 1–15. DOI: 10.1155/2020/8878300.

45. Azman, M. A.; Asyraf, M. R. M.; Khalina, A.; Petrů, M.; Ruzaidi, C. M.; Sapuan, S. M.; Wan Nik, W. B.; Ishak, M. R.; Ilyas, R. A.; Suriani, M. J. Natural Fibre Reinforced Composite Material for Product Design: A Short Review. *Polymers* **2021**, 13 (12). DOI: 10.3390/polym13121917.

46. Nazrin, A.; Sapuan, S. M.; Zuhri, M. Y. M.; Ilyas, R. A.; Syafiq, R. S.; Sherwani, S. F. K. Nanocellulose Reinforced Thermoplastic Starch (TPS), Polylactic Acid (PLA), and Polybutylene Succinate (PBS) for Food Packaging Applications. *Front. Chem.* **2020**, *8*, 213. DOI: 10.3389/fchem.2020.00213.

47. Nazrin, A.; Sapuan, S. M.; Zuhri, M. Y. M.; Tawakkal, I. S. M. A.; Ilyas, R. A. Water Barrier and Mechanical Properties of Sugar Palm Crystalline Nanocellulose Reinforced Thermoplastic Sugar Palm Starch (TPS)/Poly(Lactic Acid)(PLA) Blend Bionanocomposites. *Nanotechnol. Rev.* **2021**, *10* (1), 431–442. DOI: 10.1515/ntrev-2021-0033.

48. Shih, Y. F.; Huang, C. C.; Chen, P. W. Biodegradable Green Composites Reinforced by the Fibre Recycling from Disposable Chopsticks. *Mater. Sci. Eng. A* **2010**, *527* (6), 1516–1521. DOI: 10.1016/j.msea.2009.10.024.

49. Shih, Y. F.; Jeng, R. J. Carbon Black Containing IPNs Based on Unsaturated Polyester/ Epoxy. I. Dynamic Mechanical Properties, Thermal Analysis, and Morphology. *J. Appl. Polym. Sci.* **2002,** *86* (8), 1904–1910. DOI: 10.1002/app.11145.

50. Pearce, E. M.; Liepins, R. Public Health Implications of Components of Plastics Manufacture. Flame Retardants. *Environ. Health Perspect.* **1975,** 11, 59–69. DOI: 10.1289/ ehp.751159.

51. Lu, L.; Zhang, Q.; Wootton, D. M.; Chiou, R.; Li, D.; Lu, B.; Lelkes, P. I.; Zhou, J. Mechanical Study of Polycaprolactone-Hydroxyapatite Porous Scaffolds Created by Porogen-Based Solid Freeform Fabrication Method. *J. Appl. Biomater. Funct. Mater.* **2014,** *12* (3), 145–154. DOI: 10.5301/JABFM.5000163.

52. Misch, C. E.; Qu, Z.; Bidez, M. W. Mechanical Properties of Trabecular Bone in the Human Mandible: Implications for Dental Implant Treatment Planning and Surgical Placement. *J. Oral Maxillofac. Surg.* **1999,** *57* (6), 700–6; discussion 706. DOI: 10.1016/ s0278-2391(99)90437-8.

53. Carter, D. R.; Hayes, W. C. Bone Compressive Strength: The Influence of Density and Strain Rate. *Science* **1976,** *194* (4270), 1174–1176. DOI: 10.1126/science.996549.

54. Lalit, R.; Mayank, P.; Ankur, K. Natural Fibres and Biopolymers Characterization: A Future Potential Composite Material. *Strojnicky Casopis—J. Mech. Eng.* **2018,** *68* (1), 33–50. DOI: 10.2478/scjme-2018-0004.

55. Filippi, M.; Born, G.; Chaaban, M.; Scherberich, A. Natural Polymeric Scaffolds in Bone Regeneration. *Front. Bioeng. Biotechnol.* **2020,** *8*, 474. DOI: 10.3389/fbioe.2020.00474.

56. Khan, T.; Hameed Sultan, M. T. B.; Ariffin, A. H. The Challenges of Natural Fibre in Manufacturing, Material Selection, and Technology Application: A Review. *J. Reinf. Plast. Compos.* **2018,** *37* (11), 770–779. DOI: 10.1177/0731684418756762.

57. Fonseca, A.; Panthapulakkal, S.; Konar, S. K.; Sain, M.; Bufalinof, L.; Raabe, J.; Miranda, I. Pd. A.; Martins, M. A.; Tonoli, G. H. D. Improving Cellulose Nanofibrillation of Non-Wood Fibre Using Alkaline and Bleaching Pretreatments. *Ind. Crops Prod.* **2019,** *131*, 203–212. DOI: 10.1016/j.indcrop.2019.01.046.

58. Manimaran, P.; Sanjay, M. R.; Senthamaraikannan, P.; Jawaid, M.; Saravanakumar, S. S.; George, R. Synthesis and Characterization of Cellulosic Fibre from Red Banana Peduncle as Reinforcement for Potential Applications. *J. Nat. Fibres* **2018,** *16* (5), 768–780. DOI: 10.1080/15440478.2018.1434851.

59. Mbarki, K.; Boumbimba, R. M.; Sayari, A.; Elleuch, B. Influence of Microfibres Length on PDLA/Cellulose Microfibres Biocomposites Crystallinity and Properties. *Polym. Bull.* **2019,** *76* (3), 1061–1079. DOI: 10.1007/s00289-018-2431-x.

60. John, M. J.; Francis, B.; Varughese, K. T.; Thomas, S. Effect of Chemical Modification on Properties of Hybrid Fibre Biocomposites. *Compos. A* **2008,** *39* (2), 352–363. DOI: 10.1016/j.compositesa.2007.10.002.

61. Jacob, M.; Thomas, S.; Varughese, K. T. Mechanical Properties of Sisal/Oil Palm Hybrid Fibre Reinforced Natural Rubber Composites. *Compos. Sci. Technol.* **2004,** *64* (7–8), 955–965. DOI: 10.1016/S0266-3538(03)00261-6.

62. Bachtiar, E. V.; Kurkowiak, K.; Yan, L.; Kasal, B.; Kolb, T. Thermal Stability, Fire Performance, and Mechanical Properties of Natural Fibre Fabric-Reinforced Polymer Composites with Different Fire Retardants. *Polymers* **2019,** *11* (4), 699. DOI: 10.3390/ polym11040699.

63. Shukor, F.; Hassan, A.; Saiful Islam, M. S.; Mokhtar, M.; Hasan, M. Effect of Ammonium Polyphosphate on Flame Retardancy, Thermal Stability and Mechanical Properties of

Alkali Treated Kenaf Fibre Filled PLA Biocomposites. *Mater. Des. (1980–2015)* **2014,** *54*, 425–429. DOI: 10.1016/j.matdes.2013.07.095.

64. Chapple, S.; Anandjiwala, R. Flammability of Natural Fibre-Reinforced Composites and Strategies for Fire Retardancy: A Review. *J. Thermoplast. Compos. Mater.* **2010,** *23* (6), 871–893. DOI: 10.1177/0892705709356338.

65. Azwa, Z. N.; Yousif, B. F. Characteristics of Kenaf Fibre/Epoxy Composites Subjected to Thermal Degradation. *Polym. Degrad. Stab.* **2013,** *98* (12), 2752–2759. DOI: 10.1016/j.polymdegradstab.2013.10.008.

66. Alongi, J.; Cuttica, F.; Di Blasio, A. D.; Carosio, F.; Malucelli, G. Intumescent Features of Nucleic Acids and Proteins. *Thermochim. Acta* **2014,** *591*, 31–39. DOI: 10.1016/j.tca.2014.06.020.

67. Alongi, J.; Poskovic, M.; Visakh, P. N.; Frache, A.; Malucelli, G. Cyclodextrin Nanosponges as Novel Green Flame Retardants for PP, LLDPE and PA6. *Carbohydr. Polym.* **2012,** *88* (4), 1387–1394. DOI: 10.1016/j.carbpol.2012.02.038.

68. Kim, N. K.; Lin, R. J. T.; Bhattacharyya, D. Effects of Wool Fibres, Ammonium Polyphosphate and Polymer Viscosity on the Flammability and Mechanical Performance of PP/Wool Composites. *Polym. Degrad. Stab.* **2015,** *119*, 167–177. DOI: 10.1016/j.polymdegradstab.2015.05.015.

69. Tsobkallo, K.; Aksakal, B.; Darvish, D. Analysis of the Contribution of the Microfibrils and Matrix to the Deformation Processes in Wool Fibres. *J. Appl. Polym. Sci.* **2012,** *125* (S2), E168–E179. DOI: 10.1002/app.36535.

70. Price, D.; Horrocks, A. R. Polymer Degradation and the Matching of FR Chemistry to Degradation. In *Fire Retardancy of Polymeric Materials*; Wilkie, C. A., Morgan, A. B., Eds., 2009; pp 15–43.

71. Benisek, L. Flame Retardance of Protein Fibres. In *Flame-Retardant Polymeric Materials*; Springer: Cham, 1975; pp 137–191.

72. Nguyen, T. A.; Nguyen, T. H. Banana Fibre-Reinforced Epoxy Composites: Mechanical Properties and Fire Retardancy. *Int. J. Chem. Eng.* **2021,** 2021, 1–9. DOI: 10.1155/2021/1973644.

73. Shinoj, S.; Visvanathan, R.; Panigrahi, S.; Kochubabu, M. Oil Palm Fibre (OPF) and Its Composites: A Review. *Ind. Crops Prod.* **2011,** *33* (1), 7–22. DOI: 10.1016/j.indcrop.2010.09.009.

74. Shuit, S. H.; Tan, K. T.; Lee, K. T.; Kamaruddin, A. H. Oil Palm Biomass as a Sustainable Energy Source: A Malaysian Case Study. *Energy* **2009,** *34* (9), 1225–1235. DOI: 10.1016/j.energy.2009.05.008.

75. Bos, H. *The Potential of Flax Fibres as Reinforcement for Composite Materials*; Wageningen University and Research: Wageningen, 2004.

76. Suddell, B. C. Industrial Fibres: Recent and Current Developments. In *Proceedings of the Symposium on Natural Fibres*; Common Fund for Commodities: Amsterdam, 2008; Vol. 20; pp 71–82.

77. Huda, M. S.; Drzal, L. T.; Ray, D.; Mohanty, A. K.; Mishra, M. Natural-Fibre Composites in the Automotive Sector. In *Properties and Performance of Natural-Fibre Composites*; Woodhead Publishing: Sawston, UK, 2008; pp 221–268.

78. Sreenivas, H. T.; Krishnamurthy, N.; Arpitha, G. R. A Comprehensive Review on Light Weight Kenaf Fibre for Automobiles. *Int. J. Lightweight Mater. Manuf.* **2020,** *3* (4), 328–337. DOI: 10.1016/j.ijlmm.2020.05.003.

79. Al Rashid, A.; Khalid, M. Y.; Imran, R.; Ali, U.; Koc, M. Utilization of Banana Fibre-Reinforced Hybrid Composites in the Sports Industry. *Materials (Basel)* **2020,** *13* (14), 3167. DOI: 10.3390/ma13143167.

80. Fiore, V.; Sanfilippo, C.; Calabrese, L. Influence of Sodium Bicarbonate Treatment on the Aging Resistance of Natural Fibre Reinforced Polymer Composites Under Marine Environment. *Polym. Test.* **2019,** *80,* 106100. DOI: 10.1016/j.polymertesting.2019.106100.

81. Korniejenko, K.; Frączek, E.; Pytlak, E.; Adamski, M. Mechanical Properties of Geopolymer Composites Reinforced with Natural Fibres. *Procedia Eng.* **2016,** *151,* 388–393. DOI: 10.1016/j.proeng.2016.07.395.

82. Sá Ribeiro, R. A.; Sá Ribeiro, M. G.; Sankar, K.; Kriven, W. M. Geopolymer-Bamboo Composite—A Novel Sustainable Construction Material. *Constr. Build. Mater.* **2016,** *123,* 501–507. DOI: 10.1016/j.conbuildmat.2016.07.037.

83. Sankar, K.; Kriven, W. M. *Sodium Geopolymer Reinforced with Jute Weave. Developments in Strategic Materials and Computational Design V*; Kriven, W. M., Zhou, D., Moon, K., Hwang, T., Wang, J., Lewinssohn, C., Zhou, Y., Eds., 2015; pp 39–60.

84. Majeed, K.; Jawaid, M.; Hassan, A.; Abu Bakar, A.; Abdul Khalil, H. P. S.; Salema, A. A.; Inuwa, I. Potential Materials for Food Packaging from Nanoclay/Natural Fibres Filled Hybrid Composites. *Mater. Des.* **2013,** *46,* 391–410. DOI: 10.1016/j.matdes.2012.10.044.

85. Ngo, T. D. Natural Fibres for Sustainable Bio-composites. *Nat. Artif. Fibre-Reinf. Compos. Renew. Sources* **2018,** *3,* 107–126.

86. https://www.bcomp.ch/news/bmw-unveils-the-new-m4-gt4/

87. https://www.bcomp.ch/news/greenpadel-from-adidas-with-amplitex-flax-fibres/

CHAPTER 10

Chalcogenide Glass–Incorporated Photonic Crystal Fibers: A Game-Changer in Integrated Infrared Photonics

SOUMYA SURESH[1], AJEESH KUMAR SOMAKUMAR[2], ANUPAMA VISWANATHAN[3], and SHEENU THOMAS[1]

[1]International School of Photonics, Cochin University of Science and Technology, Cochin, Kerala, India

[2]Institute of Physics, Polish Academy of Sciences, Warsaw, Poland

[3]Sree Narayana College Nattika, Triprayar, Kerala, India

ABSTRACT

Photonic crystal fibers are microstructured optical fibers that have gained significant scientific interest over the past several years. They are widely used in various sensing applications, such as biosensors, supercontinuum sources, all-optical nonlinear devices, and so on. Their unique morphology, flexible light-guiding properties, and exceptional ability to host optical materials make them an outstanding candidate in photonics. Chalcogenide glasses are excellent infrared optical materials for active and passive photonic applications due to their remarkable optical, thermal, and electronic properties. Incorporating photonic crystal fibers with chalcogenide glasses opens a broad window into realizing integrated photonic devices. This chapter introduces photonic crystal fibers and their guiding mechanisms, and chalcogenide glass-integrated photonic crystal fibers, and their applications.

Mechanics and Physics of Porous Materials: Novel Processing Technologies and Emerging Applications.
Chin Hua Chia, Tamara Tatrishvili, Ann Rose Abraham, & A. K. Haghi (Eds.)
© 2024 Apple Academic Press, Inc. Co-published with CRC Press (Taylor & Francis)

10.1   INTRODUCTION

The invention of photonic crystal fibers (PCFs) marked an era in fiber optics technology. Conventional optical fibers have limited linear and nonlinear photonic applications due to their low index contrast between the core and the cladding. The small core size can introduce undesirable nonlinear effects. PCFs are microstructured optical fibers (MOFs) having an array of holes in their cladding along the entire length of the fiber. They are highly versatile in geometry, so the light-guiding and the optical properties can be efficiently controlled. Based on their microstructure geometry, a variety of guiding mechanisms exist in PCFs, including modified total internal reflection (MTIR), photonic bandgap (PBG) guiding, inhibited coupling guiding, anti-resonance reflecting optical waveguide (ARROW) guiding, and twist-induce guiding.[1] There also exists an exciting class of PCFs known as hybrid PCFs, in which more than one guiding mechanism coexist.[2] They find numerous applications in nonlinear fiber optics,[3] supercontinuum generation,[4] physical, chemical, and biosensing applications,[5-7] fiber lasers,[8] and so on.

The history of the evolution of PCFs begins with the PBG effect, which was first predicted by Sajeev John and Eli Yablonovitch in 1987[9,10] and gained considerable scientific attention after the experimental observations made by Yablonovitch et al. in 1989[11] and 1991.[12] PBG effect is a phenomenon in which the periodic modulation of the refractive index (RI) of material leads to the forbidden frequency ranges for the photons regardless of their polarization states. The idea was similar to forming the electronic band structure in materials due to the periodic modulation of lattice potential.

In 1991, Philip Russel realized that the periodic arrangement of subwave-length holes along the length of the fiber could introduce a PBG effect. Later, he and his coworkers demonstrated the first PCF, which consisted of an array of microholes of diameter 0.2–1.2 μm arranged in a triangular lattice with a pitch of 2.3 μm and core diameter of 4.6 μm. In 1997, they fabricated endlessly single-mode PCF to transmit single mode for the wavelength range 458–1550 nm.[13] They intended to create a MOF to guide light via the PBG effect. Still, their attempt left a solid-core PCF (SC-PCF) that guided the light via MTIR rather than the PBG effect.

In 1999, Cregan et al. successfully fabricated a hollow-core PCF (HC-PCF) for the first time that could confine the light in the hollow core with narrow transmission windows,[14] but later, HC-PCF that could transmit the light over a broad spectral range was realized.[15] Pryamikov et al. fabricated negative curvature (NC) HC-PCF with a core formed by an array of single rows of silica capillaries that could transmit light with low loss over a broad mid-infrared (MIR) spectral range.[16] This idea was further extended to fabricating nested tube antiresonant fibers in which a smaller tube is nested inside the larger one.[17] The advantage of HC-PCF is that since the light is guided through the air, low optical loss transmission is possible.

One of the key features of PCFs is that they can act as a substrate to host optical materials, gases,[18] atomic vapors,[19] and so on, inside their air holes. This feature has acquired a lot of scientific attention since incorporating active materials can alter the guiding mechanism. The properties of these incorporated materials are highly tunable with temperature,[20] applied electric/magnetic field,[21] strain,[22] pressure,[23] and so on, thereby affecting the guiding mechanism that results in various sensor fabrications.[20] Westbrook et al. fabricated the first hybrid PCF with polymer-filled holes that could tune the cladding mode resonances by varying the temperature.[24]

10.2 GEOMETRY AND THE HYBRID GUIDING MECHANISMS IN PCFS

10.2.1 MODIFIED TOTAL INTERNAL REFLECTION

Conventional single-mode fibers (SMF) have a small doped core surrounded by silica cladding. The light propagates based on the principle of total internal reflection at the core-cladding interface. The cladding is designed with a lower RI than the core.

In SC-PCF, the cross-section consists of a solid core surrounded by a periodic array of micro-hole cladding running along the fiber's entire length, which results in the average RI of the cladding lower than the core without doping, and the corresponding guiding mechanism is known as MTIR. In PCF generally, three parameters should be considered while designing: the core diameter ρ, inter-hole spacing or pitch Λ, and hole diameter d (Figure 10.1). The guiding properties of PCF entirely rely on Λ and d. Figure 10.1 shows the RI profile of conventional step-index fiber and SC-PCF. Let, n_{co} and n_{cl} are the refractive indices of the core and the cladding of the conventional fiber. n_{si}, n_{air}, and n_{eff} are the refractive indices of silica, air, and the effective RI of SC-PCF cladding, respectively.[1]

The guiding mechanism can be explained using a propagation diagram or $\beta\Lambda$ versus $\omega\Lambda/c$ plot (Figure 10.2), where β is the axial wave vector, ω is the angular frequency, and c is the velocity of light in vacuum. The diagram indicates the frequency ranges and axial wave vector component, where the light cannot propagate or is evanescent. The condition $kn = \omega n/c$ sets the maximum value of β, k is the free space wavenumber and n is the RI of the medium under consideration. Above the radiation line ($\beta < kn$), the light is free to propagate in the medium of RI n, whereas the light is evanescent below the radiation line ($\beta > kn$). Figure 10.2 can be used to explain the guiding mechanism for conventional step-index fiber and SC-PCF. In region D, the light is evanescent in both fibers; in region C, where $kn_{eff}\Lambda < \beta\Lambda < kn_{si}\Lambda$ (in the case of SMF, $kn_{cl}\Lambda < \beta\Lambda < kn_{co}\Lambda$), light is strictly confined in the core, and the propagation is turned off in photonic crystal (PC) and air (in the case of SMF, cladding, and air). The area of region C in SC-PCF is larger compared to the conventional SMF, which helps to explore the possibilities of PCFs. In region B, light can propagate in silica and the PC (core and cladding in the case of SMF) but cannot propagate in the air. In region A, light is free to propagate anywhere in the fiber. The black fingers in Figure 10.2b represent the full 2D photonic band gaps that extend to the region where $\beta < k$. Light is free to propagate in the air for regions within the black fingers but cannot penetrate the cladding due to PBG, which is the supporting data for the possibility of air guidance or hollow-core guidance. Point R represents the region where light is strictly confined inside the core in conventional step-index SMF.[25]

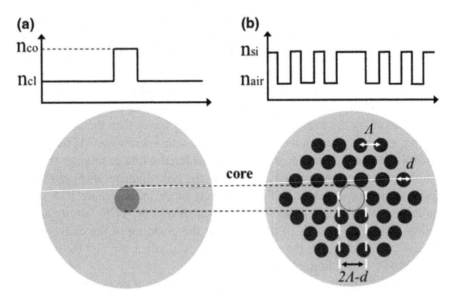

FIGURE 10.1   The cross-sections and refractive index profiles of (a) conventional step-index optical fiber and (b) SC-PCF. Gray and black regions, respectively, show silica material and air.

Source: Reproduced from Ref. [1] with the permission of the American Physical Society.

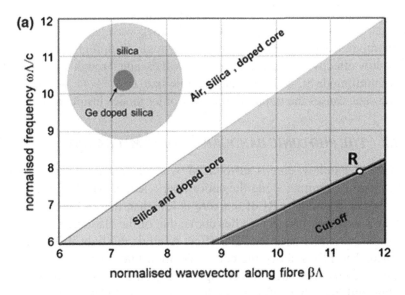

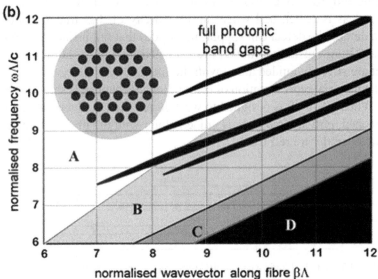

FIGURE 10.2 Propagation diagram for (a) conventional step-index single-mode fiber and (b) SC-PCF

Source: Reproduced from Ref. [1] with the permission of the American Physical Society.

The modal features of PCF are determined by the filling ratio d/Λ. If d/Λ < 0.44, PCF is an endlessly single mode (supports only a single transverse mode at all wavelengths). The fundamental mode cannot escape through the silica gaps between air holes since the effective wavelength of the fundamental mode in the transverse plane is very high. Still, higher order modes have a low effective wavelength, so they will escape through the gap. In the case of multimode SC-PCF, the relative hole size, $D/\Lambda > 0.44$, which reduces the gap size, causes the trapping of higher order modes inside the core.[25]

10.2.2 THE PHOTONIC BANDGAP GUIDING MECHANISM

In HC-PCF, a hollow core is surrounded by a periodic array of micro-holes in the cladding running along the entire fiber's length. The MTIR guidance is not possible since the RI of the core is less than that of the cladding, so the core's radiation line ($\beta = k$) lies above that of the cladding ($\beta = n_{eff}k$). In such cases, the fiber uses the PBG effect to guide the light if it is properly designed. The PBGs above the core's radiation line can be utilized for the hollow core guidance such that the light is prevented from entering the PCF cladding and strongly confined into the hollow core. In HC-PCF, the hollow core acts as a defect, and when the periodicity of the structure is broken with a defect, the defect can support the modes that fall inside the PBG. The modes falling outside the defect are refracted, and those falling inside the defect are strongly guided along with the fiber.[5,26]

The pictorial representation of the guiding mechanism is shown in Figure 10.3. When HC-PCF is designed to work in red light, and a broadband light is launched, only the red light is guided along with the fiber, and the rest of the wavelengths are refracted.[5]

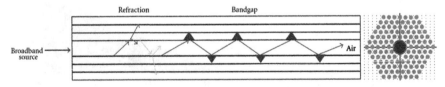

FIGURE 10.3 HC-PCF guidance via PBG (gray regions represent air holes.

Source: Reproduced from Ref. [5] with the permission of Hindawi Publishing Corporation https://creativecommons.org/licenses/by/3.0/.

The disadvantage of the abovementioned guiding mechanism in HC-PCF is that only a specific range of wavelength that falls inside PBG is guided. This feature limits the applications of HC-PCFs, but this limitation can be overcome to an extent in Kagome HC-PCF.[15,25]

10.2.3 ANTIRESONANCE REFLECTING OPTICAL WAVEGUIDE MODEL

In 1986, Duguay et al. first reported the ARROW guiding mechanism in optical waveguides for SiO_2–Si multilayer structures.[27] Later, Litchinitser et al. proposed the applicability of ARROW guiding in SC-PCF with high-index cladding layers and PBG fibers.[28] The principle of the ARROW guiding mechanism in PCF can be explained by considering the 1D planar waveguide analog (Figure 10.4), since the cladding of the PCF can be taken as an array of high and low index regions. Figure 10.4 shows a 1D waveguide in which a low index layer of RI n_1 and width a is sandwiched between two high index layers of RI n_2 and width d. The high index layers can be considered a Fabry–Perot (FP) resonator. Wavelengths satisfying the equation $kd = \pi m$, $m = 1, 2, 3, \ldots$, where k is the propagation constant, will be confined in the FP resonator. The rest of the wavelengths are reflected back to the core due to the antiresonant nature of the resonator. Let two different wavelengths propagate in the core of a waveguide. When the resonator is in the on state, the light will couple to the high index layer, resulting in transmission minima corresponding to narrow-band resonances due to the constructive interference in the FP resonator. This process is known as resonant tunneling. When the resonator is in the off state, the light is reflected to the core, and the transmission coefficients will be high, resulting in wide-band antiresonances due to the destructive interference in the FP resonator.[28]

When a PCF is incorporated with high index layers, the position of transmission minima can be predicted by the relation,[28]

$$\lambda_m = \frac{2n_1 d}{m} \sqrt{\left(\frac{n_2}{n_1}\right)^2 - 1} \quad m = 1, 2, \ldots$$

where n_1 and n_2 are the refractive indices of the host material of SC-PCF and infiltrated high index material, respectively. It is evident that the positions of transmission minima do not depend on the periodicity of the cladding. The validity of the ARROW waveguide model ceases at longer wavelengths as per the relation,[28]

$$\lambda/d > 2\sqrt{n_2^2 - n_1^2}$$

10.3 THE ROLE OF CHALCOGENIDE GLASSES

Chalcogenide glasses (ChGs) are fully amorphous semiconductor materials. They are fundamentally made up of 16th-group elements (sulfur, selenium, and tellurium) covalently bonded to network formers (germanium, arsenic, gallium, antimony, etc.).

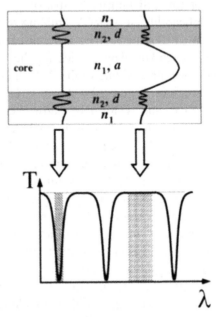

FIGURE 10.4 Schematic of the ARROW structure and the corresponding transmission spectrum.

Source: Reproduced from Ref. [28] with the permission of The Optical Society.

ChGs have potential applications in photonics because of their remarkable physical, optical, thermal, and electrical properties.[29] They have high linear RI of order $n \sim 2$–3.5, which can be exploited for light confinement applications.[30] Their excellent IR transmission window (IR cutoff is near 2–12 µm for sulfides, 2–15 µm for selenides, and 2–20 µm for tellurides)[31] can be used for designing MIR biochemical sensors since the vibrational signature of most chemical and biological species falls in the IR regime (2–25 µm).[32,33] ChGs are highly photosensitive materials that experience reversible/irreversible photostructural changes when exposed to near band-edge wavelengths.[34] The low deposition and processing temperature and high thermal and pressure stability can be used for the micro/nano-patterning on the ChG surface to realize active and passive IR photonic devices.[35] Their high optical nonlinearities and excellent rare-earth[36] solubility of specific compositions make them suitable for nonlinear optical (NLO) applications to produce IR amplifiers and lasers.[37,38] PCFs based on ChGs can harvest the advantages of both materials and open up vast opportunities, especially in IR photonics.[39,40]

Fortier et al. first observed the Brillouin and Raman scattering effect in $Ge_{15}Sb_{20}S_{65}$ PCF around 1.55 µm. They measured Brillouin and Raman gain coefficients 100 and 180 times larger than silica, respectively.[41]

The need for MIR CO_2 laser delivery fiber systems is highly desirable in medical, defense, and industrial applications.[42] Kosolapov et al. demonstrated for the first time an NC HC-PCF from the $Te_{20}As_{30}Se_{50}$ (TAS) ChG as a promising IR platform for guiding CO_2-laser radiation. The fabricated structure transmitted 10.6 µm CO_2-laser radiation with a ~11 dB/m loss. The improvements in the fabrication technology and the design optimization can reduce the optical loss below 1 dB/m.[43]

MIR is one of the most important spectral regimes where the vibrational signature of most chemical and biological species belongs. The wide IR transmission window and the high optical nonlinearities of ChGs make them ideal IR optical material for the fabrication of ChG-based PCFs for supercontinuum generation in MIR since non-ChG-based PCFs limit the supercontinuum generation for the wavelengths beyond 2.5 µm. Shaw et al. were the first to observe the broadband supercontinuum in the wavelength range 2.1–3.2 µm from a 1-m long As–Se-based PCF when pumped with 100 fs kHz pulse train at 2.5 µm.[44] This observation has gained much scientific attention and triggered scientific research for exploring supercontinuum generation in all-ChG PCFs[45–52] and in ChG-incorporated multimaterial PCFs.[53,54] Some articles point out the degradation of the optical performance of the ChG-based PCFs due to the natural aging when exposed to ambient conditions since the long-term presence of air inside the PCF holes triggers the presence of –OH and –SH absorption bands.[55–58] Mouawad et al. used methacrylate-based polymer to airproof the fiber facets to protect the fiber from natural aging by dipping the fiber ends in a polymeric solution followed by polymerization.[55] A detailed review of supercontinuum generation in ChG fibers can be found in Ref. [59].

10.3.1 *CHALCOGENIDE GLASS PCFS: FABRICATION METHODS*

The fabrication of all-ChG PCFs demands glasses with high purity, high mechanical and thermal stability, and low crystallization tendency. The first ChG MOF was proposed by Tanya M. Monro et al. from gallium lanthanum sulfide glass. Still, the lack of periodicity hindered the light guidance through the fabricated structure.[60] Frederic Desevedavy et al. were the first to design and fabricate TAS HC-PCF by stack and draw method in triangular lattice and Kagome lattice.[61] Shiryaev et al. used this method to fabricate As_2S_3 NC HC-PCF for the first time. They optimized the fiber drawing conditions with a theoretical estimated loss below 1 dB/km.[62]

There are several methods for fabricating ChG PCFs. Stack and draw technique,[41,44,60–70] mold casting method,[45,57,71–74] and mechanical drilling method[46,49,51,58,75–79] are some important methods that are discussed below.

10.3.1.1   STACK AND DRAW TECHNIQUE

The stack and draw technique is a traditional and commonly used technique for fabricating silica MOFs.[13] This method can be utilized to fabricate ChG PCFs by taking advantage of the ability of ChGs to draw into fibers.[80,81] In this process, the ChG preform is made by stacking ChG capillaries in the required manner around another ChG glass jacket and then drawing into fibers using a fiber drawing tower under pressure.[63–65,68–70] The schematic of the process is shown in Figure 10.5. Compared to other techniques, the technique is only feasible with thermally stable glasses and contains multiple-stage complex processes. Though the process enables the production of complex structures, the formation of bubbles in the interface between glass capillaries due to excessive pressure and temperature, and the in-hand manipulation,[82] and the induced RI variation can lead to scattering losses. Interstitial holes between the glass capillaries can reduce the scattering losses since their presence can reduce the effective surface area of contact between the capillaries and improve light confinement.[66]

10.3.1.2   CASTING METHOD

The mold casting method is another effective method that reduces all kinds of interface inhomogeneities, thereby considerably reducing optical losses. In this method, ChG rod and the silica mold of periodically arranged silica capillaries with negative geometry of the ChG preform are placed inside a silica ampoule, and the ampoule is heated in a rocking furnace. The molten ChG fills the silica mold which is then air-quenched. The silica ampoule is removed using a diamond tool, and the silica capillaries are etched using hydrofluoric acid (HF). The obtained structure is rinsed in deionized water and dried in an inert atmosphere to get the final ChG preform for the fiber drawing process (Figure 10.6). The process requires glasses with low viscosity so that glass can easily flow into the silica mold. The dimension of the silica capillaries has to be taken care of since the wrong choice of dimension can lead either to the breakage of silica capillaries or to the creation of mechanical stress inside the system during the molding process.[72,83]

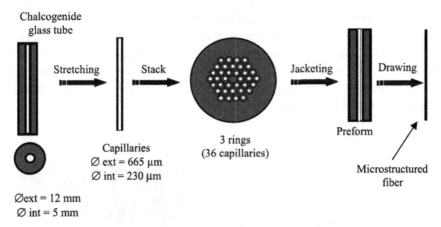

FIGURE 10.5 Stack and draw process for ChG PCFs.

Source: Reproduced from Ref. [69] with the permission of Taylor and Francis.

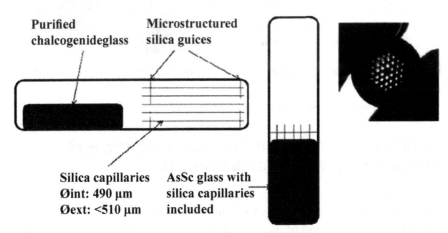

FIGURE 10.6 Schematic of the casting process. The molten glass is poured into the mold and then removed by HF treatment. The preform is shown in the right picture.

Source: Reproduced from Ref. [71] with the permission of The Optical Society.

10.3.1.3   MECHANICAL DRILLING METHOD

In the mechanical drilling method, as the name implies, the ChG preform (Figure 10.7c–e) is generated by drilling holes in ChG rod (Figure 10.7b) using mechanical drilling equipment (Figure 10.7a). The ChG rod is placed on the top of a 360° rotational stage. An x–y translational stage is used to control the position of the holes, and a water-cooling system is used to eliminate the heat generated during the process. The drilling parameters, the rotational speed and forward speed, have to be controlled to avoid cracks in the glass rod. The drilled hole's inner surface is polished to retain the optical quality. This method can reduce the interface inhomogeneities, ensures higher IR transparency, permits large-scale production, and guarantees strong light confinement in the core. Still, the process demands large quantities of glasses with high mechanical stability. [49,75,78,79]

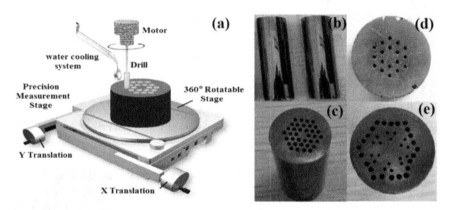

FIGURE 10.7   (a) Mechanical drilling equipment, (b) ChG rods, and (c–e) fabricated ChG preforms.

Source: Reproduced from Ref. [78] with the permission of Elsevier.

10.3.2   CHALCOGENIDE GLASS–INFILTRATED HYBRID PCFS: FABRICATION METHODS AND APPLICATIONS

The major advantage of combining silica/polymer PCFs with ChG is the possibility of modifying the guiding mechanism from MTIR to ARROW/PBG guiding. Excitation of plasmonic modes at ChG film surface can introduce an additional guiding mechanism due to the semiconducting properties of ChGs. [84,85]

10.3.2.1 *DIRECT CHALCOGENIDE GLASS–INFILTRATED HYBRID PCFS*

Granzow et al. carried out the direct infiltration of bulk $Ga_4Ge_{21}Sb_{10}S_{65}$ glass in the air holes of silica PCF by pressure-assisted melt filling technique. A strong PBG guidance is observed at the silica core, and distinct resonance bands are observed in the wavelength range 450–900 nm due to the coupling of light from the core to high index hexagonally arranged ChG strands of diameter 1.45 μm (Figure 10.8). In this process, the ChG melt is pressurized into the holes of silica PCF at a high temperature and pressure in an inert atmosphere. The method is found productive since it requires only a small sample volume compared to the fabrication of all-ChG PCFs and results in mechanically stable structures. The necessity of an inert atmosphere, high pressure and temperature, and custom-made equipment makes the method tedious.[86,87]

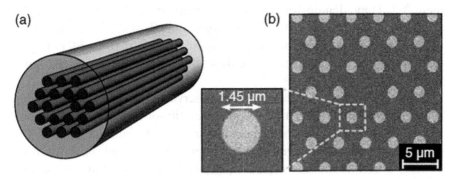

FIGURE 10.8 (a) Schematic of ChG-silica SC-PCF (red: chalcogenide strands) and (b) SEM image of the cross-section of ChG-silica SC-PCF.

Source: Reproduced from Ref. [86] with the permission of The Optical Society.

10.3.2.2 *SOLUTION-PROCESSED CHALCOGENIDE GLASS–INFILTRATED HYBRID PCFS*

The dissolution of ChGs in alkali and amine solutions is a low-cost process, resulting in the formation of nanoclusters. In 1980, Chern et al. first demonstrated the solution-derived ChG film via spin deposition using amines as solvent.[88] Solution-processed ChGs are a cost-effective and excellent choice for fabricating hybrid PCFs. They can be used for the uniform deposition of ChG nanofilms inside the holes of PCF.

The main advantages of combining PCFs with solution-processed ChGs are as follows:

1. Most chalcogenide materials dissolved in primary and secondary amines lead to homogeneous films that retain the bulk glass's optical, physical, and chemical properties.
2. The amines are volatile at low temperatures, making the film fabrication via the annealing process less tedious.
3. Solution-based deposition allows fabricating composite films by incorporating other materials, like carbon, noble metal, nanoparticles, and polymers, into the glass matrix.
4. Multilayer deposition is possible.[89]

In 2012, Christos Markos et al. reported for the first time a cost-effective method of depositing As_2S_3 ChG films inside the holes of silica PCFs by a solution-processed method without any sophisticated equipment.[90,91] They demonstrated the possibility of converting the MTIR guiding mechanism to ARROW guiding via the deposition of thin layers of ChG high index films inside the cladding holes of silica PCF. Figure 10.9 shows the SEM image of the deposited ChG films inside the holes of silica PCF.

Later, photostructural changes were observed in $As_{25}S_{75}$ ChG-infiltrated silica PCF, thereby illustrating the practicability of modifying the guiding mechanism in PCF by near bandgap illumination of the infiltrated ChG.[92] The high thermo-optic coefficient of ChGs permits the thermal tunability from 22–70 °C of transmission window (500–1300 nm) in solution-processed As_2S_3 ChG film–deposited in silica PCFs. The proposed sensing element has shown a sensitivity of 3.6 nm/°C at 1300 nm.[93] They integrated As_2S_3 ChG films in polymethylmethacrylate PCF and successfully created a transmission window in the 480–900 nm wavelength range. They observed a band edge shift of 17 nm by changing the light intensity due to the ultra-high Kerr nonlinearity of ChGs, thereby creating new routes for realizing NLO tunable devices based on polymer PCFs.[85]

10.4 CONCLUSION

This chapter aims to understand the fundamental concepts and ideas about PCFs, their guiding mechanisms, and the effect of incorporating ChGs into the PCFs toward IR photonics.

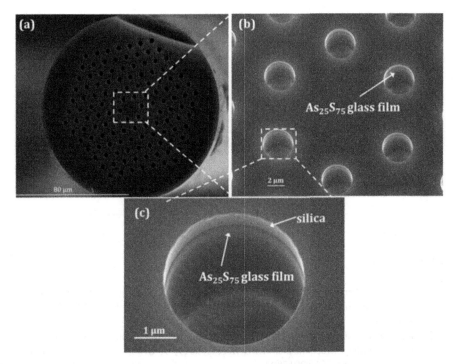

FIGURE 10.9 (a) SEM image of the cross-section of the PCF. (b) SEM image shows the ChG thin films inside silica PCF and (c) magnified SEM image of a single hole.

Source: Reproduced from Ref. [92] with the permission of AIP Publishing.

PCFs have become a vital topic of interest for the research community over several years. Their ability to manipulate light-guiding mechanisms based on their geometry is a remarkable asset for the fiber optics community. Incorporating them with an IR material is necessary to extend its possibilities to the IR regime. ChGs are an excellent platform when it comes to IR. ChG-integrated PCFs are ideal candidates for IR photonics and can be used for all-optical tunable nonlinear devices, supercontinuum generation, and sensing applications.

The fabrication of all-ChG PCFs is somewhat challenging since it demands extra thermal and pressure stability for glasses. The optical qualities can only be maintained if high-purity glasses are fabricated. Optimizing fiber drawing and design parameters can improve the fabrication of ChG PCFs with low optical losses in the MIR regime.[40] The solution-processed glasses can be an alternative for conveniently integrating ChGs with PCFs. Extensive research is necessary for further advancements in this field.

KEYWORDS

- **chalcogenide glass**
- **photonic crystal fiber**
- **ARROW model**
- **PBG guiding**
- **infrared photonics**
- **sensors**
- **integrated tunable devices**

REFERENCES

1. Markos, C.; Travers, J. C.; Abdolvand, A.; Eggleton, B. J.; Bang, O. Hybrid Photonic-Crystal Fiber. *Rev. Mod. Phys.* **2017,** *89* (4), 1–55. https://doi.org/10.1103/RevModPhys.89.045003

2. Hu, D. J. J.; Xu, Z.; Shum, P. P.; Member, S. Review on Photonic Crystal Fibers with Hybrid Guiding Mechanisms. *IEEE Access* **2019,** *7,* 67469–67482. https://doi.org/10.1109/ACCESS.2019.2917892

3. Dudley, J. M.; Taylor, J. R. Ten Years of Nonlinear Optics in Photonic Crystal Fibre. *Nat. Photonics* **2009,** *3,* 85–90. https://doi.org/10.1038/nphoton.2008.285

4. Dudley, J. M.; Genty, G.; Coen, S. Supercontinuum Generation in Photonic Crystal Fiber. *Rev. Mod. Phys.* **2006,** *78* (4), 1135–1184. https://doi.org/10.1103/RevModPhys.78.1135

5. Pinto, A. M. R.; Lopez-Amo, M. Photonic Crystal Fibers for Sensing Applications. *J. Sensors* **2012,** *2012.* https://doi.org/10.1155/2012/598178

6. Villatoro, J.; Zubia, J. New Perspectives in Photonic Crystal Fibre Sensors. *Opt. Laser Technol.* **2016,** *78,* 67–75. https://doi.org/10.1016/j.optlastec.2015.07.025

7. Algorri, J. F.; Zografopoulos, D. C.; Tapetado, A.; Poudereux, D.; Sanchez-Pena, J. M. Infiltrated Photonic Crystal Fibers for Sensing Applications. *Sensors* **2018,** *18* (12), 1–32. https://doi.org/10.3390/s18124263

8. Li, J.; Duan, K.; Wang, Y.; Zhao, W.; Zhu, J.; Guo, Y. High-Power Coherent Beam Combining of Two Photonic Crystal Fiber Lasers. *IEEE Photonics Technol. Lett.* **2008,** *20* (11), 888–890. https://doi.org/10.1109/LPT.2008.922342

9. Yablonovitch, E. Inhibited Spontaneous Emission in Solid-State Physics and Electronics. *Phys. Rev. Lett.* **1987,** *58* (20), 2059–2062. 10.1103/PhysRevLett.58.2059.

10. John, S. Strong Localization of Photons in Certain Disordered Dielectric Superlattices. *Phys. Rev. Lett.* **1987,** *58* (23). https://doi.org/10.1103/PhysRevLett.58.2486

11. Yablonovitch E.; Gmitter T. J. Photonic Band Structure: The Face-Centered-Cubic Case. *Phys. Rev. Lett.* **1989,** *63* (18), 1950–1953. https://doi.org/10.1103/PhysRevLett.63.1950

12. Yablonovitch E.; Gmitter T. J.; Leung K. M. Photonic Band Structure: The Face-Centered-Cubic Case Employing Nonspherical Atoms. *Phys. Rev. Lett.* **1991,** *67* (17). https://doi.org/10.1103/PhysRevLett.67.2295

13. Knight, J. C.; Birks, T. A.; Russell, P. S. J.; Atkin D. M. All-Silica Single-Mode Optical Fiber with Photonic Crystal Cladding: Errata. *Opt. Lett.* **1997,** *22* (7), 484–585. https:// doi.org/10.1364/OL.22.000484

14. Cregan, R. F.; Mangan, B. J.; Knight, J. C.; Birks, T. A.; Russell, P. S. J.; Roberts, P. J.; Allan, D. C. Single-Mode Photonic Band Gap Guidance of Light in Air. *Science* **1999,** *285* (5433), 1537–1539. https://doi.org/10.1126/science.285.5433.1537

15. Benabid, F.; Knight, J. C.; Antonopoulos, G.; Russell, P. S. J. Stimulated Raman Scattering in Hydrogen-Filled Hollow-Core Photonic Crystal Fiber. *Science* **2002,** *298,* 399–402. https://doi.org/10.1126/science.1076408

16. Pryamikov, A. D.; Biriukov, A. S.; Kosolapov, A. F.; Plotnichenko, V. G.; Semjonov, S. L.; Dianov, E. M. Demonstration of a Waveguide Regime for a Silica Hollow-Core Microstructured Optical Fiber with a Negative Curvature of the Core Boundary in the Spectral Region >3.5 µm. *Opt. Express* **2011,** *19* (2), 1441–1448. https://doi.org/10.1364/ OE.19.001441

17. Belardi, W.; Knight, J. C. Hollow Antiresonant Fibers with Reduced Attenuation. *Opt. Lett.* 1853 **2014,** *39* (7), 1853–1856. https://doi.org/10.1364/OL.39.001853

18. Jones, A. M.; Nampoothiri, A. V. V.; Ratanavis, A.; Kadel, R.; Wheeler, N. V.; Couny, F.; Benabid, F.; Rudolph, W.; Washburn, B. R.; Corwin, K. L. C$_2$H$_2$ Gas Laser inside Hollow-Core Photonic Crystal Fiber Based on Population Inversion. *Conf. Lasers Electro-Optics* **2010,** *19* (3), 399–402. https://doi.org/10.1364/cleo.2010.ctuu1

19. Bradley, T. D.; McFerran, J. J.; Jouin, J.; Ilinova, E.; Thomas, P.; Benabid, F. Progress towards Atomic Vapor Photonic Microcells: Coherence and Polarization Relaxation Measurements in Coated and Uncoated HC-PCF. *Adv. Slow Fast Light VI* **2013,** *8636,* 86360J. https://doi.org/10.1117/12.2013177

20. Zhang, T.; Zheng, Y.; Wang, C.; Mu, Z.; Liu, Y.; Lin, J. A Review of Photonic Crystal Fiber Sensor Applications for Different Physical Quantities. *Appl. Spectrosc. Rev.* **2018,** *53* (6), 486–502. https://doi.org/10.1080/05704928.2017.1376681

21. Mathews, S.; Farrell, G.; Semenova, Y. Directional Electric Field Sensitivity of a Liquid Crystal Infiltrated Photonic Crystal Fiber. *IEEE Photonics Technol. Lett.* **2011,** *23* (7), 408–410. https://doi.org/10.1109/LPT.2011.2107319

22. Frazao, O.; Baptista, J. M.; Santos, J. L. Temperature-Independent Strain Sensor Based on a Hi–Bi Photonic Crystal Fiber Loop Mirror. *IEEE Sens. J.* **2007,** *7* (10), 1453–1455. https://doi.org/10.1109/JSEN.2007.904884

23. Bock, W. J.; Jiahua, C.; Eftimov, T.; Urbanczyk, W. A Photonic Crystal Fiber Sensor for Pressure Measurements. *IEEE Trans. Instrument. Measure.* **2006,** 55 (4), 1119–1123. 10.1109/TIM.2006.876591.

24. Westbrook, P. S.; Eggleton, B. J.; Windeler, R. S.; Hale, A.; Strasser, T. A.; Burdge, G. L. Cladding-Mode Resonances in Hybrid Polymer-Silica Microstructured Optical Fiber Gratings. *IEEE Photonics Technol. Lett.* **2000,** *12* (5), 495–497. https://doi.org/ 10.1109/68.841264

25. Russell, P. Photonic Crystal Fibers. *Science* **2003,** *299* (5605), 358–362.

26. Russell, P. S. J. Photonic-Crystal Fibers. *J. Light. Technol.* **2006,** *24* (12), 4729–4749.

27. Duguay, M. A.; Kokubun, Y.; Koch, T. L. et al. Antiresonant Reflecting Optical Waveguides in SiO2-Si Multilayer Structures. *Appl. Phys. Lett.* **1986,** *49,* 13–15. https:// doi.org/10.1063/1.97085

28. Litchinitser, N. M.; Abeeluck, A. K.; Headley, C.; Eggleton, B. J. Antiresonant Reflecting Photonic Crystal Optical Waveguides. *Optics Lett.* **2002,** *27* (18), 1592–1594. https:// doi.org/10.1364/OL.27.001592

29. Eggleton, B. J.; Luther-Davies, B.; Richardson, K. Chalcogenide Photonics. *Nat. Photonics* **2011,** *5* (3), 141–148. https://doi.org/10.1038/nphoton.2011.309

30. Elliott, G. R.; Hewak, D. W.; Murugan, G. S.; Wilkinson, J. S. Chalcogenide Glass Microspheres; Their Production, Characterization and Potential. *Opt. Express* **2007,** *15* (26), 17542–17553. https://doi.org/10.1364/oe.15.017542

31. Savage, J. A. Optical properties of chalcogenide glasses. *J. Non-Cryst. Solids* **1982,** *47* (1), 101–115. https://doi.org/10.1016/0022-3093(82)90349-0

32. Charrier, J.; Brandily, M. L.; Lhermite, H.; Michel, K.; Bureau, B.; Verger, F.; Nazabal, V. Evanescent Wave Optical Micro-Sensor Based on Chalcogenide Glass. *Sens. Actuators, B: Chem.* **2012,** *173*, 468–476. https://doi.org/10.1016/j.snb.2012.07.056

33. Yu, C.; Ganjoo, A.; Jain, H.; Pantano, C. G.; Irudayaraj, J. Mid-IR Biosensor: Detection and Fingerprinting of Pathogens on Gold Island Functionalized Chalcogenide Films. *Anal. Chem.* **2006,** *78* (8), 2500–2506. https://doi.org/10.1021/ac051828c

34. Kohoutek, T.; Hughes, M. A.; Orava, J.; Mastumoto, M.; Misumi, T.; Kawashima, H.; Suzuki, T.; Ohishi, Y. Direct Laser Writing of Relief Diffraction Gratings into a Bulk Chalcogenide Glass. *J. Opt. Soc. Am. B* **2012,** *29* (10), 2779–2786. https://doi.org/10.1364/josab.29.002779

35. Ostrovsky, N.; Yehuda, D.; Tzadka, S.; Kassis, E.; Joseph, S.; Schvartzman, M. Direct Imprint of Optical Functionalities on Free-Form Chalcogenide Glasses. *Adv. Opt. Mater.* **2019,** *7* (19), 1–7. https://doi.org/10.1002/adom.201900652

36. Somakumar, A. K.; Upadhyay, K.; Suresh, S.; Sivasankarapillai, V. S.; Dhanusuraman, R. Photoluminescent Rare-Earth Nanocrystal-Based Characterization Methods: Advancements in Photophysical Applications. *Upconversion Nanophosphors*; Elsevier: Amsterdam, 2022; pp 1–18. https://doi.org/10.1016/B978-0-12-822842-5.00002-9

37. Yang, A.; Zhang, M.; Li, L.; Wang, Y.; Zhang, B.; Yang, Z.; Tang, D. Ga-Sb-S Chalcogenide Glasses for Mid-Infrared Applications. *J. Am. Ceram. Soc.* **2016,** *99* (1), 12–15. https://doi.org/10.1111/jace.14025

38. Zhang, M.; Yang, A.; Peng, Y.; Zhang, B.; Ren, H.; Guo, W.; Yang, Y.; Zhai, C.; Wang, Y.; Yang, Z.; Tang, D. Dy^{3+}-Doped Ga-Sb-S Chalcogenide Glasses for Mid-Infrared Lasers. *Mater. Res. Bull.* **2015,** *70*, 55–59. https://doi.org/10.1016/j.materresbull.2015.04.019

39. Shiryaev, V. S.; Churbanov, M. F. Trends and Prospects for Development of Chalcogenide Fibers for Mid-Infrared Transmission. *J. Non-Cryst. Solids* **2013,** *377*, 225–230. https://doi.org/10.1016/j.jnoncrysol.2012.12.048

40. Shiryaev, V. S. Chalcogenide Glass Hollow-Core Microstructured Optical Fibers. *Front. Mater.* **2015,** *2*. https://doi.org/10.3389/fmats.2015.00024

41. Fortier, C.; Fatome, J.; Pitois, S.; Smektala, F.; Millot, G.; Troles, J.; Desevedavy, F.; Houizot, P.; Brilland, L.; Traynor, N. Experimental Investigation of Brillouin and Raman Scattering in a $Ge_{15}Sb_{20}S_{65}$ Microstructured Chalcogenide Fiber. In *34th European Conference on Optical Communication* **2008,** *4*, 1–2. https://doi.org/10.1109/ECOC.2008.4729397

42. Pryamikov, A. D.; Kosolapov, A. F.; Plotnichenko V. G.; Dianov E. M. Transmission of CO_2 Laser Radiation through Glass Hollow Core Microstructured Fibers. *CO_2 Laser— Optimisation and Application.* 2012. https://doi.org/10.5772/37709

43. Kosolapov, A. F.; Pryamikov, A. D.; Biriukov, A. S.; Shiryaev, V. S.; Astapovich, M. S.; Snopatin, G. E.; Plotnichenko, V. G.; Churbanov, M. F.; Dianov, E. M. Demonstration of CO2-Laser Power Delivery through Chalcogenide-Glass Fiber with Negative-Curvature Hollow Core. *Opt. Express* **2011,** *19* (25), 25723–25728. https://doi.org/10.1364/oe.19.025723

44. Shaw, L. B.; Nguyen, V. Q.; Sanghera, J. S.; Aggarwal, I. D.; Thielen, P. A.; Kung, F. H. IR Supercontinuum Generation in As–Se Photonic Crystal Fiber. *Adv. Solid-State Photon. (TOPS)* **2005**, *98*, 864. https://doi.org/10.1364/assp.2005.tuc5

45. Ghosh, A. N.; Meneghetti, M.; Petersen, C. R.; Bang, O.; Brilland, L.; Venck, S.; Troles, J.; Dudley, J. M.; Sylvestre, T. Chalcogenide-Glass Polarization-Maintaining Photonic Crystal Fiber for Mid-Infrared Supercontinuum Generation. *J. Phys. Photonics* **2019**, *1* (4). https://doi.org/10.1088/2515-7647/ab3b1e

46. Mouawad, O.; Amrani, F.; Strutynski, C.; Fatome, J.; Kibler, B.; Desevedavy, F.; Gadret, G.; Jules, J. C.; Deng, D.; Ohishi, Y.; Smektala, F. Multioctave Midinfrared Supercontinuum Generation in Suspended-Core Chalcogenide Fibers. *Opt. Lett.* **2014**, *39* (9), 2684–2687. https://doi.org/10.1364/OL.39.002684

47. Saghaei, H.; Ebnali-heidari, M.; Moravvej-farshi, M. K. Midinfrared Supercontinuum Generation via As$_2$Se$_3$ Chalcogenide Photonic Crystal Fibers. *Appl. Opt.* **2015**, *54* (8), 2072–2079. https://doi.org/10.1364/AO.54.002072

48. Kalantari, M.; Karimkhani, A.; Saghaei, H. Ultra-Wide Mid-IR Supercontinuum Generation in As$_2$S$_3$ Photonic Crystal Fiber by Rods Filling Technique. *Optik* **2018**, *158*, 142–151. https://doi.org/10.1016/j.ijleo.2017.12.014

49. El-Amraoui, M.; Gadret, G.; Jules, J. C.; Fatome, J.; Fortier, C.; Desevedavy, F.; Skripatchev, I.; Messaddeq, Y.; Troles, J.; Brilland, L.; Gao, W.; Suzuki, T.; Ohishi, Y.; Smektala, F. Microstructured Chalcogenide Optical Fibers from As$_2$S$_3$ Glass: Towards New IR Broadband Sources. *Opt. Express* **2010**, *18* (25), 26655–26665. https://doi.org/10.1364/oe.18.026655

50. Cheng, T.; Kanou, Y.; Xue, X.; Deng, D.; Matsumoto, M.; Misumi, T.; Suzuki, T.; Ohishi, Y. Mid-Infrared Supercontinuum Generation in a Novel AsSe$_2$–As$_2$S$_5$ Hybrid Microstructured Optical Fiber. *Opt. Express* **2014**, *22* (19), 23019–23025. https://doi.org/10.1364/oe.22.023019

51. Savelii, I.; Mouawad, O.; Fatome, J.; Kibler, B.; Desevedavy, F.; Gadret, G.; Jules, J. C.; Bony, P. Y.; Kawashima, H.; Gao, W.; Kohoutek, T.; Suzuki, T.; Ohishi, Y.; Smektala, F. Mid-Infrared Supercontinuum Generation in Suspended-Core Sulfide and Tellurite Optical Fibers. *Optics Express* **2012**, *20* (24), 27083–27093. https://doi.org/10.1364/OE.20.027083

52. Møller, U.; Yu, Y.; Kubat, I.; Petersen, C. R.; Gai, X.; Brilland, L.; Mechin, D.; Caillaud, C.; Troles, J.; Luther-Davies, B.; Bang, O. Multi-Milliwatt Mid-Infrared Supercontinuum Generation in a Suspended Core Chalcogenide Fiber. *Opt. Express* **2015**, *23* (3), 3282–3291. https://doi.org/10.1364/oe.23.003282

53. Ben Salem, A.; Diouf, M.; Cherif, R.; Wague, A.; Zghal, M. Ultraflat-Top Midinfrared Coherent Broadband Supercontinuum Using All Normal As2S5-Borosilicate Hybrid Photonic Crystal Fiber. *Opt. Eng.* **2016**, *55* (6), 066109. https://doi.org/10.1117/1.oe.55.6.066109

54. Cheng, T.; Kawashima, H.; Xue, X.; Deng, D.; Matsumoto, M.; Misumi, T.; Suzuki, T.; Ohishi, Y. Fabrication of a Chalcogenide-Tellurite Hybrid Microstructured Optical Fiber for Flattened and Broadband Supercontinuum Generation. *J. Light Technol.* **2015**, *33* (2), 333–338. https://doi.org/10.1109/JLT.2014.2379912.

55. Mouawad, O.; Strutynski, C.; Desevedavy, F.; Gadret, G.; Jules, J.; Smektala, F. Optical Aging Behaviour Naturally Induced on As$_2$S$_3$ Microstructured Optical Fibres. *Opt. Mater. Express* **2014**, *4* (10), 2190–2203. https://doi.org/10.1364/OME.4.002190

56. Mouawad, O.; Amrani, F.; Kibler, B.; Strutynski, C.; Fatome, J.; Desevedavy, F.; Gadret, G.; Jules, J.; Heintz, O.; Lesniewska, E.; Smektala, F. Impact of Optical and Structural Aging

in As₂S₃ Microstructured Optical Fibers on Mid-Infrared Supercontinuum Generation. *Opt. Express* **2014**, *22* (20), 23912–23919. https://doi.org/10.1364/OE.22.023912

57. Toupin, P.; Brilland, L.; Mechin, D.; Adam, J. L.; Troles, J. Optical Aging of Chalcogenide Microstructured Optical Fibers. *J. Light. Technol.* **2014**, *32* (13), 2428–2432. https://doi.org/10.1109/JLT.2014.2326461

58. Mouawad, O.; Kedenburg, S.; Steinle, T.; Steinmann, A.; Kibler, B.; Desevedavy, F.; Gadret, G.; Jules, J. C.; Giessen, H.; Smektala, F. Experimental Long-Term Survey of Mid-Infrared Supercontinuum Source Based on As₂S₃ Suspended-Core Fibers. *Appl. Phys. B* **2016**. https://doi.org/10.1007/s00340-016-6453-5

59. Dai, S.; Wang, Y.; Peng, X.; Zhang, P.; Wang, X.; Xu, Y. A Review of Mid-Infrared Supercontinuum Generation in Chalcogenide Glass Fibers. *Appl. Sci.* **2018**, *8* (5). https://doi.org/10.3390/app8050707

60. Monro, T. M.; West, Y. D.; Hewak, D. W.; Broderick, N. G. R.; Richardson, D. J. Chalcogenide Holey Fibres. *Electron. Lett.* **2000**, *36* (24), 1998–2000. https://doi.org/10.1049/el:20001394

61. Desevedavy, F.; Renversez, G.; Troles, J.; Houizot, P.; Brilland, L.; Vasilief, I.; Coulombier, Q.; Traynor, N.; Smektala, F.; Adam, J. L. Chalcogenide Glass Hollow Core Photonic Crystal Fibers. *Opt. Mater.* **2010**, *32* (11), 1532–1539. https://doi.org/10.1016/j.optmat.2010.06.016

62. Shiryaev, V. S.; Kosolapov, A. F.; Pryamikov, A. D.; Snopatin, G. E.; Churbanov, M. F.; Biriukov, A. S.; Kotereva, T. V.; Mishinov, S. V.; Alagashev, G. K.; Kolyadin, A. N. Development of Technique for Preparation of As₂S₃ Glass Preforms for Hollow Core Microstructured Optical Fibers. *J. Optoelectron. Adv. Mater.* **2014**, *16* (9–10), 1020–1025.

63. Brilland, L.; Smektala, F.; Renversez, G.; Chartier, T.; Troles, J.; Nguyen, T.; Traynor, N.; Monteville, A. Fabrication of Complex Structures of Holey Fibers in Chalcogenide Glass. *Opt. Express* **2006**, *14* (3), 1280–1285. https://doi.org/10.1364/oe.14.001280

64. Desevedavy, F.; Renversez, G.; Brilland, L.; Houizot, P.; Troles, J.; Coulombier, Q.; Smektala, F.; Traynor, N.; Adam, J. L. Small-Core Chalcogenide Microstructured Fibers for the Infrared. *Appl. Opt.* **2008**, *47* (32), 6014–6021. https://doi.org/10.1364/AO.47.006014

65. Desevedavy, F.; Renversez, G.; Troles, J.; Brilland, L.; Houizot, P.; Coulombier, Q.; Smektala, F.; Traynor, N.; Adam, J. L. Te–As–Se Glass Microstructured Optical Fiber for the Middle Infrared. *Appl. Opt.* **2009**, *48* (19), 3860–3865. https://doi.org/10.1364/AO.48.003860

66. Brilland, L.; Troles, J.; Houizot, P.; Desevedavy, F.; Coulombier, Q.; Traynor, N. Interfaces Impact on the Transmission of Chalcogenides Photonic Crystal Fibres. *J. Ceram. Soc. Jpn.* **2008**, *116* (1358), 1024–1027. https://doi.org/10.2109/jcersj2.116.1024

67. Kosolapov, A. F.; Pryamikov, A. D.; Biriukov, A. S.; Vladimir, S.; Astapovich, M. S.; Snopatin, G. E.; Plotnichenko, V. G.; Churbanov, M. F.; Dianov, E. M. Demonstration of CO_2-Laser Power Delivery through Chalcogenide-Glass Fiber with Negative-Curvature Hollow Core. *Opt. Express* **2011**, 19 (25), 25723–25728. https://doi.org/10.1364/OE.19.025723

68. Smektala, F.; Desevedavy, F.; Brilland, L.; Houizot, P.; Troles, J.; Traynor, N. Advances in the Elaboration of Chalcogenide Photonic Crystal Fibers for the Mid-Infrared. *Photonic Cryst. Fibers* **2007**, *6588*, 658803. https://doi.org/10.1117/12.723259

69. Troles, J.; Brilland, L.; Smektala, F.; Houizot, P.; Desevedavy, F.; Coulombier, Q.; Traynor, N.; Chartier, T.; Nguyen, T. N.; Adam, J. L.; Renversez, G.; Brilland, L.

Chalcogenide Microstructured Fibers for Infrared Systems, Elaboration Modelization, and Characterization. *Fiber Integr. Opt.* **2009,** *28* (1), 11–26. https://doi.org/10.1080/01468030802272500

70. Liao, M.; Chaudhari, C.; Qin, G.; Yan, X.; Kito, C.; Suzuki, T.; Ohishi, Y.; Matsumoto, M.; Misumi, T. Fabrication and Characterization of a Chalcogenide-Tellurite Composite Microstructure Fiber with High Nonlinearity. *Opt. Express* **2009,** *17* (24), 21608–21614. https://doi.org/10.1364/oe.17.021608

71. Coulombier, Q.; Brilland, L.; Houizot, P.; Chartier, T.; Guyen, T. N. N.; Smektala, F.; Renversez, G.; Monteville, A.; Mechin, D.; Pain, T.; Orain, H.; Sangleboeuf, J. C.; Troles J. Casting Method for Producing Low-Loss Chalcogenide Microstructured Optical Fibers. *Opt. Express* **2010,** *18* (9), 9107–9112. https://doi.org/10.1364/OE.18.009107

72. Coulombier, Q.; Brilland, L.; Houizot, P.; Chartier, T.; Renversez, G.; Monteville, A.; Fatome, J.; Smektala, F.; Pain, T.; Orain, H.; Sangleboeuf, J. C.; Troles, J. Fabrication of Low Losses Chalcogenide Photonic Crystal Fibers by Molding Process. *Optical Comp. Mater. VII* **2010,** *7598,* 193–201. https://doi.org/10.1117/12.840868

73. Conseil, C.; Coulombier, Q.; Boussard-pledel, C.; Troles, J.; Brilland, L.; Renversez, G.; Mechin, D.; Bureau, B.; Adam, J. L.; Lucas, J. Chalcogenide Step Index and Microstructured Single Mode Fibers. *J. Non-Cryst. Solids* **2011,** *357,* 2480–2483. https://doi.org/10.1016/j.jnoncrysol.2010.11.090

74. Troles, J.; Coulombier, Q.; Canat, G.; Duhant, M.; Toupin, W. R. P.; Calvez, L.; Renversez, G.; Smektala, F.; El Amraoui, M.; Adam, J. L.; Chartier, T.; Mechin, D.; Brilland, L. Low Loss Microstructured Chalcogenide Fibers for Large Non Linear Effects at 1995 Nm. *Opt. Express* **2010,** *18* (25), 26647–26654. https://doi.org/10.1364/OE.18.026647

75. Yi, C.; Zhang, P.; Chen, F.; Dai, S.; Wang, X.; Xu, T.; Nie, Q. Fabrication and Characterization of Ge20Sb15S65 Chalcogenide Glass for Photonic Crystal Fibers. *Appl. Phys. B* **2014,** *116* (3), 653–658. https://doi.org/10.1007/s00340-013-5748-z

76. Mouawad, O.; Strutynski, C.; Desevedavy, F.; Gadret, G.; Jules, J; Smektala, F. Optical Aging Behaviour Naturally Induced on As_2S_3 Microstructured Optical Fibres. *Opt. Mater. Express* **2014,** *4* (10), 2190–2203. https://doi.org/10.1364/OME.4.002190

77. El-Amraoui, M.; Fatome, J.; Jules, J. C.; Kibler, B.; Gadret, G.; Fortier, C.; Smektala, F.; Skripatchev, I.; Polacchini, C. F.; Messaddeq, Y.; Troles, J.; Brilland, L.; Szpulak, M.; Renversez, G. Strong Infrared Spectral Broadening in Low-Loss As–S Chalcogenide Suspended Core Microstructured Optical Fibers. *Opt. Express* **2010,** *18* (5), 4547–4556. https://doi.org/10.1364/oe.18.004547

78. Zhang, P.; Zhang, J.; Yang, P.; Dai, S.; Wang, X.; Zhang, W. Fabrication of Chalcogenide Glass Photonic Crystal Fibers with Mechanical Drilling. *Opt. Fiber Technol.* **2015,** *26,* 176–179. https://doi.org/10.1016/j.yofte.2015.09.002

79. Feng, X.; Mairaj, A. K.; Hewak, D. W.; Monro, T. M. Nonsilica Glasses for Holey Fibers. *J. Light Technol.* **2005,** *23* (6), 2046. https://doi.org/10.1109/JLT.2005.849945

80. Sanghera, J. S.; Shaw, L. B.; Pureza, P.; Nguyen, V. Q.; Gibson, D.; Busse, L.; Aggarwal, I. D.; Florea, C. M.; Kung, F. H. Nonlinear Properties of Chalcogenide Glass Fibers. *Int. J. Appl. Glas. Sci.* **2010,** *1,* 296–308. https://doi.org/10.1111/j.2041-1294.2010.00021.x

81. Sanghera, J. S.; Shaw, L. B.; Aggarwal, I. D. Chalcogenide Glass-Fiber-Based Mid-IR Sources and Applications. *IEEE J. Sel. Top. Quantum Electron.* **2009,** *15* (1), 232–241. 10.1117/12.706532.

82. Fatome, J.; Jules, J. C.; Kibler, B.; Gadret, G.; Fortier, C.; Smektala, F.; Skripatchev, I.; Polacchini, C.F.; Messaddeq, Y.; Troles, J.; Brilland, L.; Szpulak, M.; Renversez,

G. Strong Infrared Spectral Broadening in Low-Loss As–S Chalcogenide Suspended Core Microstructured Optical Fibers. *Opt. Express* **2010,** *18* (5), 4547–4556. https://doi. org/10.1364/OE.18.004547

83. Coulombier, Q.; Brilland, L.; Houizot, P.; Chartier, T.; N'Guyen, T. N.; Smektala, F.; Renversez, G.; Monteville, A.; Mechin, D.; Pain, T.; Orain, H.; Sangleboeuf, J. C.; Troles, J. Casting Method for Producing Low-Loss Chalcogenide Microstructured Optical Fibers. *Opt. Express* **2010,** *18* (9), 9107–9112. https://doi.org/10.1364/oe.18.009107

84. Kuhlmey, B. T.; Pathmanandavel, K.; Mcphedran, R. C. Multipole Analysis of Photonic Crystal Fibers with Coated Inclusions. In *2006 Eur. Conf. Opt. Commun. Proceedings* 2006; pp 1–2. https://doi.org/10.1109/ECOC.2006.4801208

85. Markos, C.; Kubat, I.; Bang, O. Hybrid Polymer Photonic Crystal Fiber with Integrated Chalcogenide Glass Nanofilms. *Sci. Rep.* 2014. https://doi.org/10.1038/srep06057

86. Granzow, N.; Schmidt, M. A.; Tverjanovich, A. S.; Wondraczek, L.; Russell, P. S. J. Band-Gap Guidance in Chalcogenide-Silica Photonic Crystal Fibers. *Opt. Lett.* **2011,** *36* (13), 2432–2434. https://doi.org/10.1364/OL.36.002432

87. Schmidt, M. A.; Granzow, N.; Da, N.; Peng, M.; Wondraczek, L.; Russell, P. S. J. All-Solid Bandgap Guiding in Tellurite-Filled Silica Photonic Crystal Fibers. *Opt. Lett.* **2009,** *34* (13), 1946–1948. https://doi.org/10.1364/ol.34.001946

88. Chern, G. C.; Lauks, I. Spin-Coated Amorphous Chalcogenide Films. *J. Appl. Phys.* **1982,** *53,* 6979–6982. https://doi.org/10.1063/1.330043

89. Khan, H.; Dwivedi, P. K.; Islam, S.; Husain, M.; Zulfequar, M. Solution Processing of Chalcogenide Glasses: A Facile Path towards Functional Integration. *Opt. Mater.* **2021,** *119,* 111332. https://doi.org/10.1016/j.optmat.2021.111332

90. Markos, C.; Petersen, C. R. R. Multimaterial Photonic Crystal Fibers. *Opt. Comp. Mater. XV* **2018,** *10528,* 146–151. https://doi.org/10.1117/12.2290367

91. Markos, C.; Yannopoulos, S. N.; Vlachos, K. Chalcogenide Glass Layers in Silica Photonic Crystal Fibers. *Opt. Express* **2012,** *20* (14), 14814–14824. https://doi.org/10.1364/oe.20.014814

92. Markos, C. Photo-Induced Changes in a Hybrid Amorphous Chalcogenide/Silica Photonic Crystal Fiber. *Appl. Phys. Lett.* **2014,** *104.* https://doi.org/10.1063/1.4861374

93. Markos, C. Thermo-Tunable Hybrid Photonic Crystal Fiber Based on Solution-Processed Chalcogenide Glass Nanolayers. *Sci. Rep.* **2016,** *6.* https://doi.org/10.1038/srep31711

CHAPTER 11

Nanofibers and Nanotechnology in Porous Textiles

MARIA MATHEW[1] and RONY RAJAN PAUL[2]

[1]C-MET, Thrissur, Kerala, India

[2]Department of Chemistry, CMS College, Kottayam, Kerala, India

ABSTRACT

Nanotechnology, being the most promising technology in the 21st century, has reformed the whole world. In recent years, nanotechnology has proved to be an interdisciplinary research field, and it extended its scope of applications to versatile areas of human endeavor. Growing people's expectation for durable and functional apparel has generated a chance for materials in nanoscale to be integrated into textile substrates. The unique properties of nanomaterials can impart antibacterial, water-repellency, wrinkle free-ness, static elimination, and flame retardancy improved dyeability without changing their comfort and flexibility. Another potential application offered by nanomaterials is the production of connected garments that can sense and respond to external stimuli via electrical, color, or physiological signals. This chapter mainly focuses on the possible applications of nanotechnology in the current textile arena. The various limitations and risk factors such as nanotoxicity and their environmental impacts are also discussed.

11.1 INTRODUCTION

Recently, nanotechnology has brought immense improvement in the textile industry. Textiles and textile structures are abundantly used in high-

Mechanics and Physics of Porous Materials: Novel Processing Technologies and Emerging Applications.
Chin Hua Chia, Tamara Tatrishvili, Ann Rose Abraham, & A. K. Haghi (Eds.)

performance technical applications ranging from food packaging, furnishing, protective textiles, and medical textiles.[2] Nanotechnology is considered as a novel approach for enhancing the performance of these functional textiles. Nanotechnology, as we know is concerned with the study of materials having size in the nanoscale dimensions (1–100 nm). Nanomaterials with particle size less than 100 nm have large surface area to volume ratio which is responsible for their exceptional physical and chemical properties.[3] Nano-science-based processes are flourishing in different areas related to science and technology. Textile community is also experiencing its benefits in its diverse field of applications. Various products from nanotechnology can be comprised with textiles starting from composites in nanoscale and fibers in nanoscale to smart polymeric coating in order to provide new functionalities and improved performance.

Cotton a widely used natural fiber that exhibits high absorbency, softness, and breathability is limited in nonclassical applications due to its relatively low strength, low durability, easy creasing, and soiling and flammability.[4] Even though synthetic fibers have excellent antimicrobial and strain-resistant properties, they lack the comfort and flexibility offered by cotton.[1] Therefore, the development of a new fiber that combines the advantages of both natural and synthetic fiber as well as providing better functionalities including medical monitoring of bodily functions and metabolism,[5,6] rehabilitation,[7] and electronic and optical device integration leads to a breakthrough in the advancement of nano-engineered textiles. Such integrated textiles responsive to chemical, electrical, thermal, magnetic, and optical stimuli have potential applications in our day-to-day life. Textiles generally undergo nano-engineering to have specific functions including hydrophobicity, antibacterial properties, conductivity, antiwrinkle, antistatic properties, and UV protection. Figure 11.1 illustrates the applications of nanotechnology in textiles.

Using nanotechnology, all these properties can be achieved without affecting the texture, breathability, and flexibility of textiles. These materials may be like foams, surface coatings, voided patterning, and fillers. Emerging coating routes such as plasma, sol–gel, layer-by-layer are capable of introducing multifunction, intelligence, enhanced durability, and weather resistance to manufactured textiles. Due to all these reasons, nanotechnology has attracted researchers to the textile industry and hence a rapid progress in this field occurred.[8]

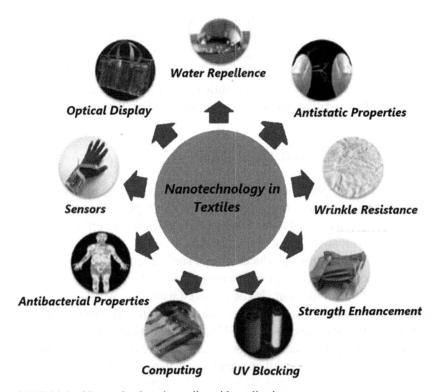

FIGURE 11.1   Nanotechnology in textiles with applications.

11.2   INFLUENCE OF NANOTECHNOLOGY IN TEXTILES

11.2.1   *NANOFIBERS*

As already mentioned, nanotechnology facilitates the development of novel fibers combining the properties of both cotton and synthetics. Figure 11.2 illustrates the textile properties enhanced by nanotechnology. Nanofibers fabricated through electrospinning technique exhibit excellent performance and good strength. In electrospinning, a polymer solution is spun at high speeds and exposed to electrostatic forces, pulling the polymers into extremely thin fibers. It is the polymer solution that determines the properties of nanofibers. Electrospun nanofibers are useful for various purposes such as to conduct electricity to resist heat for use in fire fighter's suits and to kill bacteria for applications in the medical field. Recently, three major forms of carbon nanomaterials-based nanofibrous sensors such as carbon nanotube (CNT)-coated fibrous materials, reduced graphene oxide (rGO)-coated fibrous material, and carbon fibers have been fabricated and utilized as the major sensory entities inculcated in fabrics for supervising the manufacturing procedure of fiber-filled polymeric composites (2020; Wang et al., 2018).

11.2.1.1   WATER AND OIL REPELLENCE

Water-repellant properties can be imparted to fabrics by forming nanowhiskers which are three orders of magnitude smaller than typical cotton fiber. These nanowhiskers when added to the fabrics create a peach fuzz effect.[9] The space between individual whiskers is smaller than a drop of water but larger than water molecules and thus produces a large surface tension. Due to this large surface tension, water remains on the top of the whiskers and above the fabric surface.[10,11] Breathability can also be achieved with these whiskers as they are permeable to gases. In addition to nanowhiskers, water-repellent property can be achieved through the creation of 3D surface structures on the fabric by the addition of gel-forming additives or coating the textile with nanoparticulate film.[12] Once water droplets fall on such fabrics, they bead up and roll off when the surface is slightly slopped. Therefore, the surfaces stay dry even after a heavy shower. Along with that small particles of dirt are also picked up during rolling. By varying the micro and nanoscale features of fabric surface one can achieve excellent wetting behavior. By creating roughness on the fabric, superhydrophobicity can be attained without affecting the abrasion resistance and softness of the fabric. Silica nanoparticles (SiO_2 NPs) coated over cotton in the presence of perfluorooctylated quarternary ammonium silane coupling agent (PQASCA) is excellent in producing hydrophobicity.[13] While SiO_2 NPs create roughness on the surface of cotton fibers, PQASCA lowers the surface energy. Textiles so developed exhibited water repellence at a water contact angle of 145° and oil repellency at a contact angle of 131°. Wang et al. have demonstrated that by combining the nanoparticles of hydroxylapatite, TiO_2, ZnO, and Fe_7O_3 with other organic and inorganic substances.

11.2.1.2   WRINKLE RESISTANCE

In cotton fabrics, cellulose molecules are linearly organized by themselves and passing through the crystalline and amorphous sections of the fibers.

On applying force on these fibers, hydrogen-bonded cellulose molecules get displaced from their original positions and hydrogen bonds get reformed at new locations. Conventionally, fabrics are impregnated with resins to impart wrinkle resistance. However, this approach has some limitations like decrease in

- the dyeability,
- the breathability,
- the water absorbency,
- the abrasion resistance,
- the tensile strength of fiber, and so on.

To avoid these limitations, titanium dioxide nanoparticles (TiO_2 NPs) with carboxylic acid as a catalyst are utilized to form cross-links between cellulose molecules and acid groups.[14,15] These carboxylic acid-treated fabrics were found to be softer than untreated fabrics. In addition, nanosilica with maleic acid can also successfully impart wrinkle resistance to cotton and silk fabrics.[16]

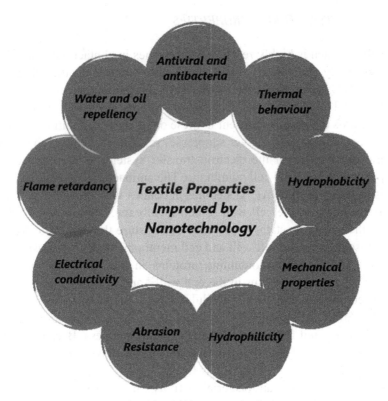

FIGURE 11.2   Textile properties enhanced by nanotechnology.

11.2.1.3   UV PROTECTION

Generally, inorganic UV blockers are nontoxic and chemically stable to high temperatures and ultraviolet radiations when compared to organic UV blockers. Nanoscale semiconductor oxides such as TiO_2, ZnO, SiO_2, and Al_2O_3 are efficient in absorbing and scattering UV radiation.[17–20] Since scattering depends upon the particle size, in order to scatter UV light having wavelength in the range 200–400 nm, the optimum particle size should be 20–40 nm.[21] Using sol–gel method a thin layer of TiO_2 can be surface coated on cotton that can retain its UV effect even after 50 launderings.[22] In order to induce scattering at a high UV protective factor rating ZnO rods have been incorporated in cotton.[23] Figure 11.3 is an SEM image of ZnO rods grown on cotton fibers. All these features can be attributed to the large surface area per unit mass and volume of these NPs, which increases the effectiveness of UV blocking.

11.2.1.4   ANTIBACTERIAL PROPERTIES

Nanosilver which is highly reactive when comes in contact with bacteria or fungi adversely affect their cellular metabolism and inhibits the cell growth. TiO_2 and ZnO NPs also exhibit similar antibacterial fungicidal properties to **textiles**.[24–26] The antibacterial property of Ag NPs are based on their reactivity towards the proteins present in these organisms. In addition to the cell growth metabolism, nano-silver suppresses respiration, reduces the activity of basal metabolism of the electron transfer system, as well as substrate transport to the bacterial cell membrane. The antibacterial mechanism of Ag NPs is as follows: when Ag NPs comes in contact with moisture or bacteria they get adhered to the cell wall or cell membrane.[27] Since silver NPs are inert in their metallic state, in presence of moisture they ionize to Ag^+ ions and diffuse through the cell wall and cell membrane to the cytoplasm. When Ag^+ ions bind to sulfur-containing proteins in the cell membrane some structural changes will occur in the cell wall[28] and it leads to the release of cellular components to the extracellular fluids, due to changes in osmotic pressure. Additionally, Ag NPs can slow down the growth and multiplication of bacteria and fungi causing bad odor and itchiness. Because of these exciting properties of Ag NPs, great research efforts have been committed to explore their applications in textile industry.

    Due to the strong antibacterial activity of nanosilver, they can be successfully applied to both natural and synthetic fibers. The antibacterial effect and durability of silver NPs are excellent if they are dispersed in a colloidal solution before they are applied to the textile fabrics. This colloidal nanosilver solution can exhibit excellent antibacterial activity in cotton and polyester fabrics even after laundering several times.[24] It has been found that Ag NPs incorporated polypropylene (PP) have superior antibacterial effect relative to PP containing micron-sized particles.[32] Fibers prepared using this, PP is found to have very good mechanical properties and better antibacterial activity against both Gram-positive and Gram-negative bacteria. The effectiveness of Ag additives for antimicrobial activity depends upon the concentration, surface area, and releasing rate of Ag^+ ions.[29-31] SEM and TEM studies on the biocidal action of silver NPs also reveal their effectiveness in antimicrobial action.

11.2.1.5 ANTISTATIC PROPERTIES

Artificial fibers such as polyester and nylon are prone to high-static charge because they are not hydrophilic. Since synthetic fibers provide poor anti-static properties, researchers are in search for new methods for improving the antistatic properties of textiles by using nanotechnology. It has been reported that nanosized particles such as TiO_2, ZnO whiskers, nano-antimony doped tin oxide (ATO), and silane nanosol can impart antistatic properties to synthetic fibers. All these materials are electrically conductive and effectively dissipate the static charge accumulated on the fabric. In addition to that, silane nanosol can enhance the antistatic property by absorbing moisture in air through hydroxyl groups.[32]

11.2.1.6 STRENGTH ENHANCEMENT

CNT-reinforced polymer composite fibers produced through melt-spinning of PP and carbon particles can improve the strength and toughness of fabrics.[33] These composite fibers can be produced through PP and carbon particles. By melt extrusion, also a wide range of nonadditive yarns with improved mechanical properties and various textures can be produced.[34] The strength of nanofibers produced via electrospinning technique can be enhanced by post-treatment approaches for example by applying heat. Highly twisted yarns have high strength, toughness, and energy-damping capability for various applications in electronic textiles like actuators, energy storage, heating, radio, and microwave absorption. It has been found out that

CNT-integrated fibers exhibited excellent strength and performance. It is noteworthy that dipping and coating methods can be utilized to immobilize CNTs on cotton.[35]

11.3 ELECTRICALLY CONDUCTIVE TEXTILE MATERIALS (E-TEXTILES)

Recently, conductive textiles are being developed from conductive nanomaterials for a wide range of smart applications such as sensors, supercapacitors, textile batteries, heating textiles, and so on. For a long run, metallic filaments (such as copper, steel, and aluminum) and intrinsically conductive polymers such as polypyrrole, polyaniline, polyacetylene, poly(3,4-ethylene dioxy thiophene)-poly(styrene sulfonate) (PEDOT:PSS), and so on are being used in conventional textiles.[36–38] However, due to limitations in these conventional conductive textiles, researchers have been looking for new alternatives that can give better electrical properties. Currently, nanotechnology has been utilized in developing conductive smart textiles by incorporating conductive nanomaterials to them.[39,40] In this regard, carbon-based nanomaterials (carbon nanotubes (CNTs) and graphene) and metallic nanomaterials have gained much attention due to their high thermal and electrical conductivity.

11.3.1   CONDUCTIVE METALLIC NANOMATERIALS

Conductive metals (silver, copper, gold, aluminum, and tin) and metal oxide (TiO_2, ZnO) nanomaterials offer a better alternative to conventional conductive materials.[41–43] These conductive nanomaterials (NPs or nanowires) can be incorporated with conductive polymer nanocomposites, which can be used for making conductive fibers or coated textiles. Similarly, antistatic fabrics can be developed from conductive NPs such as Ag, ZnO, TiO_2 that have the ability to dissipate the static charge of synthetic fibers because of their good electro-conductive properties.[44,45]

11.3.2   CARBON-BASED NANOMATERIALS

Even though metal NPs can impart excellent conductive properties to textiles, there are some drawbacks in using them for textile applications, since the processing of metal NPs is very complex and involves toxic chemicals. Thus, there is a growing interest for carbon-based functional materials from biological sources. In recent days, carbon-based conductive materials such as graphene, graphene oxide (GO), rGO, CNTs, carbon black (CB), and activated carbon (AC) are receiving great importance due to their extraordinary mechanical stability and electrochemical performance.[46] For application in textiles, these materials are suspended in liquid solvents to form conductive ink or printing paste.[47,48] These textiles can be used for various applications depending on the purity of materials, morphology of the particles, ink viscosity, surface properties of textiles, and coating procedures. Spun coating, inkjet printing, wet transfer monolayer, brush coating, dip coating, screen printing methods are the different methods used for coating carbon-based conductive materials to the textile substrates.[49–54]

The electrical conductivity of CNT is as high as 10^6–10^7 S/m and for pure graphene; it can be up to 10^5 S/m. The recent trend of using CNTs for fabricating electronic textiles is because of their high electrical conductivity, excellent thermal stability (up to 4000 K), high tensile strength (63 GPa), and low density (1.3–1.4 g/cm^3).[55] Additionally, due to the high aspect ratio CNTs can create percolation network on rough textile substrates. CNTs are generally synthesized by arc discharge,[56] laser ablation,[57] and chemical vapor deposition.[55,58] CNTs thus synthesized can be deposited on the textile substrate via dip coating, flexographic printing, inkjet printing, and doctors' blade technology.[49,59–61] CNT deposited cotton fabric is then treated with acid for hydrogen bond formation with cotton by imparting acidic group.[36] Thus, cotton fabric with high electrical conductivity, flexibility, and foldability can be produced. These unique properties are achieved because of the strong adhesion of CNTs with cotton textiles due to high van der Waals and hydrogen bonding between cotton and acid-treated CNTs.[48] These CNT-coated textiles possess excellent properties like flame retardancy, UV absorption capability, and water repellency.[62]

Another most important carbon-based conductive material used for the fabrication of E-textiles is CB.[62] Compared to other carbon-based materials for E-textiles, the production and fabrication of CB is simple and cost-effective.[63] Combustion at high temperature, gasification, pyrolysis, and thermal cracking are some of the conventional methods for preparing CB.[64] Due to the large particle size and porous structure CB particles are not easily adsorbed to the cotton fabrics and hence CB-coated fabrics exhibit low conductivity. However, screen printing of CB can impart good conductivity because the inter-yarn spaces of fabrics are filled with CB particles. In screen printing, a thick layer of CB is coated over the fabric surface which increases the electron mobility through the fabrics. However, one of the major issues with CB coated textiles is their low wash durability compared to other carbon-based materials. Chen et al.[65] have studied the effect of different carbon-based nanomaterials and their aspect ratios on the thermal conductivity of epoxy nanocomposites and observed that increasing filler concentration and aspect ratio increased the thermal conductivity.

11.4 APPLICATIONS OF CONDUCTIVE TEXTILES

Conductive nanomaterials incorporated textiles offer high-performance applications such as electroconductive textiles, EMI shield textiles, anti-static textiles, textile sensors, textile batteries, supercapacitors, wearable antenna photonic textiles, and many others.[40,66,67] These are summarized in Figure 11.3. Recently, E-textiles have a huge demand in various fields like healthcare, military and defense, sports, workwear, and many other technical fields. In military and defense sector, conductive textiles are used for protection against extreme weather conditions, sweat management, integration of high-tech materials ballistic impact, nuclear, biological, and chemical threats. In healthcare sector, conductive textiles are used for both clinical and nonclinical applications to monitor and communicate the conditions of patients as they can detect, acquire and transmit physiological signals. Increasing health awareness among youngsters led to a higher demand for conductive textiles to monitor sleep, heart rate, calories consumed, blood pressure, and so on.

11.5 TOXICITY OF TEXTILE NANOMATERIALS

Exposure to nano-engineered textiles will definitely have certain risks to health, environment, and sustainability. During washing or abrasive exposure, NPs from these commercial textiles are released and get accumulated in environment. Nanotextile materials mainly use nanofibers, nanocomposites, and nanometal oxides. Many studies have shown that release of silver nanoparticles that impart antibacterial properties to textiles during washing and release of sweat can cause various environmental issues. Amount of silver NPs released may vary depending on the quantity of silver coating, quality of fabric, and sweat formulation. Chemistry of these NPs also vary when these textiles are washed.[68] These silver NPs released into wastewater get accumulated on sewage sludge and get deposited on agricultural fields, causing the depletion of soil fertility.

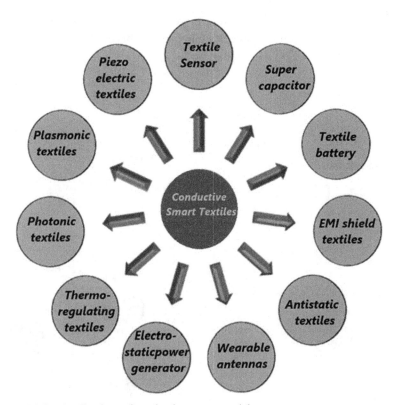

FIGURE 11.3 Applications of conductive nanomaterials.

Titanium oxide (TiO_2) NPs, which are used in several functional textiles due to its applications in UV absorption, degradation of dyes and as a pigment is also hazardous to human beings and our surrounding environment. NPs of anatase TiO_2 of sizes 10–20 nm may cause oxidative DNA damage and micronuclei formation.

There are various other direct and indirect health effects for textile nanomaterials. The direct exposure of NPs can come through wearing of nanotextiles. They may cause harmful effects when these materials start degradation. Indirect exposure to NPs occurs through air, water, soil, and food chain. It has been reported that concentration of <50 µg/m³ or lower can easily enter the human body.[69] Once these NPs enter our respiratory tract through the air, they spread to different parts of the lungs by diffusional forces. Gastrointestinal tract also gets effected, while they are moving through the respiratory tract. Rezic[70] reported that when NPs are administered in human body through intravenous, subcutaneous, transdermal, inhalation, or oral means, they get accumulated in lungs and brain and then move to heart and blood leading their way to liver, kidneys, spleen, and even embryo. Figure 11.4 shows a model of transport of NPs in human body.

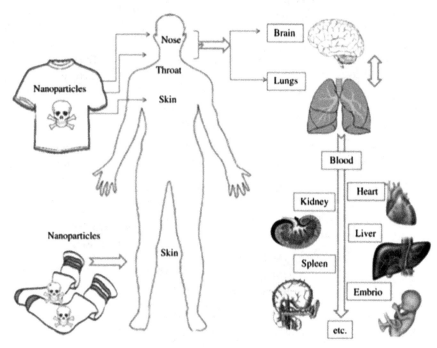

FIGURE 11.4   Model of transport of nanoparticles in human body.

11.6   CONCLUSION

Development of nano-engineered textiles has revolutionized the textile industry and these smart textiles have the ability to revolutionize the processing of our clothing and fabrics. Incorporation of nanotechnology in the textile industry can improve the functionalities for intelligent textiles, such as self-cleaning, antibacterial properties, UV protection, water repellence, wrinkle resistance, antistatic properties, and strength enhancement. The recent advancements in E-textiles and their applications in various sectors also have been discussed. These conductive nanomaterials have a huge potential for enhancing the conductive properties in textiles and will drive the futuristic development of smarter conductive textiles for multidirection applications.

Along with the development of nanotextiles, life-cycle assessments and toxicity of released nanomaterials from textiles also have to be evaluated to an extent. Nanotechnology-based products will definitely continue to emerge with new applications; however, manufacturers and regulatory agencies must ensure that these technologies have no negative impact on human health and the environment during their processing and life cycle.

KEYWORDS

- **textiles**
- **nanofibers**
- **carbon nanotubes**
- **silver nanoparticles**
- **conductive textiles**
- **nanotoxicity**

REFERENCES

1. Mukhopadhyay, S. *Text. Rev.* **2007,** *October,* 85–99.
2. Matsuo, T. Advanced Technical Textile Products. *Text. Prog.* **2008,** *40* (3), 123–181. DOI: 10.1080/00405160802386063.
3. Zhao, Q. Q.; Boxman, A.; Chowdhry, U. *J. Nanopart. Res.* **2003,** *5*, 567–572.
4. Cherenack, K.; van Pieterson, L. Smart Textiles: Challenges and Opportunities. *J. Appl. Phys.* **2012,** *112* (9), 091301. DOI: 10.1063/1.4742728.
5. Paradiso, R.; Loriga, G.; Taccini, N.; Gemignani, A.; Ghelarducci, B. Wealthy—A Wearable Healthcare System: New Frontier on e-Textile. *J. Telecomm. Inf. Technol.* **2005,** *4*, 105–113.
6. Cheng, M. H.; Chen, L. C.; Hung, Y. C.; Yang, C. M. A Realtime Maximum-Likelihood Heart-Rate Estimator for Wearable Textile Sensors. In *Engineering in Medicine and Biology Society, 2008. EMBS 2008. 30th Annual International Conference of the IEEE*; IEEE, 2008; pp 254–257.
7. Mattmann, C.; Amft, O.; Harms, H.; Tröster, G.; Clemens, F. Recognizing Upper Body Postures Using Textile Strain Sensors. In *11th IEEE International Symposium on Wearable Computers*, Boston, MA, 2007, pp 1–8. DOI: 10.1109/ISWC.2007.4373773.
8. Patra, K.; Gouda, S. Application of Nanotechnology in Textile Engineering: An Overview. *J. Eng. Technol. Res.* **2013,** *5* (5), 104–111. DOI: 10.5897/JETR2013.0309.
9. Russell, E. Nanotechnologies and the Shrinking World of Textiles. *Text. Horiz.* **2002,** *9*, 7–9.

10. Marmur, A. The Lotus Effect: Superhydrophobicity and Metastability. *Langmuir* **2004,** *20* (9), 3517–3519. DOI: 10.1021/la036369u.

11. Gao, L.; McCarthy, T. J. The "Lotus Effect" Explained: Two Reasons Why Two Length Scales of Topography Are Important. *Langmuir* **2006,** *22* (7), 2966–2967. DOI: 10.1021/la0532149.

12. El-Khatib, E. M. Antimicrobial and Self-Cleaning Textiles Using Nanotechnology. *Res. J. Text. Apparel* **2012,** *16* (3), 156–174. DOI: 10.1108/RJTA-16-03-2012-B016.

13. Yu, M.; Gu, G.; Meng, W.-D.; Qing, F.-L. Superhydrophobic Cotton Fabric Coating Based on a Complex Layer of Silica Nanoparticles and Perfluorooctylated Quaternary Ammonium Silane Coupling Agent. *Appl. Surf. Sci.* **2007,** *253* (7), 3669–3673. DOI: 10.1016/j.apsusc.2006.07.086.

14. Chien, H.; Chen, H.; Wang, C. The Study of Non-formaldehyde Crease-Resist Finishing Fabrics Treated with the Compound Catalyst of Nanometer Grade TiO$_2$ under UV Light and Different Polycarboxylic Acid. *J. Hwa Gang Text.* **2003,** *10*, 104–114.

15. Wang, C. C.; Chen, C. C. Physical Properties of Crosslinked Cellulose Catalyzed with Nano Titanium Dioxide. *J. Appl. Polym. Sci.* **2005,** *97* (6), 2450–2456. DOI: 10.1002/app.22018.

16. Song, X.; Liu, A.; Ji, C.; Li, H. The Effect of Nano-Particle Concentration and Heating Time in the Anti-Crinkle Treatment of Silk. *Jilin Gongxueyuan Xuebao* **2001,** *22*, 24–27.

17. Xin, J. H.; Daoud, W. A.; Kong, Y. Y. A New Approach to UV-Blocking Treatment for Cotton Fabrics. *Text. Res. J.* **2004,** *74* (2), 97–100. DOI: 10.1177/004051750407400202.

18. Yang, H.; Zhu, S.; Pan, N. Studying the Mechanisms of Titanium Dioxide as Ultraviolet-Blocking Additive for Films and Fabrics by an Improved Scheme. *J. Appl. Polym. Sci.* **2004,** *92* (5), 3201–3210. DOI: 10.1002/app.20327.

19. Saito, M. Antibacterial, Deodorizing, and UV Absorbing Materials Obtained with Zinc Oxide (ZnO) Coated Fabrics. *J. Ind. Text.* **1993,** *23*, 150–164.

20. Xiong, M.; Gu, G.; You, B.; Wu, L. Preparation and Characterization of Poly(Styrene Butylacrylate) Latex/Nano-ZnO Nanocomposites. *J. Appl. Polym. Sci.* **2003,** *90* (7), 1923–1931. DOI: 10.1002/app.12869.

21. Burniston, N.; Bygott, C.; Stratton, J. Nanotechnology Meets Titanium Dioxide. *Surf. Coat. Int. A* **2004,** *87*, 179–184.

22. Daoud, W. A.; Xin, J. H. Low Temperature Sol–Gel Processed Photocatalytic Titania Coating. *J. Sol Gel Sci. Technol.* **2004,** *29* (1), 25–29. DOI: 10.1023/B:JSST.0000016134. 19752.b4.

23. Wang, R.; Xin, J. H.; Tao, X. M.; Daoud, W. A. ZnO Nanorods Grown on Cotton Fabrics at Low Temperature. *Chem. Phys. Lett.* **2004,** *398* (1–3), 250–255. DOI: 10.1016/j. cplett.2004.09.077.

24. Yeo, S. Y.; Lee, H. J.; Jeong, S. H. Preparation of Nanocomposite Fibers for Permanent Antibacterial Effect. *J. Mater. Sci.* **2003,** *38* (10), 2143–2147. DOI: 10.1023/A:1023767828656.

25. Lee, H. J.; Yeo, S.; Jeong, S. Antibacterial Effect of Nanosized Silver Colloidal Solution on Textile Fabrics. *J. Mater. Sci.* **2003,** *38* (10), 2199–2204. DOI: 10.1023/A:1023736416361.

26. Yeo, S. Y.; Jeong, S. H. Preparation and Characterization of Polypropylene/Silver Nanocomposite Fibers. *Polym. Int.* **2003,** *52* (7), 1053–1057. DOI: 10.1002/pi.1215.

27. Klasen, H. J. Historical Review of the Use of Silver in the Treatment of Burns. I. Early Uses. *Burns* **2000,** *26* (2), 117–130. DOI: 10.1016/s0305-4179(99)00108-4.

28. Feng, Q. L.; Wu, J.; Chen, G. Q.; Cui, F. Z.; Kim, T. N.; Kim, J. O. A Mechanistic Study of the Antibacterial Effect of Silver Ions on *Escherichia coli* and *Staphylococcus aureus*. *J. Biomed. Mater. Res.* **2000,** *52* (4), 662–668. DOI: 10.1002/1097-4636(20001215) 52:4<662::aid-jbm10>3.0.co;2-3.

29. Jeong, S. H.; Yeo, S. Y.; Yi, S. C. The Effect of Filler Particle Size on the Antibacterial Properties of Compounded Polymer/Silver Fibers. *J. Mater. Sci.* **2005,** *40* (20), 5407–5411. DOI: 10.1007/s10853-005-4339-8.

30. Nowack, B.; Krug, H. F.; Height, M. 120 Years of Nanosilver History: Implications for Policy Makers. *Environ. Sci. Technol.* **2011,** *45* (4), 1177–1183. DOI: 10.1021/es103316q.

31. Kumar, R.; Howdle, S.; Münstedt, H. Polyamide/Silver Antimicrobials: Effect of Filler Types on the Silver Ion Release. *J. Biomed. Mater. Res. B Appl. Biomater.* **2005,** *75* (2), 311–319. DOI: 10.1002/jbm.b.30306.

32. Pal, S.; Tak, Y. K.; Song, J. M. Does the Antibacterial Activity of Silver Nanoparticles Depend on the Shape of the Nanoparticle? A Study of the Gram-Negative Bacterium *Escherichia coli*. *Appl. Environ. Microbiol.* **2007,** *73* (6), 1712–1720. DOI: 10.1128/AEM.02218-06.

33. Xu, P.; Wang, W.; Chen, S. Application of Nanosol on the Antistatic Property of Polyester. *Melliand Int.* **2005,** *11*, 56–59.

34. Kumar, S.; Doshi, H.; Srinivasarao, M.; Park, J. O.; Schiraldi, D. A. Fibers from Polypropylene/Nanocarbon Fiber Composites. *Polymer* **2002,** *43* (5), 1701–1703. DOI: 10.1016/S0032-3861(01)00744-3.

35. Schaerlaekens, M. Melt Extrusion with Nano-additives. *Chem. Fib. Int.* **2003,** *53*, 100.

36. Liu, Y.; Wang, X.; Qi, K.; Xin, J. H. Functionalization of Cotton with Carbon Nanotubes. *J. Mater. Chem.* **2008,** *18* (29), 3454–3460. DOI: 10.1039/b801849a.

37. Kaur, G.; Adhikari, R.; Cass, P.; Bown, M.; Gunatillake, P. Electrically Conductive Polymers and Composites for Biomedical Applications. *RSC Adv.* **2015,** *5* (47), 37553–37567. DOI: 10.1039/C5RA01851J.

38. Maity, S.; Chatterjee, A. Textile/Polypyrrole Composites for Sensory Applications. *J. Compos.* **2015,** *2015*, 1–6. DOI: 10.1155/2015/120516.

39. Ouyang, J. Recent Advances of Intrinsically Conductive Polymers. *Acta Phys. Chim. Sin.* **2018,** *34* (11), 1211–1220. DOI: 10.3866/PKU.WHXB201804095.

40. Joshi, M.; Adak, B. Advances in Nanotechnology Based Functional, Smart and Intelligent Textiles: A Review. In *Comprehensive Nanoscience and Nanotechnology*; Andrews, D. L., Lipson, R. H., Nann, T., Eds.; Elsevier: Amsterdam, 2019; pp 253–290.

41. Coyle, S.; Wu, Y.; Lau, K. T.; De Rossi, D.; Wallace, G.; Diamond, D. Smart Nanotextiles: A Review of Materials and Applications. *MRS Bull.* **2007,** *32* (5), 434–442. DOI: 10.1557/mrs2007.67.

42. Naghdi, S.; Rhee, K. Y.; Hui, D.; Park, S. J. A Review of Conductive Metal Nanomaterials as Conductive, Transparent, and Flexible Coatings, Thin Films, and Conductive Fillers: Different Deposition Methods and Applications. *Coatings* **2018,** *8* (8), 278. DOI: 10.3390/coatings8080278.

43. Xue, C. H.; Chen, J.; Yin, W.; Jia, S. T.; Ma, J. Z. Superhydrophobic Conductive Textiles with Antibacterial Property by Coating Fibers with Silver Nanoparticles. *Appl. Surf. Sci.* **2012,** *258* (7), 2468–2472. DOI: 10.1016/j.apsusc.2011.10.074.

44. Karimi, L.; Yazdanshenas, M. E.; Khajavi, R.; Rashidi, A.; Mirjalili, M. Using Graphene/TiO_2 Nanocomposite as a New Route for Preparation of Electroconductive, Self-Cleaning, Antibacterial and Antifungal Cotton Fabric without Toxicity. *Cellulose* **2014,** *21* (5), 3813–3827. DOI: 10.1007/s10570-014-0385-1.

45. Zhang, F.; Yang, J. Preparation of Nano-ZnO and Its Application to the Textile on Antistatic Finishing. *Int. J. Chem.* **2009,** *1* (1). DOI: 10.5539/ijc.v1n1p18.

46. Joshi, M.; Adak, B. Nanotechnology-Based Textiles: A Solution for Emerging Automotive Sector. *Rubber Nanocomposites and Nanotextiles*; De Gruyter: Berlin, 2018; pp 207–266.

47. Hansora, D. P.; Shimpi, N. G.; Mishra, S. Performance of Hybrid Nanostructured Conductive Cotton Materials as Wearable Devices: An Overview of Materials, Fabrication, Properties and Applications. *RSC Adv.* **2015,** *5* (130), 107716–107770. DOI: 10.1039/C5RA16478H.

48. Weng, W.; Chen, P.; He, S.; Sun, X.; Peng, H. Smart Electronic Textiles. *Angew. Chem. Int. Ed. Engl.* **2016,** *55* (21), 6140–6169. DOI: 10.1002/anie.201507333.

49. Shahariar, H. *Process Engineering & Materials Characterization for Printing Fexible and Durable Passive Electronic Devices on Nonwoven*, 2017. https://repository.lib.ncsu.edu/handle/1840.20/34966.

50. Hu, L.; Pasta, M.; Mantia, F. L.; Cui, L.; Jeong, S.; Deshazer, H. D.; Choi, J. W.; Han, S. M.; Cui, Y. Stretchable, Porous, and Conductive Energy Textiles. *Nano Lett.* **2010,** *10* (2), 708–714. DOI: 10.1021/nl903949m.

51. Yapici, M. K.; Alkhidir, T.; Samad, Y. A.; Liao, K. Graphene-Clad Textile Electrodes for Electrocardiogram Monitoring. *Sens. Actuators B* **1474,** *221* (2015), 469.

52. Jost, K.; Stenger, D.; Perez, C. R.; McDonough, J. K.; Lian, K.; Gogotsi, Y.; Dion, G. Knitted and Screen Printed Carbon-fiber Supercapacitors for Applications in Wearable Electronics. *Energy Environ. Sci.* **2013,** *6* (9), 2698–2705. DOI: 10.1039/c3ee40515j.

53. Majee, S.; Liu, C.; Wu, B.; Zhang, S.-L.; Zhang, Z.-B. Ink-Jet Printed Highly Conductive Pristine Graphene Patterns Achieved with Water-Based Ink and Aqueous Doping Processing. *Carbon* **2017,** *114*, 77–83. DOI: 10.1016/j.carbon.2016.12.003.

54. Khan, Z. U.; Kausar, A.; Ullah, H.; Badshah, A.; Khan, W. U. A Review of Graphene Oxide, Graphene Buckypaper, and Polymer/Graphene Composites: Properties and Fabrication Techniques. *J. Plast. Film Sheeting* **2016,** *32* (4), 336–379. DOI: 10.1177/8756087915614612.

55. Kim, I.; Shahariar, H.; Ingram, W. F.; Zhou, Y.; Jur, J. S. Inkjet Process for Conductive Patterning on Textiles: Maintaining Inherent Stretchability and Breathability in Knit Structures. *Adv. Funct. Mater.* **2019,** *29* (7), 1807573. DOI: 10.1002/adfm.201807573.

56. Purohit, R.; Purohit, K.; Rana, S.; Rana, R. S.; Patel, V. Carbon Nanotubes and Their Growth Methods. *Proc. Mater. Sci.* **2014,** *6*, 716–728. DOI: 10.1016/j.mspro.2014.07.088.

57. Shi, Z.; Lian, Y.; Liao, F. H.; Zhou, X.; Gu, Z.; Zhang, Y.; Iijima, S.; Li, H.; Yue, K. T.; Zhang, S. L. Large Scale Synthesis of Single-Wall Carbon Nanotubes by Arc-Discharge Method. *J. Phys. Chem. Solids* **2000,** *61* (7), 1031–1036. DOI: 10.1016/S0022-3697(99)00358-3.

58. Che, G.; Lakshmi, B. B.; Martin, C. R.; Fisher, E. R.; Ruoff, R. S. Chemical Vapor Deposition Based Synthesis of Carbon Nanotubes and Nanofibers Using a Template Method. *Chem. Mater.* **1998,** *10* (1), 260–267. DOI: 10.1021/cm970412f.

59. Loffredo, F.; Mauro, A. D. G. D.; Burrasca, G.; La Ferrara, V. L.; Quercia, L.; Massera, E.; Di Francia, G.; Sala, D. D. Ink-Jet Printing Technique in Polymer/Carbon Black Sensing Device Fabrication. *Sens. Actuators B* **2009,** 143 (1), 421–429. DOI: 10.1016/j.snb.2009.09.024.

60. Besra, L.; Liu, M. A Review on Fundamentals and Applications of Electrophoretic Deposition (EPD). *Prog. Mater. Sci.* **2007,** *52* (1), 1–61. DOI: 10.1016/j.pmatsci.2006.07.001.

61. Bøggild, P. The War on Fake Graphene. *Nature* **2018**, 562 (7728), 502–503. DOI: 10.1038/d41586-018-06939-4.

62. Islam, R.; Khair, N.; Ahmed, D. M.; Shahariar, H. Fabrication of Low Cost and Scalable Carbon-Based Conductive Ink for E-Textile Applications. *Mater. Today Commun.* **2019**, *19*, 32–38. DOI: 10.1016/j.mtcomm.2018.12.009.

63. Pahalagedara, L. R.; Siriwardane, I. W.; Tissera, N. D.; Wijesena, R. N.; De Silva, K. M. N. Carbon Black Functionalized Stretchable Conductive Fabrics for Wearable Heating Applications. *RSC Adv.* **2017**, *7* (31), 19174–19180. DOI: 10.1039/C7RA02184D.

64. Fabry, F.; Fulcheri, L. *Synthesis of Carbon Blacks and Fullerenes from Carbonaceous Wastes by 3-Phase AC Thermal Plasma to Cite This Version*; Horticulture Australia, 2016. To cite this version: HAL Id: hal-01328472.

65. Chen, J.; Gao, X.; Song, W. Effect of Various Carbon Nanofillers and Different Filler Aspect Ratios on the Thermal Conductivity of Epoxy Matrix Nanocomposites. *Results Phys.* **2019**, *15*, 102771. DOI: 10.1016/j.rinp.2019.102771.

66. Atwa, Y.; Maheshwari, N.; Goldthorpe, I. A. Silver Nanowire Coated Threads for Electrically Conductive Textiles. *J. Mater. Chem. C* **2015**, *3* (16), 3908–3912. DOI: 10.1039/C5TC00380F.

67. Fu, K. K.; Padbury, R.; Toprakci, O.; Dirican, M.; Zhang, X. Conductive Textiles. In *Engineering of High-Performance Textiles*; Miao, M., Xin, J. H., Eds.; Woodhead Publishing: Sawston, UK, 2018, pp 305–334.

68. Lombi, E.; Donner, E.; Scheckel, K. G.; Sekine, R.; Lorenz, C.; Von Goetz, N.; Nowack, B. Silver Speciation and Release in Commercial Antimicrobial Textiles as Influenced by Washing. *Chemosphere* **2014**, *111*, 352–358. DOI: 10.1016/j.chemosphere.2014.03.116.

69. Karakoti, A. S.; Hench, L. L.; Seal, S. The Potential Toxicity of Nanomaterials—The Role of Surfaces. *JOM* **2006**, *58* (7), 77–82. DOI: 10.1007/s11837-006-0147-0.

70. Rezić, I. Determination of Engineered Nanoparticles on Textiles and in Textile Wastewaters. *Trends Anal. Chem.* **2011**, *30* (7), 1159–1167. DOI: 10.1016/j.trac.2011.02.017.

CHAPTER 12

Metal/Metal-Oxide-Decorated Carbon-Based Nanofillers for Energy Storage Applications

POOJA MOHAPATRA, LIPSA SHUBHADARSHINEE, and
ARUNA KUMAR BARICK

*Department of Chemistry, Veer Surendra Sai University of Technology,
Siddhi Vihar, Burla, Sambalpur, Odisha, India*

ABSTRACT

Energy storage devices such as lithium-ion batteries, supercapacitors, electronic double-layer capacitors, and pseudocapacitors are going through incessant modifications in their conduction as there is an enduring requirement for energy storage devices, which are rechargeable having very large specific energy and getting quickly charged. In this book chapter, electrochemical studies like galvanic charge and discharge, cyclic voltammetry, and electrochemical impedance spectroscopy are discussed for various metal and metal-oxide-embedded carbon nanofillers (e.g., reduced graphene oxide, graphene oxide, and graphene-based hybrid nanomaterials for energy storage and conversion applications. Basically, metals like Au and Ag and metal oxides like ZnO, CuO, TiO_2, MnO_2, and so on are incorporated into the carbon nanofillers to prepare nanohybrids. The electrochemical properties such as capacity, cycling stability, and rate capability are significantly improved, which have a current necessity in the industry and feasibility for future installation in an environmentally friendly way. Studies are also going on about the future prospects and challenges for the advancement in this field by the implementation of conducting polymers.

Mechanics and Physics of Porous Materials: Novel Processing Technologies and Emerging Applications.
Chin Hua Chia, Tamara Tatrishvili, Ann Rose Abraham, & A. K. Haghi (Eds.)
© 2024 Apple Academic Press, Inc. Co-published with CRC Press (Taylor & Francis)

12.1   INTRODUCTION

Due to the increase in global economic expansion, the population of the world, and the reliance of human beings on gadgets that consume high energy, the stipulation of global energy is accelerating, resulting in the scarcity of it in the near future.[1-3] According to the World Energy Council, we will need twice the amount of energy from the current scenario by 2050. Therefore, energy storage is pivotal for countering the future worldwide challenges, as a lack of energy may pose a threat to the environment, human health, security, and so on. Hence, nowadays, in the regime of nanoscales, energy storage is considered the center of attention not only for the majority of the scientific communities but also for the whole world.

As one of the most sensitive issues of society, the energy issue must be resolved through methods that are cost-effective, environmentally friendly, and efficient energy depository devices that can be helpful in energy-demanding areas while keeping an eye on the current development of economy.[4] As it is said that necessity is the mother of invention, this future threat forces us to invent some high-power systems for energy conversion and energy storage that must be power sustainable and have very low exhaust emissions.[2,3] So, keeping these key points in mind, researchers are focusing on utilities, manufacturers, and project developers to identify, test, evaluate, and certify systems that will amalgamate smoothly with today's grid while planning for tomorrow. On the basis of recent research, new technologies offer new materials and procedures for energy storage devices. Graphene, GO, rGO single-walled and multiwalled carbon nanotubes (SWCNTs and MWCNTs), nanodiamonds, carbon nanosheets, activated carbons, and so on, which are known as carbon-based nanomaterials that individually or embedded with metal/metal oxides play some significant roles in energy storage devices to increase their efficiency.[1,5,6] In other words, we can say that these new materials introduced by the current technologies that act as frontiers against the energy shortage in the near future. Carbon-based nano-materials are frequently used electrodes due to cost-effectiveness and their abundance presence. Due to their high storage capacity, these are the most suitable nanocomposites for electrochemical electrodes.[7-9]

The graphene consists of a two-dimensional (2D) honeycomb structure of sp^2 hybridized carbon. For carbon nanomaterials, it has been considered to be the mother material as a building block.[10-13] Having surprising physicochemical properties such as strong mechanical strength, large specific surface area, extraordinary optical transmittance, excellent electron transport capabilities, and thermal conductivity, it shows a profound impact in the electrochemistry field.[11,12,14,15] As previously mentioned, graphene, with all its efficiency, is at the vanguard of the condensed matter physics and materials science research.[13,17–19] As the thinnest known material, it has the advantage of being used for basic constituent for many new carbon nanofiller derivatives, because it can be piled up into three-dimensional (3D) graphite and also rolled into one-dimensional (1D) carbon nanotubes (CNTs). It is also enfolded into a spherical fullerene complemented with pentagons. Above all, it can be represented as the mother of all graphitic materials.[18] Hence, theoretical works on graphene and its derivatives have been studied for years. Isolated graphene with other 2D atomic layers has the disadvantage of being thermodynamically unstable.[20] It is a zero-gap semiconductor having distinctive electronic properties. As a reverberation of its flawless crystal structure, low-energy quasi-particles enclosed in it, go for a dispersion relation that is linear, similar to massless relativistic particles. Many peculiar electronic properties have been led by this essential characteristic of a gapless semiconductor,[18,21–23] including transport, ballistic, pseudospin chirality on the basis of a normal temperature half-integral chiral quantum hall effect, berry phase, and conductivity in the absence of charge carriers, which build it a propitious option for advanced electronic products, both as an interconnect and a device.[24] It is conjectured that the graphene overshadows metals, CNTs, semiconductors, and graphite in many applications due to its unique properties whether as an individual material or a component in a composite or hybrid material.[24]

The carboxylic groups, as well as the plentiful hydroxyl and epoxide groups, dominate the chemical structure of graphene oxide (GO).[25] The GO is fascinated by many physicochemical properties due to the presence of functional groups containing oxygen. Those properties comprise thermal, electronic, mechanical, optical, and electrochemical properties including chemical reactivity.[26,27] There are benefits of oxygen functional groups as it can be served as an effective site to immobilize various electroactive species for functionalization or chemical modification of GO.[25,28–30] Therefore, in chemical, thermal, or electrochemical ways, modulation of physicochemical properties happens.[28,31–33] Since stable dispersion is formed in solvents either aqueous or nonaqueous, regular methods via solution processing namely drop-casting, dip-coating, spray-coating, or spin-coating procedures are frequently used for the fabrication of thin films based on GO.[34] The GO can be effectively conveyed as an amphiphilic substance due to the presence of carboxylic groups, which are hydrophilic in nature, and unoxidized benzene rings, which are hydrophobic, having a polyaromatic network.[35] This amphiphilicity increases the likelihood of GO self-assembly at various interfaces where aqueous GO dispersion is typically used as a pairing of one phase with another phase that can be air, liquid, or solid.[36] To get controlled microstructured and adaptable properties of GO-based materials, the self-assembly methods are the best with simple and effective strategies,[36] which makes them efficient in various applications for catalytic support, energy storage, transparent conductors, and polymer composites.[34] Apparently, over the past decades, considerable effort has been devoted to the development of renewable as well as green energy sources because of the global concerns about the upcoming depletion of fossil fuels and the ever-increasing environmental problems.[37] GO and its derivatives, along with composites embedded with metal/metal oxides, are being focused on demandable uses of electrochemical energy storage, comprising batteries, capacitors, and fuel cells due to their excellent properties as well as structural diversity. Behaving as oxidants the reduced oxygen functional groups of GO form closely packed composites with the values of reduced graphene oxide (rGO).[10]

The magnificent specific surface area, including the changeable interlayer distance of the thin layered nanostructures of GO, permits to take place the electrochemical reactions while taking note of size or volume according to the change of products.[10] By using free p-electrons on the surface, GO could be used as an insulating dielectric spacer or an electrically conductive substrate, depending on the extent of reduction. The GO has shown great potential in energy systems because of the flexible functions resulting from its characteristic nanostructure, which contributed to electrodes, electrocatalysts, protecting layers, printing inks, reinforcements, and membranes, among other things. In this section, the applications of GO in energy storage and conversion systems are summarized, as well as discussed the roles and benefits of GO in depth.[10]

Priority must be imposed to explore large-scale production of high-grade graphene with a low-cost and easy synthesis route to advance the end uses of 2D graphene-based products, which show enormous potential for several functions owing to their unique properties.[18,38] Until now, various strategies have been used to produce graphene, primarily thermal/chemical reduce precursor materials into the GO,[40] mechanical or ultrasonic exfoliation,[41] chemical intercalation,[42] chemical vapor deposition (CVD)/plasma-enhanced CVD,[43,44] electric arc discharge,[45] epitaxial growth,[46] and so on. Zhao et al.[47] went through a sol–gel polymerization and thermal treatment process for the synthesis of a Fe_3O_4/Fe/carbon composite, which exhibits a steady and reversible capacity of >600 mA h/g. He et al. clearly mentioned that Fe_3O_4 nanoparticles engulfed by carbon through an *in-situ* method embedded in 2D porous graphitic nanosheets obtained a capacity of ~900 mA h/g.[46]

GO as a water-soluble nanomaterial[48] is created by chemically treating graphite powder to incorporate oxygen defects in the graphite stack.[49] This results in complete solid abrasion into nanosheets with a thickness of atoms, aided by thermal, chemical, or mechanical treatments.[50–54] The rGO thin films can act as transparent conductors with the same level of optical clarity and electrical conductivity of CNTs networks. The primary advantages of GO over other carbon-based nanomaterials are easy synthesis routes, excellent solubility and economic processing, fine adjustable conductivity, huge surface area, excellent biocompatibility, and availability of low-cost resources.[55] The capacitance of GO film is lower than rGO. The elimination of oxygen comprising polar functional groups improves conductivity of GO film, ultimately increasing its capacitive behavior.[56] Surface-enhanced 3D rGO networks were also investigated as a high-performance electrode materials for energy storage devices such as RBs, asymmetric, pseudo, and hybrid supercapacitors.

Metal or metal oxides and conductive polymers divulge long cycle life as well as high-power density as they are used as common supercapacitor electrode materials. Among the composites, graphene with metal oxides exhibits characteristics that combine graphene's exceptional cycle stability with metal oxides' high-capacity properties, significantly improving the overall properties of nanocomposites.[57] Until now, the advancements of metal/metal-oxide-decorated graphene (MGr) composites in electrochemical capacitors have been interpreted based on their collective properties. It has been demonstrated that MGr composites have significant improvements in rate capability, cycling stability, and capacity when compared to their individual impacts. The morphology, structure, and bonding of the nano-hybrid are usually analyzed by various instruments such as atomic force microscopy, scanning electron microscopy (SEM), X-ray diffraction, X-ray photoelectron spectroscopy, and so on.[39]

At present, green energy and portable electronic devices have appeared with tremendous mechanical properties; they are easy to assemble, flexible, light weight, and so on.[58] The primary requirements for electronic gadgets that make them appealing for various high-tech applications are the fast and efficient power supply.[58–61] To get this uninterrupted power supply, we need sustainable energy sources that can be derived from various already utilized methodologies as described here.

12.2 METAL/METAL-OXIDE-DECORATED CARBON-BASED HYBRID NANOFILLERS

12.2.1 METAL/METAL-OXIDE-DECORATED GRAPHENE-BASED HYBRID NANOFILLERS

Nandi et al.[57] explicitly explain that lofty power density along with prolonged cycle life can be obtained by using metal or metal oxide and some conductive polymers as supercapacitor electrode materials. Due to their distinctive electronic structure and extensive surface area, the capacity to transfer charge could be enhanced, but because of the presence of van der Waals force, the graphene nanosheets get easily agglomerated by reducing their specific surface area and volume.[62] Hence, by introducing metal or metal oxides, they get stuffed between the adjacent graphene layers by enriching their charge capacity as supercapacitor electrode materials.[63–69] Wu et al.[24] provided some authentic data for various comparisons to give us a better opinion in Table 12.1.

Metal/Metal-oxide-embedded graphene composites can be synthesized by various methods like the solution mixing method,[70] atomic layer deposition (ALD) method,[71] sol–gel method,[72] solvothermal method,[73,74] self-assembly method,[75–77] electrochemical deposition method,[78,79] microwave method,[80] reduction method,[81,82] coprecipitation method,[83] photochemical synthesis,[84] green conversion methods,[85] and so on. Their current developments in the area of electrochemical capacitors are clarified on the basis of their synergistic properties. Based on their morphology, structure, and reactions on the electrode surface, these composites obtained large capacitances. Comparison of energy density and power density among various metal/MGr supercapacitors are listed in Table 12.2.

Juan et al.[16] represented that 3D graphene can be synthesized by a simple solvent treatment[86] and any modified Hummers method.[87,88] Due to the scalability, that is, having a good capacity to change size and scale, graphite oxide is the commonly employed route for synthesizing graphene.[89] Regardless of having a very small surface area, that is, only 81.7 m^2/g, excellent electrochemical behavior in aqueous as well as organic electrolytes is shown by 3D graphene. It achieves capacitance of 341 and 166 F/g in aqueous and nonaqueous electrolytes and shows energy densities of 16.2 and 52.5 W h/kg, respectively. As per the literature survey, although in aqueous solutions, the specific capacitance value of graphene-based supercapacitors is very promising but still below expectations. Theoretically, the capacitances

TABLE 12.1 Benefits and Drawbacks of Graphene, Metal Oxides (MO), and Graphene/MO Composites in LIBs and ECs.[24]

Benefits of graphene	Drawbacks of graphene	Benefits of MO	Drawbacks of MO	Benefits of graphene/MO composites
Superior electrical conductivity	Serious agglomeration	Very large capacity/capacitance	Poor electrical conductivity	Synergistic effects
Abundant surface functional groups	Restacking	High packing density	Large volume change	Suppressing the volume change of MO
Thermal and chemical stability	Large irreversible capacity	High-energy density	Severe aggregation/agglomeration	Suppressing agglomeration of MO and restacking of graphene
Large surface area	Low initial coulombic efficiency	Rich resources	Large irreversible capacity	Uniform dispersion of MO
Ultrathin thickness	No clear lithium storage mechanism	—	Poor rate capability	High capacity/capacitance, good rate capability
Structural flexibility	No obvious voltage plateau	—	Poor cycling stability	Improved cycling stability
Broad electrochemical window	Large voltage hysteresis	—	—	Improved energy/power densities

can be obtained up to 550 F/g, if the entire surface area of graphene can be utilized.[90] This height can be achieved by introducing pseudo-capacitance, which will slow the charge/discharge rate[91] by adding metal/metal oxide nanoparticles[92] like RuO_2,[93] MnO_2,[94] Fe_2O_3,[95] Ni^{2+}, Al^{3+}, and polymers such as polyaniline,[87,96,97] and so on. For the interpretation of electrochemical performances, electrochemical impedance spectroscopy (EIS), cyclic voltammetry (CV), and galvanostatic charge–discharge measurement (GCD) are generally employed. Accordingly, in organic and alkaline electrolytes, over 86% and 96% of the original capacitance can be retained after 100 cycles of GCD tests.

TABLE 12.2 Energy and Power Density of Various Metal/Metal-Oxide-Decorated Graphene Supercapacitors.[57]

Samples	C_{sp} (F/g)	Energy density (W h/kg)	Power density (kW/kg)
$Ni_3(PO_4)_2$/Gr	125	49.2	0.5
MnO_2/Gr	30	30.4	5.0
Co–Ni/Gr	156	23.9	2.5
MnO_2–textile/Gr	315	12.5	110
Ru/Gr	270	15	76.4
Au/Gr	174	29.4	0.2
Ag/Gr	472	41	1.5

Ejigu et al.[68] specifically demonstrated the importance of transition metals during the exfoliation of graphite, which can also go through a redox reaction by taking metal/metal oxide as the starting materials that enhance the electrochemical properties of the composite and also regulate the structure of the graphene.

12.2.2 METAL/METAL-OXIDE-DECORATED GRAPHENE OXIDE-BASED HYBRID NANOFILLERS

Tian et al.[10] elucidated the potential use of graphene and its derivatives in the area of electrochemistry due to their exceptionally good physicochemical properties, including satisfactory optical transmittance, high mechanical strength, appreciable thermal conductivity, electron transport capabilities, and high surface area.[27–30] They prioritized the impact of GO and it embedded with metal/metal oxide on the energy storage devices. The GO could be treated as a predecessor for the preparation of graphene by a thermal or chemical process through reduction.[33] GO and its derivatives can be used in various kinds of batteries like redox flow, lithium ion, lithium–sulfur, lithium–air, and so on, and also in supercapacitors, dielectric capacitors, and fuel cells as energy storage devices. Their versatile utilizations of electrochemical energy storage are shown in Figure 12.1 which includes capacitors, fuel cells, and batteries.

Ramaraju et al.[4] supported for the metal-oxide-deposited GO as a good application for storing energy. Accordingly, taking advantages of their light weight, long cycle lives, and high-energy density, RBs show effective promise for energy storage devices.[105,106] The increased resistance can be partially dependent on decreased cell resistance and electrochemical milling effects that can diminish the particle size and can also affect a material's morphology.

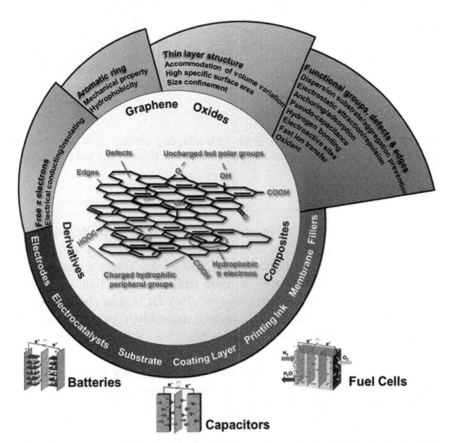

FIGURE 12.1 Pictorial illustration of applications of graphene oxide in electrochemical energy storage.[10]

Source: Reprinted with permission from Ref. [10]. Copyright 2021. Elsevier

12.2.3   *METAL/METAL-OXIDE-DECORATED CARBON NANOTUBES BASED HYBRID NANOFILLERS*

Jyoti et al.[58] reported 3D graphene/carbon nanotubes (3D-GCNTs) hybrid materials with extraordinary properties that could be considered to develop sustainable energy technology. Figure 12.2a and b represent the different applications of the 3D-GCNTs and relationship between power density and energy density, respectively.

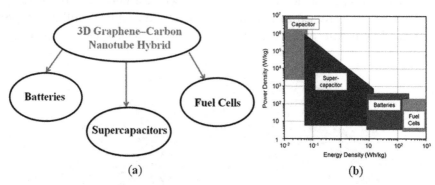

(a)   (b)

FIGURE 12.2   (a) Graphene/CNTs nanohybrids exhibiting different applications[58] and (b) assessment of power density of capacitors, supercapacitors, batteries, and fuel cells based on the energy density.[107]

Source: Reprinted with permission from Ref. [58] and Ref. [107].

The 2D carbon-based nanostructures are widely used depending on their outstanding properties like fine thermal conductivity (3000–5000 W/m K), large specific surface area (2630 m^2/g), Young's modulus ($\approx$1 TPa), superior charge mobility at room temperature ($\approx$10,000 cm^2/V S), and satisfactory optical transparency ($\approx$97.7%). GCNTs can be synthesized by various means like layered-by-layered assembly, CVD, surface modification using acid treatment, vacuum filtration (VF) techniques, *in-situ* technique, and so on.[58]

Owing to their long-life, high-power density, better charge–discharge behavior, and environmentally friendly nature, supercapacitors have drawn significant research attention. The achievement of the supercapacitors relies upon some of the electrodes and electrolytes. They are recorded in Table 12.3.

To gain a higher power and energy density, the organic electrolyte was carried out at a wide voltage range. At discrete current densities, the GCD curve of r[GO-CNTs] at various current densities are proposed in Figure

TABLE 12.3 Various GCNTs and Doped GCNTs Hybrid Composites Showing Electrochemical Performance in Supercapacitor.[58]

Sl. no.	Electrode material	Electrolyte	Preparation technique	C_{sp} (F/g)	Energy (W h/kg)	Power (W/kg)
1	GCNTs	KOH	Layered-by-layered assembly	120	–	–
2	r[GO–CNTs]	Organic 1 M tetraethylammonium tetrafluoroborate in polycarbonate	Surface modification technique using the acid	160	4.3	400
3	$NiCo_2O_4$ nitrogen-doped graphene CNTs	6 M KOH aqueous	Solution mixing using vacuum suction	2292.7 at 5 A/g	24.69	15,485
4	3D graphene form/CNTs/MnO_2 hybrid	0.5 M Na_2SO_4	Hydrothermal technique	–	22.8	860
5	Honeycomb $NiCo_2O_4$@ $Ni(OH)_2$ supported 3D GCNTs sponge	6 M KOH aqueous	Homogenous solution mixing with the help of stirring	138 at 5 A/g	42.3	476
6	Ruthenium oxide (RuO_2) anchored graphene and CNT hybrid foam	2 M Li_2SO_4	GCNT hybrids were synthesized using CVD	502.78 at 1 mA/cm³	39.28	128.01

Source: Reprinted with permission from Ref. [58].

12.3a. The less contorted the triangular curves, better the performance of electronic double-layer capacitors. The voltage drop across the system was recorded to be 3.0–2.8 V. By scrutinizing slope of the discharge curve, the value of specific capacitance was found, which was the same between 0.5 and 1 A/g and decreased from 110 to 78 F/g. Figure 12.3b reported the volumetric capacitance of two composites r[GO-CNT] and rPM[GO/oCNTs] to be 165 and 82.7 F/m³, respectively, at 1 A/g.[108] Due to the graphene composites having less effective surface area, the value reduces, and the reason behind this is the accumulation of CNTs and the restacking issue of graphene. The CV plot in Figure 12.3c illustrates the rapid ion charge transport properties of graphene-based composite electrodes. In the span of 0.0–3.0 V, the integrated area of the CV curve presented by r[GO-CNTs] is lower in comparison with rPM[GO/oCNT]. Figure 12.3d demonstrates that r[GO-CNTs] has the highest activity at a high energy density of 34.3 W h/kg at 4000 W/kg, whereas gradually decreases from 26.6 to 12.8 kW h/kg for rPM [GO/oCNTs].

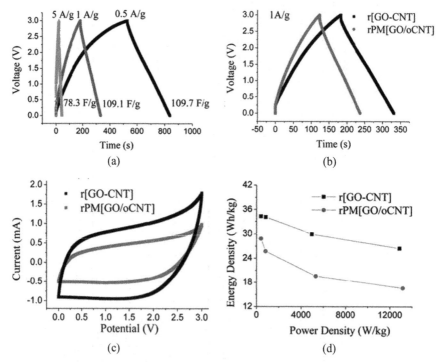

FIGURE 12.3   (a) Galvanostatic charging/discharging curves at applied current density, (b) galvanostatic charging/discharging curves, (c) CV plots, and (d) Ragone plots.[108]

Source: Reprinted with permission from Ref. [108]. Copyright 2013. John Wiley and Sons.

Zhang et al.[109] delineated how numerous carbon nanofillers like graphene, CNTs, graphite, and GO/graphite/CNT (GGCC) perform electrochemical behavior. In comparison to individual carbon nanofillers, the GGCC composite showed excellent electrochemical performance as it improves the electron transfer properties. Figure 12.4a, b, e depicts the morphology of graphite, CNTs, and GGCC as seen through SEM within a diameter range of 20–50 nm. The graphite sheet is uniformly covered by GO sheets and turned smooth after hybridization by a rough surface. SEM and transmission electron microscope (TEM) images of GO sheets can be seen in Figure 12.4c and d, respectively. Figure 12.4f displays TEM images of CNTs and graphite covered with GO sheets. Under high reversible capacity, the GGCC hybrid composite evinced a stable cyclic performance, and this cyclic performance of composites is depicted in Figure 12.4g. The significant improvement in electrochemical performance of the GGCC electrode is represented in Figure 12.4h. As shown in Figure 12.4i for different carbon fillers, the GO electrode showed the lower value decreasing from 353.1 mA h/g at 0.5 C to 19.7 mA h/g at 10 C based on their rate capability performance. Figure 12.4j depicts the rate capability performance of the GO/graphite and GGCC electrodes at various current rates.

In Table 12.4, the electrochemical performance is compared for various GCNTs hybrids along with their doping materials in lithium-ion batteries (LIBs) is compared.

12.2.4 METAL/METAL-OXIDE-DECORATED REDUCED GRAPHENE-OXIDE-BASED HYBRID NANOFILLERS

Kumar et al.[110] reported various Li-ion capacitors and batteries, which have been tried for highly performed energy storage devices.[111–115] These capacitors outperform the batteries in terms of charge/discharge cycle and specific power, whereas the later has high specific energy storage capacity. The surface-endowed 3D rGO hybrids are used in various hybrid, pseudo, and asymmetric supercapacitors, and also in RBs as high-achieving electrode materials for energy storage devices. There are huge comparisons between metal/metal-oxide-decorated reduced graphene-based hybrid nanofillers with other hybrid nanofillers on the basis of CV, GCD, and EIS (Table 12.5).

The electrochemical properties of a simple copper oxide deposited on GO where Cu_{ox}–rGO (CuO/Cu_2O hollow polyhedrons on rGO) was obtained, which was used in LIBs and sodium-ion batteries (NIBs) as they show excellent electrochemical properties. The brisk use of rechargeable batteries

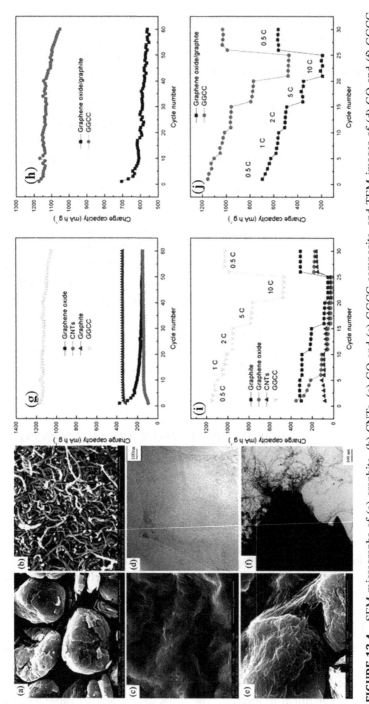

FIGURE 12.4 SEM micrographs of (a) graphite, (b) CNTs, (c) GO and (e) GGCC composite, and TEM images of (d) GO and (f) GGCC composite, (g) and (h) electrochemical cyclic performance, and (i) and (j) rate capabilities of electrodes.[109]

Source: Reprinted with permission from Ref. [109]. Copyright 2014. Elsevier.

TABLE 12.4 Various GCNTs and Doped GCNTs Hybrid Showing Electrochemical Performance in LIBs.[58]

Sl. no.	Material and doping material	Charge/Discharge capacity (mA h/g)	Rate (mA/g)	Voltage (V)	CE (%)	Capacity of retention
1	GCNTs	526.26/900	100	0.01–3	100	99%
2	Nitrogen-doped graphene–CNTs–Mn_3O_4	1394.6/1572.54	250	0.005–3	99.6	–
3	SnO_2–graphene–carbon nanotube	823/1043	100	0.005–3	–	85.5%
4	MoS_2/graphene foam/carbon nanotubes (GF/CNT)	823/1368	200	0.01–3	99	81.3%
5	$CoSnO_3$/graphene/CNTs	1363.9/1830	100	0.005–3	–	99%

TABLE 12.5 Various Applications of MO-rGO-Coupled Composites.[110]

Sl. no.	Nanocomposite composition	Applications
1	Fe_3O_4/rGO composite	Supercapacitor
2	Co_3O_4-reduced graphene oxide	High-performance supercapacitor electrodes
3	Heterostructure of MoS_2–rGO	High-performance supercapacitors
4	3D MoS_2@CNT/rGO network composites	Advanced flexible supercapacitors
5	MoS_2 nanosheet arrays rooted on hollow rGO spheres	Anode material for enhanced lithium-storage performance
6	MoS_2 nanosheet arrays rooted on hollow rGO spheres	As electrode in supercapacitors
7	Porous Nb_4N_5/rGO nanocomposites	Ultrahigh-energy-density lithium-ion hybrid capacitor
8	Mesoporous H–Nb_2O_5/rGO nanocomposites	Latest Li-ion hybrid supercapacitors

(RBs) is heavily reliant on the synthesis of nanomaterials with outstanding storage properties. Metal–organic frameworks (MOFs) have shown attentions for others as new kinds of porous material caused by their probable uses in nonlinear optics, ion exchange, and gas storage capabilities.[116–120] As illustrated in Figure 12.5a, the CV was implemented to analyze the reaction mechanism in electrochemical cycling in half-cell assembly (Li/Cu_{ox}–rGO) between 0.005 and 3 V against Li at a scan rate of 0.1 mV/s. It was subjected to GCD tests at a constant current density of 200 mA/g, as observed in Figure 12.5b. In Figure 12.5c, the cycling performance of several electrodes (GO, MOF, CuO–rGO blank, and Cu_{ox}–rGO) is mentioned at a current density of

200 mA/g. It can be observed that the electrode with Cu$_{ox}$–rGO exhibits the best electrochemical performance with a reversible capacity of 1490 mA h/g after 220 cycles as compared to other samples. Figure 12.5d shows that, in the interim, the Cu$_{ox}$–rGO electrode shows exceptional rate capability at various current densities of 200–4000 mA/g. Furthermore, moving back to the current density of 200 mA/g, the cell also shows good retention of capacity.

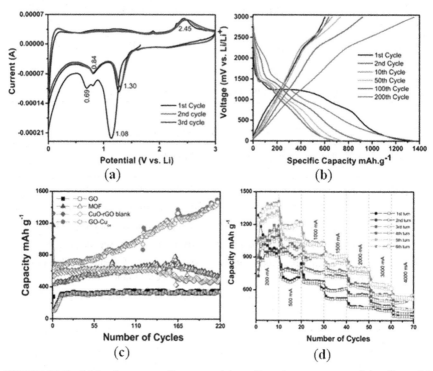

FIGURE 12.5 Li-ion battery performance: (a) cyclic voltammograms of the Cu$_{ox}$–rGO composite electrode, (b) discharge–charge curves of the Cu$_{ox}$–rGO composite electrode, (c) cycle performance of various electrodes, (d) cycle performance of the Cu$_{ox}$–rGO composite electrode at applied current rates.[4]

Source: Reprinted with permission from Ref. [4]. Copyright 2016. Royal Society of Chemistry.

As shown in Figure 12.6, from the EIS, it is evident that as an anode material in batteries, there are benefits if Cu$_{ox}$–rGO is used. The plot of Cu$_{ox}$–rGO in addition to the other electrode systems is represented in Figure 12.6a and is known as the Nyquist plot. Nyquist plot demonstrates the presence of both diffusion and kinetic components, as well as the presence of two profound semicircles in the regions of middle- and upper-frequency, which could be authorized to the charge transfer resistance and resistance of the solid electrolyte interface, respectively, whereas diffusion resistance may be assigned in the lower frequency region by a straight line. To inquest the change in the EIS spectra, EIS spectra of the electrode before and after the 1st, 15th, 50th, 100th, 150th, 200th, 250th, and 300th cycles are represented in Figure 12.6b. The EIS analysis reveals the fact that the charge resistance is directly proportional to cycling, which can be obvious in the lower frequency region where the peaks disappear.

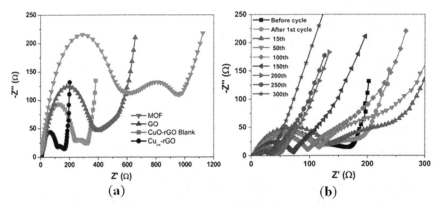

FIGURE 12.6 Li-ion battery performance in terms of electrochemical impedance spectra: (a) for various electrodes, (b) for Cu_{ox}–rGO before cyclic performance and after applied cycles.[4]

Source: Reprinted with permission from Ref. [4]. Copyright 2016. Royal Society of Chemistry.

12.3 CONCLUSIONS

Carbon-based nanofillers decorated with metal/metal oxide nanoparticles and conducting polymers are felicitously an integral part of energy storage devices. Still, a lot of developments are needed in terms of high-power output, long charge–discharge cycle life, and high-power output of batteries showing good permanence without any appreciable decrease in specific energy. Although in the areas of asymmetric supercapacitors, hybrid supercapacitors, pseudo-supercapacitors, and Li-ion RBs, tremendous achievement has been attained, in terms of structural stability of electrode material, requirement of fast electron transfer, highly conducting electrolyte, steady cyclic stability up to 4000 charge–discharge cycles, light in weight, soaring specific energy and charge storage capacity, and so on. There is still much to be achieved. We need to initiate the production of energy storage devices that can inhibit the energy density of a battery without compromising cycling stability and power density, which persist as a basic obstacle to making energy storage devices such as supercapacitors as discrete power sources. It will help us in the field of energy sources for our future advancement, along with the safekeeping of the sources for longer period.

KEYWORDS

- **graphene**
- **energy storage**
- **electrochemical properties**
- **lithium-ion batteries**
- **supercapacitor**

REFERENCES

1. Dubey, R.; Guruviah, V. *Ionics* **2019,** *25* (4), 1419–1445.
2. Conway, B. E. *Electrochemical Supercapacitors*; Springer: Massachusetts, USA, 1999.
3. Burke, A. *J. Power Sources* **2000,** *91*, 37–50.
4. Ramaraju, B.; Li, C. H.; Prakash, S.; Chen, C. C. *Chem. Commun.* **2016,** *52* (5), 946–949.
5. Dulyaseree, P.; Yordsri, V.; Wongwiriyapan, W. *Jpn. J. Appl. Phys.* **2016,** *55* (2S), 02BD05.
6. Frackowiak, E.; Delpeux, S.; Jurewicz, K.; Szostak, K.; Cazorla-Amoros, D.; Beguin, F. *Chem. Phys. Lett.* **2002,** *361* (1–2), 35–41.
7. Poizot, P. L. S. G.; Laruelle, S.; Grugeon, S.; Dupont, L.; Tarascon, J. M. *Nature* **2000,** *407* (6803), 496–499.
8. Cabana, J.; Monconduit, L.; Larcher, D.; Palacin, M. R. *Adv. Mater.* **2010,** *22* (35), E170-E192.
9. Reddy, M. V.; Subba Rao, G. V.; Chowdari, B. V. R. *Chem. Rev.* **2013,** *113* (7), 5364–5457.
10. Tian, Y.; Yu, Z.; Cao, L.; Zhang, X. L.; Sun, C.; Wang, D. W. *J. Energy Chem.* **2021,** *55*, 323–344.
11. Allen, M. J.; Tung, V. C.; Kaner, R. B. *Chem. Rev.* **2010,** *110*, 132–145.
12. Choi, W.; Lahiri, I.; Seelaboyina, R.; Kang, Y. S. *Crit. Rev. Solid State Mater. Sci.* **2010,** *35* (1), 52–71.
13. Novoselov, K. S.; Geim, A. K.; Morozov, S. V.; Jiang, D. E.; Zhang, Y.; Dubonos, S. V.; Grigorieva, I. V.; Firsov, A. A. *Science* **2004,** *306* (5696), 666–669.
14. Neto, A. C.; Guinea, F.; Peres, N. M.; Novoselov, K. S.; Geim, A. K. *Rev. Mod. Phys.* **2009,** *81* (1), 109.
15. Brownson, D. A.; Kampouris, D. K.; Banks, C. E. *J. Power Sources* **2011,** *196* (11), 4873–4885.
16. Hu, J.; Kang, Z.; Li, F.; Huang, X. *Carbon* **2014,** *67*, 221–229.

17. Novoselov, K. S.; Geim, A. K.; Morozov, S. V.; Jiang, D.; Katsnelson, M. I.; Grigorieva, I.; Dubonos, S.; Firsov, A. *Nature* **2005**, *438* (7065), 197–200.

18. Geim, A. K.; Novoselov, K. S. The rise of graphene. In Nanoscience and Technology: A Collection of Reviews from Nature Journals, World Scientific Publishing, New Jersey, USA, 2009; pp 11–19.

19. Geim, A. K. *Science* **2009**, *324* (5934), 1530–1534.

20. Dreyer, D. R.; Ruoff, R. S.; Bielawski, C. W. *Angew. Chem. Int. Ed.* **2010**, *49* (49), 9336–9344.

21. Fal'Ko, V. I.; Geim, A. K. *Eur. Phys. J.: Spec. Top* **2007**, *148* (1), 1–4.

22. Zhang, Y.; Tan, Y. W.; Stormer, H. L.; Kim, P. *Nature* **2005**, *438* (7065), 201–204.

23. Kopelevich, Y.; Esquinazi, P. *Adv. Mater.* **2007**, *19* (24), 4559–4563.

24. Wu, Z. S.; Zhou, G.; Yin, L. C.; Ren, W.; Li, F.; Cheng, H. M. *Nano Energy* **2012**, *1* (1), 107–131.

25. Dreyer, D. R.; Park, S.; Bielawski, C. W.; Ruoff, R. S. *Chem. Soc. Rev.* **2010**, *39* (1), 228–240.

26. Compton, O. C.; Cranford, S. W.; Putz, K. W.; An, Z.; Brinson, L. C.; Buehler, M. J.; Nguyen, S. T. *ACS Nano* **2012**, *6* (3), 2008–2019.

27. Saxena, S.; Tyson, T. A.; Shukla, S.; Negusse, E.; Chen, H.; Bai, J. *Appl. Phys. Lett.* **2012**, *99* (1), 013104.

28. Ganguly, A.; Sharma, S.; Papakonstantinou, P.; Hamilton, J. *J. Phys. Chem. C* **2011**, *115* (34), 17009–17019.

29. Wang, S.; Chia, P. J.; Chua, L. L.; Zhao, L. H.; Png, R. Q.; Sivaramakrishnan, S.; Zhou, M.; Goh, R. G. S.; Friend, R. H.; Wee, A. T. S.; Ho, P. K. H. *Adv. Mater.* **2008**, *20* (18), 3440–3446.

30. Yang, Y.; Wang, J.; Zhang, J.; Liu, J.; Yang, X.; Zhao, H. *Langmuir* **2009**, *25* (1814), 11808–11811.

31. Eda, G.; Fanchini, G.; Chhowalla, M. *Nat. Nanotechnol.* **2008**, *3* (5), 270–274.

32. Mattevi, C.; Eda, G.; Agnoli, S.; Miller, S.; Mkhoyan, K. A.; Celik, O.; Mastrogiovanni, D.; Cranozzi, C.; Carfunkel, E.; Chhowalla, M. *Adv. Funct. Mater.* **2009**, *19*, 2577–2583.

33. Chen, D.; Feng, H.; Li, J. *Chem. Rev.* **2012**, *112* (11), 6027–6053.

34. Kim, F.; Cote, L. J.; Huang, J. *Adv. Mater.* **2010**, *22* (17), 1954–1958.

35. Kim, J.; Cote, L. J.; Kim, F.; Yuan, W.; Shull, K. R.; Huang, J. *J. Am. Chem. Soc.* **2010**, *132* (23), 8180–8186.

36. Shao, J. J.; Lv, W.; Yang, Q. H. *Adv. Mater.* **2014**, *26* (32), 5586–5612.

37. Li, F.; Jiang, X.; Zhao, J.; Zhang, S. *Nano Energy* **2015**, *16*, 488–515.

38. Geim, A. K. *Rev. Mod. Phys.* **2011**, *83* (3), 851–862.

39. Rai, S.; Bhujel, R.; Khadka, M.; Chetry, R. L.; Swain, B. P.; Biswas, J. *Mater. Today Chem.* **2021**, *20*, 100472.

40. Mao, S.; Pu, H.; Chen, J. *RSC Adv.* **2012**, *2* (7), 2643–2662.

41. Hernandez, Y.; Nicolosi, V.; Lotya, M.; Blighe, F. M.; Sun, Z.; De, S.; McGovern, I. T.; Holland, B.; Byrne, M.; Gun'Ko, Y. K.; Boland, J. J.; Niraj, P.; Duesberg, G.; Krishnamurthy, S.; Goodhue, R.; Hutchison, J.; Scardaci, V.; Ferrari, A. C.; Coleman, J. N. *Nat. Nanotechnol.* **2008**, *3* (9), 563–568.

42. Malik, S.; Vijayaraghavan, A.; Erni, R.; Ariga, K.; Khalakhan, I.; Hill, J. P. *Nanoscale* **2010**, *2* (10), 2139–2143.

43. Bo, Z.; Yu, K.; Lu, G.; Wang, P.; Mao, S.; Chen, J. *Carbon* **2011**, *49* (6), 1849–1858.

44. Reina, A.; Jia, X.; Ho, J.; Nezich, D.; Son, H.; Bulovic, V.; Dresselhaus, M. S.; Kong, J. *Nano Lett.* **2009**, *9* (1), 30–35.

45. Sun, Z.; Yan, Z.; Yao, J.; Beitler, E.; Zhu, Y.; Tour, J. M. *Nature* **2010,** *468* (7323), 549–552.
46. Sutter, P. W.; Flege, J. I.; Sutter, E. A. *Nat. Mater.* **2008,** *7*, 406–411.
47. Hu, C.; Zhai, X.; Liu, L.; Zhao, Y.; Jiang, L.; Qu, L. *Sci. Rep.* **2013,** *3*, 2065.
48. Becerril, H. A.; Mao, J.; Liu, Z.; Stoltenberg, R. M.; Bao, Z.; Chen, Y. *ACS Nano* **2008,** *2* (3), 463–470.
49. William, S.; Hummers J.; Offeman R. E. *J. Am. Chem. Soc.* **1958,** *80*, 1339.
50. Schniepp, H. C.; Li, J. L.; McAllister, M. J.; Sai, H.; Herrera-Alonso, M.; Adamson, D. H.; Prud'homme, R. K.; Car, R.; Saville, D. A.; Aksay, I. A. *J. Phys. Chem. B* **2006,** *110* (17), 8535–8539.
51. Stankovich, S.; Dikin, D. A.; Piner, R. D.; Kohlhaas, K. A.; Kleinhammes, A.; Jia, Y.; Wu, Y.; Nguyen, S. T.; Ruoff, R. S. *Carbon* **2007,** *45* (7), 1558–1565.
52. McAllister, M. J.; Li, J. L.; Adamson, D. H.; Schniepp, H. C.; Abdala, A. A.; Liu, J.; Herrera-Alonso, M.; Milius, D. L.; Car, R.; Prud'homme, R. K.; Aksay, I. A. *Chem. Mater.* **2007,** *19* (18), 4396–4404.
53. Szabó, T.; Berkesi, O.; Forgó, P.; Josepovits, K.; Sanakis, Y.; Petridis, D.; Dékány, I. *Chem. Mater.* **2006,** *18* (11), 2740–2749.
54. Zhao, J.; Liu, L.; Li, F. *Graphene Oxide: Physics and Applications*; Springer: Berlin, Heidelberg, Germany, 2015.
55. Lahaye, R. J. W. E.; Jeong, H. K.; Park, C. Y.; Lee, Y. H. *Phys. Rev. B* **2009,** *79* (12), 125435.
56. Yang, J.; Gunasekaran, S. *Carbon* **2013,** *51*, 36–44.
57. Nandi, D.; Mohan, V. B.; Bhowmick, A. K.; Bhattacharyya, D. *J. Mater. Sci.* **2020,** *55* (15), 6375–6400.
58. Jyoti, J.; Gupta, T. K.; Singh, B. P.; Sandhu, M.; Tripathi, S. K. *J. Energy Storage* **2022,** *50*, 104235.
59. Adusei, P. K.; Kanakaraj, S. N.; Gbordzoe, S.; Johnson, K.; DeArmond, D.; Hsieh, Y. Y.; Fang, Y.; Mishra, S.; Phan, N.; Alvarez, N. T.; Shanov, V. *Electrochim. Acta* **2019,** *312*, 411–423.
60. Niu, Z.; Zhang, Y.; Zhang, Y.; Lu, X.; Liu, J. *J. Alloys Compounds* **2020,** *820*, 153114.
61. Jyoti, J.; Singh, B. P.; Tripathi, S. K. *J. Energy Storage* **2021,** *43*, 103112.
62. Meyyappan, M. *J. Vac. Sci. Technol. A* **2013,** *31* (5), 050803.
63. Zhu, L.; Zhang, S.; Cui, Y.; Song, H.; Chen, X. *Electrochim. Acta* **2013,** *89*, 18–23.
64. Roy, N.; Sengupta, R.; Bhowmick, A. K. *Prog. Polym. Sci.* **2012,** *37* (6), 781–819.
65. Xiao, Y.; Huang, L.; Zhang, Q.; Xu, S.; Chen, Q.; Shi, W. *Appl. Phys. Lett.* **2015,** *107* (1), 013906.
66. Mondal, T.; Bhowmick, A. K.; Krishnamoorti, R. *RSC Adv.* **2014,** *4* (17), 8649–8656.
67. Purushothaman, K. K.; Saravanakumar, B.; Babu, I. M.; Sethuraman, B.; Muralidharan, G. *RSC Adv.* **2014,** *4* (45), 23485–23491.
68. Ejigu, A.; Fujisawa, K.; Spencer, B. F.; Wang, B.; Terrones, M.; Kinloch, I. A.; Dryfe, R. A. *Adv. Funct. Mater.* **2018,** *28* (48), 1804357.
69. Liu, Y.; Ying, Y.; Mao, Y.; Gu, L.; Wang, Y.; Peng, X. *Nanoscale* **2013,** *5* (19), 9134–9140.
70. Paek, S. M.; Yoo, E.; Honma, I. *Nano Lett.* **2009,** *9* (1), 72–75.
71. Sun, X.; Zhou, C.; Xie, M.; Sun, H.; Hu, T.; Lu, F.; Scott, S. M.; George, S. M.; Lian, J. J. Mater. Chem. A, 2014 *2* (20), 7319–7326.
72. Du, J.; Lai, X.; Yang, N.; Zhai, J.; Kisailus, D.; Su, F.; Wang, D.; Jiang, L. *ACS Nano* **2011,** *5* (1), 590–596.
73. El-Hout, S. I.; Chen, C.; Liang, T.; Yang, L.; Zhang, J. *Mater. Chem. Phys.* **2017,** *198*, 99–106.

74. Huang, X.; Zhou, X.; Zhou, L.; Qian, K.; Wang, Y.; Liu, Z.; Yu, C. *ChemPhysChem* **2011,** *12* (2), 278–281.

75. Kim, I. Y.; Lee, J. M.; Kim, T. W.; Kim, H. N.; Kim, H. I.; Choi, W.; Hwang, S. J. *Small* **2012,** *8* (7), 1038–1048.

76. Rai, A. K.; Anh, L. T.; Gim, J.; Mathew, V.; Kang, J.; Paul, B. J.; Singh, N. K.; Song, J.; Kim, J. *J. Power Sources* **2013,** *244,* 435–441.

77. Hu, T.; Xie, M.; Zhong, J.; Sun, H. T.; Sun, X.; Scott, S.; George, S. M.; Liu, C. S.; Lian, J. *Carbon* **2014,** *76,* 141–147.

78. Madhu, R.; Dinesh, B.; Chen, S. M.; Saraswathi, R.; Mani, V. *RSC Adv.* **2015,** *5* (67), 54379–54386.

79. Yin, Z.; Wu, S.; Zhou, X.; Huang, X.; Zhang, Q.; Boey, F.; Zhang, H. *Small* **2010,** *6* (2), 307–312.

80. Sun, W.; Li, H.; Wang, Y. *Rep. Electrochem.* **2015,** *5,* 1–19.

81. Hurtado, R. B.; Cortez-Valadez, M.; Aragon-Guajardo, J. R.; Cruz-Rivera, J. J.; Martínez-Suárez, F.; Flores-Acosta, M. *Arab. J. Chem.* **2020,** *13* (1), 1633–1640.

82. Zhang, H.; Hines, D.; Akins, D. L. *Dalton Trans.* **2014,** *43* (6), 2670–2675.

83. Zhang, J.; Jiang, J.; Zhao, X. S. *J. Phys. Chem. C* **2011,** *115* (14), 6448–6454.

84. Zhang, Y.; Yuan, X.; Wang, Y.; Chen, Y. *J. Mater. Chem.* **2012,** *22* (15), 7245–7251.

85. Al-Marri, A. H.; Khan, M.; Shaik, M. R.; Mohri, N.; Adil, S. F.; Kuniyil, M.; Alkhathlan, H. Z.; Al-Warthan, A.; Tremel, W.; Tahir, M. N.; Khan, M.; Siddiqui, M. R. H. *Arab. J. Chem.* **2016,** *9* (6), 835–845.

86. Gao, Z.; Wang, J.; Li, Z.; Yang, W.; Wang, B.; Hou, M.; He, Y.; Liu, Q.; Mann, T.; Yang, P.; Zhang, M.; Liu, L. *Chem. Mater.* **2011,** *23* (15), 3509–3516.

87. Zhang, K.; Zhang, L. L.; Zhao, X. S.; Wu, J. *Chem. Mater.* **2010,** *22* (4), 1392–1401.

88. Hummers, Jr., W. S.; Offeman, R. E. *J. Am. Chem. Soc.* **1958,** *80* (6), 1339–1339.

89. Segal, M. *Nat. Nanotechnol.* **2009,** *4* (10), 612–614.

90. El-Kady, M. F.; Strong, V.; Dubin, S.; Kaner, R. B. *Science* **2012,** *335* (6074), 1326–1330.

91. Wang, G.; Zhang, L.; Zhang, J. *Chem. Soc. Rev.* **2012,** *41* (2), 797–828.

92. Anwar, A. W.; Majeed, A.; Iqbal, N.; Ullah, W.; Shuaib, A.; Ilyas, U.; Bibi, F.; Rafique, H. M. *J. Mater. Sci. Technol.* **2015,** *31* (7), 699–707.

93. Wu, Z. S.; Wang, D. W.; Ren, W.; Zhao, J.; Zhou, G.; Li, F.; Cheng, H. M. *Adv. Funct. Mater.* **2010,** *20* (20), 3595–3602.

94. Yan, J.; Fan, Z.; Wei, T.; Qian, W.; Zhang, M.; Wei, F. *Carbon* **2010,** *48* (13), 3825–3833.

95. Zhu, X.; Zhu, Y.; Murali, S.; Stoller, M. D.; Ruoff, R. S. *ACS Nano* **2011,** *5* (4), 3333–3338.

96. Wu, Q.; Xu, Y.; Yao, Z.; Liu, A.; Shi, G. *ACS Nano* **2010,** *4* (4), 1963–1970.

97. Chen, K.; Chen, L.; Chen, Y.; Bai, H.; Li, L. *J. Mater. Chem.* **2012,** *22* (39), 20968–20976.

98. Ji, L.; Rao, M.; Zheng, H.; Zhang, L.; Li, Y.; Duan, W.; Guo, J.; Cairns, E. J.; Zhang, Y. *J. Am. Chem. Soc.* **2011,** *133* (46), 18522–18525.

99. Chabot, V.; Higgins, D.; Yu, A.; Xiao, X.; Chen, Z.; Zhang, J. *Energy Environ. Sci.* **2014,** *7* (5), 1564–1596.

100. Song, M. K.; Park, S.; Alamgir, F. M.; Cho, J.; Liu, M. *Mater. Sci. Eng.: R: Rep.* **2011,** *72* (11), 203–252.

101. Leung, P. K.; Li, X.; de León, C. P.; Berlouis, L.; Low, C. T. J.; Walsh, F. C. *RSC Adv.* **2012,** *2* (27), 10125–10156.

102. Wang, W.; Luo, Q.; Li, B.; Wei, X.; Li, L.; Yang, Z. *Adv. Funct. Mater.* **2013,** *23* (8), 970–986.

103. Weber, A. Z.; Mench, M. M.; Meyers, J. P.; Ross, P. N.; Gostick, J. T.; Liu, Q. *J. Appl. Electrochem.* **2011,** *41* (10), 1137–1164.

104. Blasi, O. D.; Briguglio, N.; Busacca, C.; Ferraro, M.; Antonucci, V.; Blasi, A. D. *Appl. Energy* **2015,** *147*, 74–81.

105. Kazda, T.; Vondrák, J.; Di Noto, V.; Sedlaříková, M.; Čudek, P.; Omelka, L.; Šafaříková, L.; Kašpárek, V. J. *Solid State Electrochem.* **2015,** *19* (6), 1579–1590.

106. Pagot, G.; Bertasi, F.; Nawn, G.; Negro, E.; Carraro, G.; Barreca, D.; Maccato, C.; Polizzi, S.; Noto, V. D. *Adv. Funct. Mater.* **2015,** *25* (26), 4032–4037.

107. Meng, C.; Gall, O. Z.; Irazoqui, P. P. Biomed. *Microdevices* **2013,** *15* (6), 973–983.

108. Jung, N.; Kwon, S.; Lee, D.; Yoon, D. M.; Park, Y. M.; Benayad, A.; Choi, J. Y.; Park, J. S. *Adv. Mater.* **2013,** *25* (47), 6854–6858.

109. Zhang, J.; Xie, Z.; Li, W.; Dong, S.; Qu, M. *Carbon* **2014,** *74*, 153–162.

110. Kumar, H.; Sharma, R.; Yadav, A.; Kumari, R. *J. Energy Storage* **2021,** *33*, 102032.

111. Zeng, Y.; Yu, M.; Meng, Y.; Fang, P.; Lu, X.; Tong, Y. *Adv. Energy Mater.* **2016,** *6* (24), 1601053.

112. Simon, P.; Gogotsi, Y. *Philos. Trans. Royal Soc. A: Math., Phys. Eng. Sci.* **2010,** *368* (1923), 3457–3467.

113. Xie, K.; Li, J.; Lai, Y.; Lu, W.; Liu, Y.; Zhou, L.; Huang, H. *Electrochem. Commun.* **2011,** *13* (6), 657–660.

114. Zheng, S.; Li, X.; Yan, B.; Hu, Q.; Xu, Y.; Xiao, X.; Xue, H.; Pang, H. *Adv. Energy Mater.* **2017,** *7* (18), 1602733.

115. Liu, J.; Guan, C.; Zhou, C.; Fan, Z.; Ke, Q.; Zhang, G.; Liu, C.; Wang, J. *Adv. Mater.* **2016,** *28* (39), 8732–8739.

116. Yaghi, O. M.; Li, G.; Li, H. *Nature* **1995,** *378* (6558), 703–706.

117. Kitaura, R.; Kitagawa, S.; Kubota, Y.; Kobayashi, T. C.; Kindo, K.; Mita, Y.; Matsuo, A.; Kobayashi, M.; Chang, H. C.; Ozawa, T. C.; Suzuki, M.; Sakata, M.; Takata, M. *Science* **2002,** *298* (5602), 2358–2361.

118. Chae, H. K.; Siberio-Perez, D. Y.; Kim, J.; Go, Y.; Eddaoudi, M.; Matzger, A. J.; O'Keeffe, M.; Yaghi, O. M. *Nature* **2004,** *427* (6974), 523–527.

119. Hasegawa, S.; Horike, S.; Matsuda, R.; Furukawa, S.; Mochizuki, K.; Kinoshita, Y.; Kitagawa, S. *J. Am. Chem. Soc.* **2007,** *129* (9), 2607–2614.

120. Dincă, M.; Long, J. R. *Angew. Chem. Int. Ed.* **2008,** *47* (36), 6766–6779.

PART III
Porous Materials for Biomedical Applications

CHAPTER 13

Porous Composites and Nanofibers with Enhanced Biomedical Applicability

ADRIJA GHOSH[1], ARUNA KUMAR BARICK[2], and
DIPANKAR CHATTOPADHYAY[1,3]

[1]Department of Polymer Science and Technology, University of Calcutta, Kolkata, West Bengal, India

[2]Department of Chemistry, Veer Surendra Sai University of Technology, Siddhi Vihar, Burla, Sambalpur, Odisha, India

[3]Center for Research in Nanoscience and Nanotechnology, Acharya Prafulla Chandra Roy Sikhsha Prangan, University of Calcutta, Kolkata, West Bengal, India

ABSTRACT

In recent years, polymeric nanofibers have gained great interest because of their multifaceted functions. There are several ways of preparing polymeric nanofibers which include sonochemical method, centrifugal jet spinning, blow spinning, melt blowing, and electrospinning. Polymeric nanofibers have a wide range of biomedical applications owing to their unique characteristics such as high porosity, large aspect ratio, high permeability, easy functionalization methods, and good mechanical strength. This chapter provides an outlook on the various methods of fabricating polymeric nanofibers and their biomedical uses, for example, tissue engineering, drug/ gene delivery, wound dressing, medical prostheses, and immunosensing. In addition to that, it concisely discusses the challenges and future perspectives of the prepared polymeric nanofibers and using them for practical biomedical applications.

Mechanics and Physics of Porous Materials: Novel Processing Technologies and Emerging Applications.
Chin Hua Chia, Tamara Tatrishvili, Ann Rose Abraham, & A. K. Haghi (Eds.)
© 2024 Apple Academic Press, Inc. Co-published with CRC Press (Taylor & Francis)

13.1 INTRODUCTION

Nanofibers, as the name suggests, are one-dimensional fiber-like structures having diameter in the range of nanometers. Such features provide them with unique characteristics such as high porosity, large aspect ratio, high permeability, ease of functionalization methods, and good mechanical strength. Nanofibers can be prepared from both natural and man-made polymers. Due to its various unique properties, nanofibrous mats have applications in the field of wastewater treatment,[1] protective devices,[2] conducting materials,[3] and energy storing devices.[4] Other than these, it also has numerous biomedical applications.[5]

The similarity in structure of the majority of organs and tissues with nanofibers drives the passion of most researchers to explore the applications of nanofibers in the field of biomedical. Fibrous proteins like collagen, laminin, and elastin, which constitute the extracellular matrix (ECM) have structures and properties resembling to that of nanofibers. ECM helps in cell attachment, its growth, and movement. It also facilitates the restoration of damaged tissues. Similarly, nanofibers promote the proliferation of cells, migration, and other aspects. Nanofibers can be used for tissue engineering, wound dressings, drug delivery, and medical prostheses. It also works as biosensors. Biocompatibility of the polymers is an important aspect for their use in the biomedical field. Incorporation of nanoparticles in the fiber matrix helps in targeted drug delivery. It might help in imparting antimicrobial and antioxidant properties to the fiber. Drugs are loaded into nanofibers for wound-healing purposes. Inclusion of drugs in the matrix of the polymer leads to its sustained release in the wound area. This helps in prolonging the therapeutic effect of the drugs. Thus, easy encapsulation of particles in the fibrous matrix enhances the remedial efficacy of the fibers.

Recently, efforts have been made to develop smart nanofibers that can function in response to external stimuli. Smart polymers or stimuli-responsive polymers are a class of polymers that mimic the nature of living beings to react to external stimuli. Thus, these polymers can respond to stimuli like pH, temperature, light, glucose, and so on. This feature is extremely important in the field of nanotheranostics, as it aids in both tissue-specific drug deliveries along with the intake of drugs in adequate amounts that is necessary for the body. For example, while treating diseases like cancer targeting tumor cells and not healthy cells is significant. Thus, tissue-specific drug delivery is a crucial need. Since, tumor cells have lower pH than normal cells, pH-responsive drug delivery would lead to effective cancer treatment. Also, over intake of drugs like insulin might result in hypoglycemic conditions in diabetic patients. Thus, glucose-responsive drug delivery systems can solve this issue. In recent years, nanofiber-based wound dressing materials have been developed that deliver drugs only in response to certain stimuli. In upcoming years, we might witness the development of more smart polymeric nanofibers and practical use of them in biomedical field.

13.2 METHODS OF PREPARATION OF POLYMERIC NANOFIBERS

13.2.1 CENTRIFUGAL JET SPINNING

In centrifugal jet spinning, the fluid intended to be spun is taken in a rotating spin head, which has several nozzles near the side walls. On reaching an optimum value of rotating speed, centrifugal force prevails over the surface tension of spinning liquid leading to the formation of liquefied jets from the nozzle ends of spinning head. Simultaneous application of centrifugal and air frictional force results in the elongation of jets forming nanofibers ultimately. While other forces like gravitational, surface tension, and rheological might also play a vital role in its fabrication. The elongated jets are accumulated on a collector in the form of non-oven nanofibers.[6]

13.2.2 MELT BLOWING

In melt blowing process, the molten polymer is transferred into a die head by an extruder. Highly compressed and particulate-free hot air is pumped into the chambers at constant pressure. It then passes within slits, positioned on both ends of die tips and also by the side of its segment at extremely large velocity. These air streams converge together at a particular position on axis of each orifice, forming a stream of air having high velocity alongside the axis. As the molten polymer extrudate leaves the orifice, it is attenuated by this stream of hot air, which transforms into fibers on solidification under ambient conditions. Conclusively, a layer of fibers is deposited on the surface of collector.[7]

13.2.3 BLOW SPINNING

This process involves the preparation of polymer solution by dissolving them in a volatile solvent and placing it into a needle tip with the help of a pump. Highly paced stream of air is directed along the needle forming a thin jet of liquid. This leads to evaporation of the solvent and the subsequent formation of fibers, which are collected. Change in pressure during the process is converted into kinetic energy of the solution.[7]

13.2.4 ELECTROSPINNING

In this process, a polymer solution is prepared by stirring it with a solvent that dissolves the polymer. Then it is poured into a syringe, which is suspended and a metallic collector is placed below it. On application of adequate voltage, polymeric solution gets charged and an electric field is generated between the fluid and the metallic plate that results in formation of a droplet at the tip of the syringe. When field is large enough to overcome the surface tension of fluid, it forms the "Taylor cone" transforming into a jet. While the jet proceeds toward the collector, it endures bending instabilities and evaporation of the solvent takes place. This leads to the fabrication of longer fibers with smaller diameters.[8]

There are various other methods for fabricating nanofibers which have been summarized in Table 13.1.

TABLE 13.1   Different Methods of Fabricating Polymeric Nanofibers.

Methods for fabricating nanofibers	Features
Drawing	• Simple machinery
	• Discontinuous process of fabrication
Template synthesis	• Constant process
	• Easy fabrication of fibers with variable dimensions
Phase separation	• Simple machinery
	• Handy
	• Easy fabrication of fibers with varying mechanical strength
	• Restricted to specific polymers
Molecular self-assembly	• Complex process
	• Fabrication of fibers of very small diameter and length
	• Expensive
Freeze drying	• Costeffective
	• Simple process
	• Nonuniform porosity

13.3 POLYMERIC NANOFIBERS FOR TISSUE ENGINEERING

Prevalence of tissue injuries and organ failure has increased the demand for organ transplantation. Tissue engineering offers a lucrative alternative to it, by focusing on the regeneration of damaged tissues and curbing out the after-effects of organ transplantation. It aims at the preparation of 3D structured tissues, which mimic actual cells along with ECM molecules. These ECM molecules are primarily interwoven nanoscaled fibrous structures.[9]

13.3.1 VASCULAR TISSUE ENGINEERING

Vascular tissue engineering aims at fabrication of functional tissue scaffolds that can interact with cells, resulting in the formation of new blood vessels. Kong et al.[10] electrospun chondroitin sulfate incorporated gelatin (Gt)/poly-caprolactone (PCL) nanofibers, which functioned as tissue scaffolds that facilitated regeneration of blood vessels. The fiber had good anticoagulation properties and exhibited cell attachment and proliferation. Jia et al.[11] cova-lently attached hyaluronic acid with collagen and electrospun the product thus obtained. The reaction involved is represented in Figure 13.1. The scaffold fabricated was proposed for vascular tissue engineering applications.

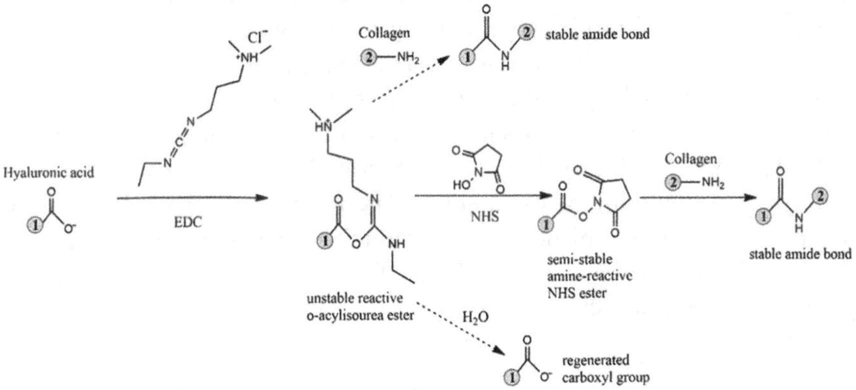

FIGURE 13.1 Schematic representation of cross-linking hyaluronic acid with collagen using EDC/NHS.[11]

Source: Reprinted with permission from Ref. [11]. Copyright 2019. Springer Nature.

13.3.2 BONE TISSUE ENGINEERING

Bone healing involves numerous osteogenic processes making the whole procedure an intricate one. Scaffolds meant to be used for bone tissue engineering should promote regeneration of collagen and cell ingrowth. Mousa et al.[12] fabricated iron acetate incorporated cellulose acetate nanofibers. The fibers were formed via electrospinning. The scaffolds thus prepared exhibited good proliferation when tested against human fetal-osteoblast cells. Sedgi et al.[13] prepared graphene oxide (GO) and zinc–curcumin-laden coaxial nanofibers having a diameter of 153 nm. These scaffolds ameliorated osteogenic functions and inhibited bacterial growth reducing the risks of infections. Awasthi et al.[14] electrospun 2D MoS_2 nanosheet-loaded PCL/zein nanofibers, which possessed good proliferation, biocompatibility, and cell adhesion.

13.3.3   NEURAL TISSUE ENGINEERING

The intricate architecture of axonal tract changes as a result of traumatic brain injury, spinal cord injury, peripheral nerve injury, or neurodegenerative illness, which inhibits growth and impairs long distance guidance. The neural tissue engineering theory has arisen to address the drawbacks of only cell-based therapy. Sadeghi et al.[15] prepared scaffolds comprising PCL, chitosan, and polypyrrole via electrospinning. The researchers performed *in-vitro* study using PC12 cells and discovered that the scaffolds had the potential for cell attachment and proliferation making it apposite for substituting neural tissues. Nune et al.[16] electrospun melanin-loaded silk fibroin nanofibrous scaffolds and tested its potential for tissue engineering by employing human neuroblastoma cells. It facilitated proliferation and adhesion to neural cells. Hu et al.[17] chemically modified polyglycerol sebacate (PGS) by copolymerizing it with methyl methacrylate for easy electrospinning of the polymer. It was further blended with gelatin for enhancing its hydrophilicity. The copolymer possessed the potential for nerve tissue regeneration and promoted cell proliferation.

13.3.4   CARTILAGE TISSUE ENGINEERING

Treatment of cartilage tissues is a complex process due to its intricate structure that comprises of special chondrocytes possessing characteristics variant to regions. Tissue engineering can be a possible solution to this. Cao et al.[18] prepared GO nanosheet incorporated polyvinyl alcohol (PVA)/chitosan composite nanofibers. The scaffolds promoted the growth of ATDC5 cells and supported cell proliferation. Thus, it could be used for cartilage tissue engineering. Silva et al.[19] prepared PGS/PCL nanofibers loaded with kartogenin (KGN) molecule, which facilitates chondrogenesis. On culturing with human bone marrow mesenchymal stem cells, the scaffolds showed good cell proliferation. Also, the co-axially electrospun nanofibers exhibited more sustained release of KGN than mono-axially electrospun scaffolds. Thus, the prepared scaffold could be used for cartilage tissue engineering.

13.4 POLYMERIC NANOFIBERS AS WOUND DRESSING MATERIALS

13.4.1 *ANTIBACTERIAL WOUND DRESSING*

Invasion by bacteria causes chronic infections, which worsens existing wounds at a higher pace. High surface area of nanofibers helps to impregnate them with antibacterial agents and thus can be used as antibacterial wound dressing. Yang et al.[20] prepared a Janus nanofiber comprising polymers polyvinylpyrrolidone (PVP) and ethyl cellulose. Ciprofloxacin and silver nanoparticles were loaded into it for imparting antibacterial properties to the prepared wound dressing. It inhibited the development of Gram-positive and Gram-negative bacteria together. Amiri et al.[21] prepared teicoplanin-loaded chitosan/polyethylene oxide (PEO) nanofibers. Nanofibers loaded with 4% of the drug showed faster wound contraction.

13.4.2 *DIABETIC WOUND HEALING*

A major part of the population is affected by diabetes and its related health complications such as nephropathy, retinopathy, and diabetic foot ulcer. Diabetic wounds take longer time period for healing than other because of inability of the diabetic patients to suppress the higher concentration of reactive oxygen species formed due to wounds. It is accompanied by a delayed inflammatory response and collapse of angiogenesis process. High porosity, good oxygen exchange capability, and superior absorbance of humidity make nanofibers ideal candidates for diabetic wound dressing materials.

Cam et al.[22] fabricated nanofibers from gelatin and bacterial cellulose. It was further loaded with two antidiabetic drugs named metformin (Gel-BC-Met) and glybenclamide (Gel-BC-Gb). The nanofibrous mat thus obtained could be used for the treatment of diabetic wounds. While nanofibers containing glybenclamide showed a faster epithelization rate than that loaded with metformin as shown in Figure 13.2.

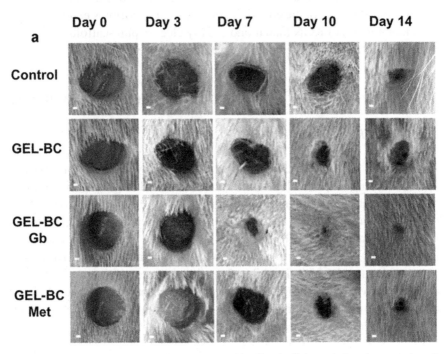

FIGURE 13.2   Representation of *in-vivo* wound healing in diabetes-induced rats on days 0, 3, 7, 10, and 14 after incision (scale bar of 10 mm).[22]

Source: Reprinted with permission from Ref. [22]. Copyright 2020. Elsevier

13.4.3   *BURN WOUND HEALING*

Bayat et al.[23] fabricated bromelain incorporated chitosan nanofibers via electrospinning. The nanofibrous film was effective as a burn wound dressing material. It healed burn wounds in rats within 21 days. Abid et al.[24] prepared a double-layered nanofiber containing two different drugs in each for treatment of burn wounds. The authors electrospun a contact layer of PEO loaded with a nervous pain relief drug gabapentin and a sodium alginate layer containing acetaminophen (gentle pain killer). The dressing quickly released gabapentin while releasing acetaminophen in a sustained manner. Thus, it had the potential to relieve pain in burn patients. Guo et al.[25] electrospun α-lactalbumin with PCL and assessed its efficacy in treating burn scars. The fibrous mats ameliorated the formation of collagen and fastened the wound-healing process.

13.5 POLYMERIC NANOFIBERS FOR MEDICAL PROSTHESES

Nanofibers are preferred for the fabrication of prostheses owing to their soft texture. Zhai et al.[26] prepared heparin-laden poly(L-lactide)-*co*-poly(ε-caprolactone) co-axial fibers (P(LLA-*co*-PCL)/heparin) via electrospinning. Tests were conducted on model dogs. The patency rate of the P(LLA-*co*-PCL)/heparin graft was found to be 100% in early period; while it was 25% in the long term. Thus, it had the potential to substitute femoral artery.

Li et al.[27] fabricated a double-layered vascular graft consisting of an inner layer made of microfibrous poly(L-lactide-ε-caprolactone) and an outer layer of PCL nanofibers. *In-vivo* tests were conducted for 18 months and results showed that the implant does not exhibit aneurysm or burst. Also, the grafts promoted vascular regeneration. Similarly, Jirofti et al.[28] prepared vascular graft from polyethylene terephthalate and polytetrafluoroethylene via co-electrospinning. *In-vivo* tests were carried out on rats and sheep models. The results revealed no aneurysm and complete patency following a period of eight months. It confirmed the use of a prepared scaffold as small-diameter vascular graft.

13.6 POLYMERIC NANOFIBERS FOR DRUG DELIVERY

Nanofibers have been extensively used as drug delivery system. Many smart nanofibers have been developed that deliver drug in response to external stimuli as represented in Table 13.2.

13.6.1 TRANSDERMAL DRUG DELIVERY

Cui et al.[38] prepared PVA/chitosan nanofibers and loaded them with drug. It was then subsequently cross-linked through glulataraldehyde. It released drug at a slow rate making it apposite for sustained delivery of drug through skin. Mendes et al.[39] fabricated hybrid nanofibers composed of chitosan blended with phospholipids as shown in Figure 13.3. Its biocompatibility was confirmed by cytotoxicity study. The nanofibrous mat had the potential of delivering several drugs transdermally including curcumin, vitamin B12, and diclofenac.

TABLE 13.2 Stimuli-Responsive Drug Delivery by Nanofibers.

Nanofiber	Stimuli	Drug	Application	Reference
Gold nanorods loaded poly(N-isopropylacrylamide)	Near-infrared	Fluorescein	Cancer treatment through chemotherapy	[29]
Poly(N-isopropylacrylamide)-co-poly(acrylic acid)/polyurethane	Thermal and pH	Nifedipine	Treatment of hypertension or angina	[30]
5,11,17,23-tetra-tert-butyl-25,27-bis(3-aminomethyl-pyridineamido)-26,28-dihydroxycalix[4]arene	pH	Doxorubicin	Tumor therapy	[31]
Poly(N-isopropylacrylamide-co-acrylamide-co-vinylpyrrolidone)	pH	Doxorubicin	Cancer treatment	[32]
Cellulose nanocrystal-poly[2-(dimethylamino)ethyl methacrylate] reinforced poly(3-hydroxybutyrate-co-3-hydroxy valerate)	Temperature and pH	Tetracycline hydrochloride	Treatment of bacterial infections	[33]
Poly(vinylidene fluoride-trifluroethylene)	Mechanoresponsive	Crystal violet	Mechanoresponsive drug delivery	[34]
Polyethyleneimine-N-isopropylacrylamide grafted cellulose	Temperature and pH	Doxorubicin	Cancer treatment	[35]
Hyperbranched polyethylenimine modified with isobutyramide/carboxylated cellulose	Temperature and pH	Doxorubicin	Cancer treatment	[36]
Polydopamine coated poly(ε-caprolactone)	pH	Doxorubicin	Cancer treatment	[37]

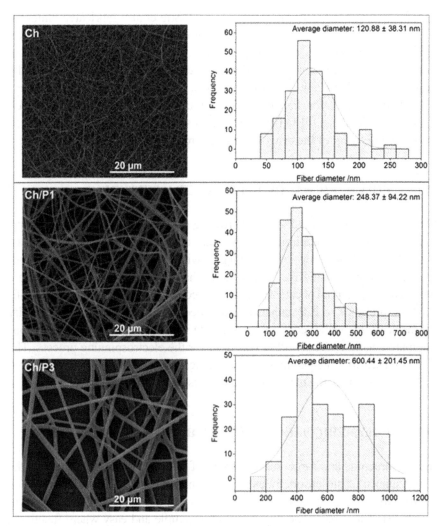

FIGURE 13.3 SEM images and along with histograms representing the diameter of the prepared chitosan (Ch), 1:1 chitosan/phospholipids (Ch/P1), and 1:3 chitosan/phospholipids (Ch/P3) nanofibers.[39]

Source: Reprinted with permission from Ref. [39]. Copyright 2016. Elsevier.

13.6.2 TRANSMUCOSAL DRUG DELIVERY

Tuğcu-Demiröz et al.[40] fabricated PVP nanofibers loaded with metronidazole. The formulation was found to be suitable for delivery of drug through vaginal mucosa. Lancina et al.[41] prepared chitosan/PEO nanofibers and encapsulated insulin within it. The nanofibers had the potential for delivering insulin through transbuccal route. Similarly, Stie et al.[42] fabricated chitosan/PEO-based nanofibers and assessed the effect of degree of deacetylation on the mucoadhesivity of the film prepared. It was concluded that with an increase in the degree of deacetylation of chitosan had no influence on the mucoadhesive behavior of nanofibers. Also, Stie et al.[43] prepared a nanofiber-film hybrid patch for transmucosal delivery of desmopressin. The patch consisted of chitosan/PEO nanofiber and a saliva-repelling film. It exhibited sustained delivery of the drug via mucosal route.

13.7   POLYMERIC NANOFIBERS FOR CELL CULTURE

Cell culture is the method used for growing cells under optimized conditions, outside their usual environment. Abazari et al.[44] prepared nanofiber mats of polyvinylidene fluoride and collagen with and without incorporation of platelet-rich plasma (PRP). Following its characterization, the scaffolds were tested for its potential of osteoinductivity. It could be concluded from the results that the presence of PRP enhanced the biocompatibility and osteoinductivity of the scaffold. Majidi et al.[45] fabricated 3D scaffolds comprising nanofibrous alginate/gelatin hydrogel. PEO was used for the electrospinning process. The prepared scaffolds helped in proliferating mesenchymal stem cells and favored cell motility. Thus, the nanofiber could be used for cell culture.

13.8   POLYMERIC NANOFIBERS FOR BIOSENSORS

Small structure and large surface areas of nanofibers help to improve the performance of biosensors by enhancing their working speed, selectivity, and sensitivity. There are mainly two approaches toward fabricating biosensors from polymeric nanofibers. The first approach is simple and easy where functional polymers are made into nanofibers having inducing function. The produced nanofibers are utilized as inducing components of the biosensors. The second approach involves synthesizing sensitive materials and depositing it on the surface of synthesized template nanofibers; followed by chemical modification.

Khalid et al.[46] fabricated nanodiamond-incorporated silk fibroin nanofibers. The negatively charged nitrogen-vacancy centres present in nanodiamond could sense the temperature and thus could assess the temperature distinctions of wounded or inflammated areas of body. It had good biocompatibility and promoted wound contraction. Finally, it could work as both biosensor and wound dressing material.

Apetrei and Camurlu[47] electrospun montmorillonite (MMT)-loaded polyacrylonitrile (PAN) nanofibers (PAN/MMT). The clay was further functionalized with dioctadecyl dimethyl ammonium chloride and methylene blue (PAN/DDAC-MMT). The prepared nanofiber was used as glucose biosensor. Modification and incorporation of MMT clay into the nanofibers ameliorated its biosensing performance and sensitivity. Figure 13.4 represents the amperometric response of the biosensors synthesized.

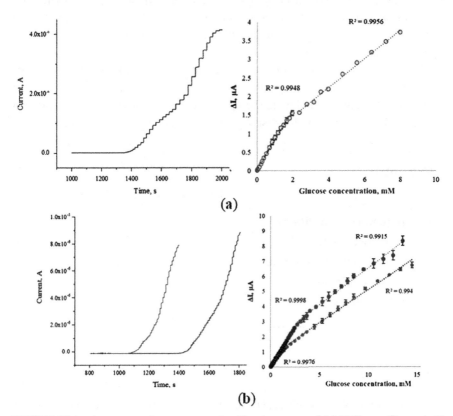

FIGURE 13.4 Amperometric response and calibration curve of (a) PAN nanofibers at pH 6.5 and 0.7 V and (b) PAN/MMT (red line, pink markers) and PAN/DDAC-MMT (black line, purple markers) nanofibers at pH 7.0 and 0.65 V against Ag/AgCl.[47]

Source: Reprinted with permission from Ref. [47]. Copyright 2020. Elsevier.

13.9 CONCLUSION AND FUTURE PERSPECTIVES

Nanofibers have various applications and can be made through multiple techniques. Among these, electrospinning has gained the most popularity as it is noncomplex and flexible in terms of fabricating fibers with different morphology. Nanofibers can be widely used for biomedical applications by correctly choosing the polymer to be used for the delivery of drugs. Despite the various advantages of the electrospinning process, difficulty to produce nanofibers in large-scale manner with desirable physical and chemical characteristics poses hurdle in advancement of its biomedical application. Clogging of spinnerets during the process of electrospinning is an undesirable part of the process, which needs to be overcome. Most of the experimental tests conducted with nanofibers are done *in vitro*. *In-vivo* assessments are necessary to confirm the practicability of using nanofibers for biomedical applications.

More research works can be directed toward developing smart nanofibrous patches that can be used for tissue engineering and explore their use for regenerative medicine and gene therapy. Also, further advancement is needed in the field of 3D printing for discovering its potential to fabricate nanofibers.

KEYWORDS

- **polymeric nanofibers**
- **electrospinning**
- **biomedical applications**
- **tissue engineering**
- **drug delivery**

REFERENCES

1. Cui, J.; Li, F.; Wang, Y.; Zhang, Q.; Ma, W.; Huang, C. *Sep. Purif. Technol.* **2020,** *250,* 117116.
2. Blachowicz, T.; Hütten, A.; Ehrmann, A. *Fibers* **2022,** *10* (6), 47.
3. Baker, C. O.; Huang, X.; Nelson, W.; Kaner, R. B. *Chem. Soc. Rev.* **2017,** *46* (5), 1510–1525.
4. Liang, J.; Zhao, H.; Yue, L.; Fan, G.; Li, T.; Lu, S.; Chen, G.; Gao, S.; Asiri, A. M.; Sun, X. *J. Mater. Chem. A* **2020,** *8* (33), 16747–16789.
5. Leung, V. Ko, F. *Polym. Adv. Technol.* **2011,** *22* (3), 350–365.
6. Zhang, X.; Lu, Y. *Polym. Rev.* **2014,** *54* (4), 677–701.
7. Yarin, A. L.; Pourdeyhimi, B.; Ramakrishna, S. Melt- and Solution Blowing. In *Fundamentals and Applications of Micro- and Nanofibers*; Cambridge University Press, Cambridge, UK, 2014; pp 89–178.

8. Ghosh, A.; Orasugh, J. T.; Chattopadhyay, D.; Ghosh, S. *Groundw. Sustain. Dev.* **2022**, *16*, 100716.

9. Nemati, S.; Kim, S. J.; Shin, Y. M.; Shin, H. *Nano Converg.* **2019**, *6*, 36.

10. Kong, X.; He, Y.; Zhou, H.; Gao, P.; Xu, L.; Han, Z.; Yang, L.; Wang, M. *Nanoscale Res. Lett.* **2021**, *16*, 62.

11. Jia, W.; Li, M.; Kang, L.; Gu, G.; Guo, Z.; Chen, Z. *J. Mater. Sci.* **2019**, *54* (15), 10871–10883.

12. Mousa, H. M.; Hussein, K. H.; Sayed, M. M.; El-Rahman, A.; Mohamed, K.; Woo, H. M. *Polymers* **2021**, *13* (8), 1339.

13. Sedghi, R.; Sayyari, N.; Shaabani, A.; Niknejad, H.; Tayebi, T. *Polymer* **2018**, *142*, 244–255.

14. Awasthi, G. P.; Kaliannagounder, V. K.; Maharjan, B.; Lee, J. Y.; Park, C. H.; Kim, C. S. *Mater. Sci. Eng.: C* **2020**, *116*, 111162.

15. Sadeghi, A.; Moztarzadeh, F.; Mohandesi, J. A. *Int. J. Biol. Macromol.* **2019**, *121*, 625–632.

16. Nune, M.; Manchineella, S.; Govindaraju, T.; Narayan, K. S. *Mater. Sci. Eng.: C* **2019**, *94*, 17–25.

17. Hu, J.; Kai, D.; Ye, H.; Tian, L.; Ding, X.; Ramakrishna, S.; Loh, X. J. *Mater. Sci. Eng.: C* **2017**, *70* (2), 1089–1094.

18. Cao, L.; Zhang, F.; Wang, Q.; Wu, X. *Mater. Sci. Eng.: C* **2017**, *79*, 697–701.

19. Silva, J. C.; Udangawa, R. N.; Chen, J.; Mancinelli, C. D.; Garrudo, F. F.; Mikael, P. E.; Cabral, J. M.; Ferreira, F. C.; Linhardt, R. J. *Mater. Sci. Eng. C* **2020**, *107*, 110291.

20. Yang, J.; Wang, K.; Yu, D. G.; Yang, Y.; Bligh, S. W. A.; Williams, G. R. *Mater. Sci. Eng.: C* **2020**, *111*, 110805.

21. Amiri, N.; Ajami, S.; Shahroodi, A.; Jannatabadi, N.; Darban, S. A.; Bazzaz, B. S. F.; Pishavar, E.; Kalalinia, F.; Movaffagh, J. *Int. J. Biol. Macromol.* **2020**, *162*, 645–656.

22. Cam, M. E.; Crabbe-Mann, M.; Alenezi, H.; Hazar-Yavuz, A. N.; Ertas, B.; Ekentok, C.; Ozcan, G. S.; Topal, F.; Guler, E.; Yazir, Y.; Parhizkar, M.; Edirisinghe, M. *Eur. Polym. J.* **2020**, *134*, 109844.

23. Bayat, S.; Amiri, N.; Pishavar, E.; Kalalinia, F.; Movaffagh, J.; Hashemi, M. *Life Sci.* **2019**, *229*, 57–66.

24. Abid, S.; Hussain, T.; Nazir, A.; Zahir, A.; Khenoussi, N. *Polym. Bull.* **2019**, *76* (12), 6387–6411.

25. Guo, X.; Liu, Y.; Bera, H.; Zhang, H.; Chen, Y.; Cun, D.; Foderà, V.; Yang, M. *ACS Appl. Mater. Interfaces* **2020**, *12* (41), 45702–45713.

26. Zhai, W.; Qiu, L. J.; Mo, X. M.; Wang, S.; Xu, Y. F.; Peng, B.; Liu, M.; Huang, J. H.; Wang, G. C.; Zheng, J. H. *J. Biomed. Mater. Res. B: Appl. Biomater.* **2019**, *107* (3), 471–478.

27. Li, W.; Wu, P.; Zhang, Y.; Midgley, A. C.; Yuan, X.; Wu, Y.; Wang, L.; Wang, Z.; Zhu, M.; Kong, D. *ACS Appl. Bio Mater.* **2019**, *2* (10), 4493–4502.

28. Jirofti, N.; Mohebbi-Kalhori, D.; Samimi, A.; Hadjizadeh, A.; Kazemzadeh, G. H. *Biomed. Mater.* **2018**, *13* (5), 055014.

29. Singh, B.; Shukla, N.; Kim, J.; Kim, K.; Park, M. H. *Pharmaceutics* **2021**, *13* (8), 1319.

30. Lin, X.; Tang, D.; Yu, Z.; Feng, Q. *J. Mater. Chem. B* **2014**, *2* (6), 651–658.

31. Ozcan, F.; Cagil, E. M. *J. Appl. Polym. Sci.* **2021**, *138* (11), 50041.

32. Salehi, R.; Irani, M.; Rashidi, M. R.; Aroujalian, A.; Raisi, A.; Eskandani, M.; Haririan, I.; Davaran, S. *Des. Monomers Polym.* **2013**, *16* (6), 515–527.

33. Chen, Y.; Abdalkarim, S. Y. H.; Yu, H. Y.; Li, Y.; Xu, J.; Marek, J.; Yao, J.; Tam, K. C. *Int. J. Biol. Macromol.* **2020**, *155*, 330–339.

34. Jariwala, T.; Ico, G.; Tai, Y.; Park, H.; Myung, N. V.; Nam, J. *ACS Appl. Bio Mater.* **2021,** *4* (4), 3706–3715.

35. Liang, Y.; Zhu, H.; Wang, L.; He, H.; Wang, S. *Carbohydr. Polym.* **2020,** *249*, 116876.

36. He, H.; Shi, X.; Chen, W.; Chen, R.; Zhao, C.; Wang, S. *J. Agric. Food Chem.* **2020,** *68* (28), 7425–7433.

37. Jiang, J.; Xie, J.; Ma, B.; Bartlett, D. E.; Xu, A.; Wang, C. H. *Acta Biomater.* **2014,** *10* (3), 1324–1332.

38. Cui, Z.; Zheng, Z.; Lin, L.; Si, J.; Wang, Q.; Peng, X.; Chen, W. *Adv. Polym. Technol.* **2018,** *37* (6), 1917–1928.

39. Mendes, A. C.; Gorzelanny, C.; Halter, N.; Schneider, S. W.; Chronakis, I. S. *Int. J. Pharm.* **2016,** *510* (1), 48–56.

40. Tuğcu-Demiröz, F.; Saar, S.; Tort, S.; Acartürk, F. *Drug Dev. Ind. Pharm.* **2020,** *46* (6), 1015–1025.

41. Lancina III, M. G.; Shankar, R. K.; Yang, H. *J. Biomed. Mater. Res. A* **2017,** *105* (5), 1252–1259.

42. Stie, M. B.; Gätke, J. R.; Wan, F.; Chronakis, I. S.; Jacobsen, J.; Nielsen, H. M. *Carbohydr. Polym.* **2020,** *242*, 116428.

43. Stie, M. B.; Gätke, J. R.; Chronakis, I. S.; Jacobsen, J.; Nielsen, H. M. *Int. J. Mol. Sci.* **2022,** *23* (3), 1458.

44. Abazari, M. F.; Soleimanifar, F.; Faskhodi, M. A.; Mansour, R. N.; Mahabadi, J. A.; Sadeghi, S.; Hassannia, H.; Saburi, E.; Enderami, S. E.; Khani, M. M.; Karizi, S. Z. *J. Cell. Physiol.* **2020,** *235* (2), 1155–1164.

45. Majidi, S. S.; Slemming-Adamsen, P.; Hanif, M.; Zhang, Z.; Wang, Z.; Chen, M. *Int. J. Biol. Macromol.* **2018,** *118*, 1648–1654.

46. Khalid, A.; Bai, D.; Abraham, A. N.; Jadhav, A.; Linklater, D.; Matusica, A.; Nguyen, D.; Murdoch, B. J.; Zakhartchouk, N.; Dekiwadia, C.; Reineck, P.; Simpson, D.; Vidanapathirana, A. K.; Houshyar, S.; Bursill, C. A.; Ivanova, E. P.; Gibson, B. C. *ACS Appl. Mater. Interfaces* **2020,** *12* (43), 48408–48419.

47. Apetrei, R. M.; Camurlu, P. *Electrochim. Acta* **2020,** *353*, 136484.

CHAPTER 14

Porous Nanocomposites for Anticancer Therapy: Comparative Study in Water and Hank's Solution by UV–Vis Spectroscopy

YULIIA KUZIV, PAVLO VIRYCH, VADIM PAVLENKO, and
NATALIYA KUTSEVOL

Taras Shevchenko National University of Kyiv, Kyiv, Ukraine

ABSTRACT

Nanoscience is an approach to generate polymer-based nanocomposite dye for PDT or photodynamic therapy. Polymer molecules can be in preference, retained in tumor, and prevent photosensitizer aggregation. Moreover, stimuli-sensitive polymers can be used for controlled drug release.

Poly(N-isopropylacrylamide) is extensively studied as a promising nano-carrier for biomedical application because of its conformational transition within the physiological temperature. Star-like shape of dextran-graft-poly-N-isopropylacrylamide (D-PNIPAM) used for the creation of multicomponent nanosystem D-PNIPAM/AuNPs/Ce6 nanosystems. These nanosystems were tested by optical absorption in water and in Hank's buffer solution within a temperature range of 25–40C and by dynamic-like scattering at 25 and 37C. It has been proved that no drastic changes are accompanied by an aggregation process in the nanosystem in Hank's saline solution in comparison with one prepared in water.

In-vitro examination of the PDT activity of the D-PNIPAM/AuNPs/Ce6 nanosystem on the culture of MT-4 human malignant lymphocytes revealed the absence of cytotoxicity without red-light irradiation. The death of lymphocytes pre-incubated with nanocomposite after laser irradiation was around of 40%.

Mechanics and Physics of Porous Materials: Novel Processing Technologies and Emerging Applications.
Chin Hua Chia, Tamara Tatrishvili, Ann Rose Abraham, & A. K. Haghi (Eds.)
© 2024 Apple Academic Press, Inc. Co-published with CRC Press (Taylor & Francis)

14.1 INTRODUCTION

Photodynamic therapy (PDT) in the last decades focuses the attention of researchers as an alternative therapy for cancer diseases. PDT is based on photochemical reactions with photosensitizing drug (PS). Most often porphyrins with subsequent light irradiation at the proper wavelength lead to the photochemical conversion of molecular oxygen ($^3O_2$) into very reactive singlet oxygen ($^1O_2$) which is a cytotoxic agent provoking the damage of tumor cells via apoptosis or necrosis. Oxidative process causes damage of nucleic acids, proteins, and lipids in the cells.[1-4] The classic PDT treatment of cancer based on an accumulation of a photosensitizer drug in tumor tissue, then illuminated with light which the cell death is activated. For efficient PDT: (1) photosensitizer should demonstrate high absorption in the 650–850-nm range where tissue penetration is high, (2) PS should have high singlet oxygen quantum yield for the photochemical reaction, (3) should have low dark toxicity, and (4) be soluble in water and should not form aggregates. That is why, it is important to create PS with improved properties. Nowadays, despite several clinically approved PS, the highly efficient and selective PS agent is still not found.[1]

Nanotechnology is a promising way for creation of highly efficient composites for PDT therapy. Polymer nanoparticles can be used as a nano-platform for drug delivery.[5] Current research and reviews have established that polymer molecules can be preferentially accumulated in tumors due to passive or receptor-mediated active targeting.[6] However, the use of traditional polymers leads to uncontrollable drug release. To overcome this problem, the stimuli-sensitive polymer as thermosensitive one can be used for controlled release of drugs.[7,8] However, for future development of stimuli-responsive nanoplatforms, it should be deep knowledge on nanocarrier biocompatibility, drug-loading capacity, nanocarrier stability, and toxicity.

It is known that poly(*N*-isopropylacrylamide) (PNIPAM) can undergo a tunable phase transition in the region of physiological temperature.[9] Several reviews related to the employment of PNIPAM with the scope of biomedicine have been reported in recent years. The most explored areas in the applied biomedical science are

- bioimaging,
- biosensors, and
- drug delivery.

However, PNIPAM is widely studied because of its solubility in water and also its "LCST" or lower critical solution temperature. This is of course very near to human body temperature according to reference.[10] The unpredictability of the hydrophilic/hydrophobic state is expected by diversified temperature (this means to be either less or greater than the critical value). Below LCST, polymers are soluble. When the temperature goes over the LCST, these polymers at the first stage undertake a phase transition; then in the second stage, it is expected to collapse and form aggregates. It is worth mentioning that the linear PNIPAM has an LCST value of approximately 32°C. Meanwhile, PNIPAM is in fact soluble at room temperature, but its phase separates at physiological temperature (37°C) and therefore can be used for controlled release of the drug for cancer therapy.[11]

We have synthesized star-shaped temperature-sensitive copolymers of dextran-poly-*N*-isopropylacrylamide (D-PNIPAM), which have a phase transition in the region of physiological temperatures.[12] It was shown that it is possible to shift the temperature of the conformational for the D-PNIPAM by 3–4°C compared to the linear PNIPAM. Thus, LCST becomes closer to human-body temperature.

The D-PNIPAM copolymer has demonstrated the high *in-vitro* efficiency as nanocarriers for anticancer drug doxorubicin.[13] Also, the nanosystems consisting gold nanoparticles (AuNPs) into D-PNIPAM water solution were synthesized and studied.[14–16] It is known that nanosized gold is practically nontoxic and has an affinity with biological factors that support vascular growth. Gold-based nanomaterials enhance the therapeutic effect of the drugs for photodynamic and photothermal therapy.[17] The ability of gold nanoparticles to penetrate the endothelium of tumor vessels provides higher selectivity for the accumulation of anticancer agents in the tumor.[18] However, the size characteristics of gold nanoparticles and the aggregative stability are very important.[18] For this purpose, polymer matrices are used, since it is the macromolecules of polymers affect the processes of formation of metal particles, controlling their size and shape and are like a steric barrier for nanoparticle aggregation.[19]

Recently, we have reported on the high efficiency of nanocomposites for PDT, consisting of gold nanoparticles (AuNPs), a photosensitizer of chlorine e6 (Ce6) incorporated into the anionic star-shape copolymer dextran-polyacrylamide-*co*-polyacrylic acid.[20,21]

The aim of the present work was to investigate the behavior of the multi-component nanosystem D-PNIPAM/AuNPs/Ce6 in water and in Hank's buffer solution used for cell-culture assays. The study was carried out at physiological temperatures; the nanocomposite was tested for its photodynamic efficacy.

14.2   MATERIALS AND METHODS

14.2.1   *THERMOSENSITIVE POLYMER*

Dextran with molecular weights $M_w = 7 \times 10^5$ g/mol from Fluka, N-isopropylacrylamide (NIPAM) from Aldrich was used for the synthesis of dextran-graft-PNIPAM copolymer. The number of grafting sites per dextran backbone was predetermined by a molar ratio of acrylamide to cerium ions, and it was equal to 15. The sample was designated as D-PNIPAM. The synthesis and characterization of D-PNIPAM with 15 grafts were reported in Ref. [12]. The molecular characteristics of D-PNIPAM are given in Table 14.1 (M_w—the weight average molecular weight of the copolymer, M_w/M_n—polydispersity of copolymer, and D_h—hydrodynamic diameter of macromolecule).

TABLE 14.1   Molecular Parameters of D-PNIPAM Copolymers.

Sample	$M_w \times 10^{-6}$ (g/mol)	M_w/M_n	$D_h 25C$ (nm)
D-PNIPAM	1.03	1.52	40

14.2.2   *NANOSYSTEMS*

14.2.2.1   *REAGENTS USED FOR NANOSYSTEMS SYNTHESIS*

Photosensitizer Ce6 was obtained from Santa Cruz Biotechnology (USA); doxorubicin hydrochloride was purchased from Sigma-Aldrich (USA); cisplatin was obtained from "EBEVE" (Austria); $NaBH_4$, $AgNO_3$, and $HAuCl_4$ were purchased from Sigma-Aldrich (USA).

Hank's balanced salt solution was purchased from Sigma-Aldrich (USA). Hank's solution was phenol red-free and consisted of two components: glucose (1.0 g/L) and $NaHCO_3$ (0.35 g/L); dimethyl sulfoxide (DMSO) was obtained from Serva (Germany).

14.2.2.2   *NANOSYSTEM: POLYMER/AUNPS*

Tetrachloroauric acid (Au precursor) and $NaBH_4$ (reductant) have been used for *in-situ* AuNPs synthesis into polymer solution. The synthesis process was reported in Refs. [14–16].

14.2.2.3 NANOSYSTEM: POLYMER/AUNPS/CE6 IN WATER AND HANK'S SOLUTION

A stock solution of Ce6 (C = 0.182 mg/mL) was prepared in DMSO. Then, 0.55 mL of this solution was added to 0.30 mL of distilled water. This mixture was added to 1.15 mL of polymer/AuNPs sol under constant stirring.

Eight milliliters of distilled water or Hank's buffer solution was added to 2 mL of obtained three-component D-PNIPAM/AuNPs/Ce6 system. The reaction mixture was stirred.

Three-component polymer/AuNPs/Ce6 and four-component nanosystems were prepared by the method described above and then were centrifuged. The absorption spectra of the supernatant showed the absence of Ce6 in the solution, thus indicating its total incorporation into the polymer nanocarrier. These nanocomposites were used for UV–vis and *in-vitro* anticancer examination.

14.2.3 METHODS

14.2.3.1 UV–VIS SPECTROSCOPY AND TRANSMISSION ELECTRON MICROSCOPY (TEM)

The UV absorption spectra of D-PNIPAM copolymer, D-PNIPAM/AuNPs, and D-PNIPAM/AuNPs/Ce6 in water and Hank's solution were obtained using a Varian Cary 50 scan UV spectrophotometer (Palo Alto, CA, USA) in the range of 200–800 nm in the temperature region of 25–40°C in 1°C. The studied systems were kept for 2 min at each temperature before measurement. *Transmission electron microscopy* measurements were carried out as reported in Ref. [12].

14.2.4 IN-VITRO EXPERIMENTS WITH MALIGNANT CELLS

Cells of the MT-4 line (human T-cell leukemia) were obtained from the cell culture bank of the R. E. Kavetsky Institute of Experimental Pathology, Oncology and Radiobiology, NAS of Ukraine. The cells were maintained in RPMI-1640 medium containing 10% fetal bovine serum at a temperature of 37°C in an incubator with 95% humidity and 5% CO_2 in air. For photodynamic treatment, cell suspensions in the logarithmic growth phase (0.5 Å ~ 10^6 mL^{-1}) were prepared in Hank's balanced salt solution. After 1.5 h of incubation of the cells at 37°C with Ce6 or with its nanocomposite, the samples were washed twice with a 10× volume of Hank's solution and exposed to a red laser (wavelength 658 nm, specific power—1.1 mW/cm², doses—1 J/cm²). After irradiation, the cells were transferred to the culture medium and incubated at 37°C for 18 h to complete the photodynamically induced apoptosis process. The cell viability was then determined by Trypan blue dye exclusion test. Each experiment was repeated at least five times.

A semiconductor laser (PMNP Photonik Plus, Cherkasy) with a wavelength of 658 nm, which coincides with one of the Ce6 absorption maxima, was used as a light source for photodynamic damage to cells.

14.3   RESULTS AND DISCUSSION

The typical TEM image of D-g-PNIPAM/AuNPs nanosystem obtained at a temperature of 25°C is presented in Figure 14.1. It is seen that AuNPs have size of $d = (8 \pm 3)$ nm and are spherical in shape. AuNPs were synthesized in the dilute aqueous solution of D-g-PNIPAM (polymer concentration was below the concentration of crossover). Thus, the distance between the neighboring PNIPAM macromolecules is larger than the macromolecule size by several times.

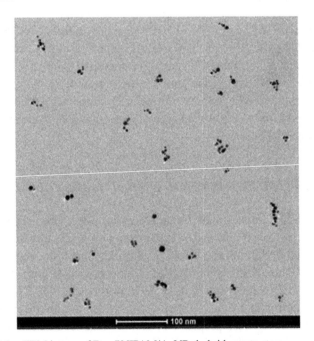

FIGURE 14.1   TEM image of D-g-PNIPAM/AuNPs hybrid nanosystem.

Figure 14.2(a, b) demonstrates the UV spectra of the D-PNIPAM copolymer solutions prepared in water and in Hank's solution, respectively. The optical absorption spectra of PNIPAM exhibit maximum absorption (λ_{max}) at 228 nm corresponding to π- and n-electron transition localized on C=O group of PNIPAM.[22] The small peak at 250–275 nm obviously corresponds to some internal molecular structures formed by C=O groups caused by hydrogen bonding. Upon heating, absorption band intensity at 250–275 nm slight increases. It testifies to the fact of some changes in the studied system in the region of LCST. It is known that the macromolecule PNIPAM becomes partially hydrophobic after LCST that leads to macrocoil conformational transition.

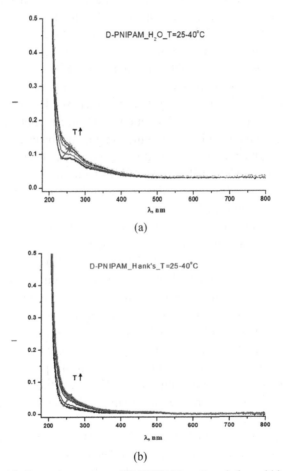

(a)

(b)

FIGURE 14.2 (a) Absorption spectra of D-PNIPAM water solution within a temperature range of 25–40°C. (b) Absorption spectra of D-PNIPAM copolymer in Hank's solution within a temperature range of 25–40°C.

Similar changes were observed for D-PNIPAM in Hank's balanced salt solution. However, these changes are less pronounced. It is not surprising as Hank's solution consists of the addition of salts which can affect the formation of internal hydrogen bond in the system.

Absorption spectra of Au sols synthesized *in situ* in the D-PNIPAM, in water, and with the addition of Hank's solution are shown in Figure 14.3(a, b). In both systems, the surface plasmon resonance (SPR) peak of Au nanoparticles at 530 nm is registered. This peak does not shift when heated. However, a slight change in the intensity of the SPR band is observed. That may also be caused by the conformational changes of the macromolecule within the LCST region leading to the change in the distance between AuNPs. In Hank's solution, the change in the intensity of the SPR under heating is less pronounced than in the aqueous solution.

It is known that the Ce6 absorption spectrum has two characteristic bands: in the region of 470–670 nm (first electronic transition) and in the region 350–440 nm (second electronic transition), the so-called Sore band.[23]

As can be seen, for the nanosystem polymer/AuNPs/Ce6 (Figure 14.4), two components appear in the absorption spectrum: a photosensitizer (Sore band at 400 nm and one Q-band 671 nm) and gold nanoparticles (530 nm). As it is noted above, the polymer matrix does not absorb in this wavelength range.

The nanocomposite D-PNIPAM/AuNPs/Ce6 in Hank's solution was studied by DLS to estimate the size of nano-object in nanosystems which can be used for *in-vitro* PDT examination.

The experiment was performed at 25 and 37°C, as the nanocomposites are prepared at room temperature, but biological testing is carried out at 37 °C. We have reported[14–16] that for D-PNIPAM/AuNPs in water solution, few scattering objects are registered: the AuNPs of 10 nm in size, D-PNIPAM macromolecules with incorporated AuNPs of 40 nm in size, and some aggregates of macromolecules. Dimensional characteristics of D-PNIPAM/AuNPs/Ce6 in Hank's solution at 25 and 37 °C are represented in Figure 14.5. It is seen that AuNPs of 10 nm in size in registered, and bimodal distribution of scattering objects of 100 and 300 nm are present in nanosytem at 25°C. Increasing the temperature to 37°C caused the decrease in size of aggregates to 100 nm. Moreover, the individual macromolecules of 20–30 nm in size are registered.

Thus, it has been proved that no drastic changes accompanied by the aggregation process in the nanosystem in Hank's saline solution, the nanosystem remains stable with heating to physiological temperature. Thus, this ternary nanosystem can be recommended to test for PDT activity.

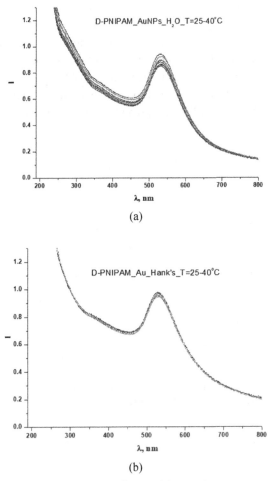

FIGURE 14.3 (a) Absorption spectra of Au sol into polymer water solution within a temperature range of 25–40°C. (b) Absorption spectra of Au sol into Hank's solution within a temperature range of 25–40°C.

For the *in-vitro* examination, the photodynamic activity of the D-PNIPAM/ AuNPs/Ce6 nanosystem on the culture of MT-4 human malignant lympho-cytes was performed (Figure 14.6) in comparison to Ce6. *In-vitro* study of the toxicity of hybrid gold–polymer composites was reported in Ref. [24] and proved low toxicity of examined compounds even in high doses.

Figure 14.6 demonstrates that in the absence of red irradiation and the D-PNIPAM/AuNPs/Ce6 does not show cytotoxicity. At the same time, the mortality of lymphocytes pre-incubated with nanocomposite after laser irradiation demonstrated 40% of the mortality of tumor cells. This result is comparable with efficacy of Ce6. But, the nanocomposite can prevent the aggregation of photosensitizer *in-vivo* experiments as well as can provide target delivery of drug to tumor cell. Moreover, for nanosystem polymer/ AuNPs under illumination with light with $\lambda = 650$ nm, the local heating in the small volume near the laser beam was registered.[25] It means that nanosystems with AuNPS during PDT can provide also hyperthermia effect.

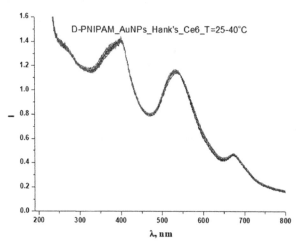

FIGURE 14.4 Absorption spectra of nanosystems D-PNIPAM/AuNPs/Ce6 in Hank's solution within a temperature range of 25–40°C.

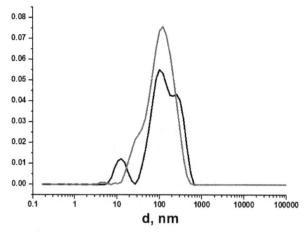

FIGURE 14.5 The dimensional characteristics on hydrodynamic diameter for D-PNIPAM/ AuNPs/Ce6 nanocomposites in Hank's solution at 25°C (black) and 37°C (red).

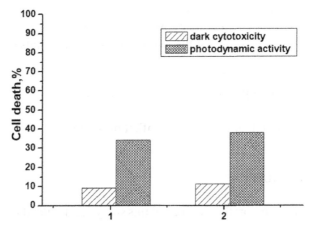

FIGURE 14.6 Dark toxicity and PDT efficacy: 1—Ce6; 2—D-PNIPAM/AuNPs/Ce6 nanocomposite.

14.4 CONCLUSIONS

It was shown that for multicomponent nanosystems D-PNIPAM/AuNPs/Ce6 in Hank's solution, it was not observed the increasing of aggregation process. The nanosystem remains stable when heated to physiological temperature, although the *in-vitro* study shows that the photodynamic activity of the D-PNIPAM/AuNPs/Ce6 nanosystem is approximately equal in efficiency to Ce6. However, the nanocomposite can prevent the aggregation of photosensitizer *in-vivo* experiments and can provide target delivery of drug to tumor cell. Thus, this ternary nanosystem can be recommended for further research for PDT therapy.

FUNDING STATEMENT

This study was supported in part by the Ministry of the Education and Science of Ukraine, Project 0122U001818 "Hybrid nanosystems based on 'smart' polymers for biotechnology and medicine" (2022–2023).

ACKNOWLEDGMENTS

The authors are grateful to Dr. I. Shton and Dr. E. Shyshko from R. E. Kavetsky Institute of Experimental Pathology, Oncology and Radiobiology for *in-vitro* testing of nanocomposite for PDT. The authors thank the French PAUSE program for an emergency welcome of Ukrainian scientists and the brave defenders of Ukraine for the opportunity to finalize this publication.

CONFLICT OF INTEREST

The authors declare that there is no conflict of interest.

DATA AVAILABILITY

The data used to support the findings of this study are included in this chapter.

KEYWORDS

- **nanoscience**
- **polymers**
- **comparative study**
- **water**
- **anticancer**

REFERENCES

1. Paszko, E.; Ehrhardt, C.; Senge, M. O.; Kelleher, D. P.; Reynolds, J. V. Nanodrug Applications in Photodynamic Therapy. *Photodiagn. Photodyn. Ther.* **2011,** *8* (1), 14–29. DOI: 10.1016/j.pdpdt.2010.12.001.
2. Dolmans, D. E.; Fukumura, D.; Jain, R. K. Photodynamic Therapy for Cancer. *Nat. Rev. Cancer* **2003,** *3* (5), 380–387. DOI: 10.1038/nrc1071.
3. Wang, H.; Xu, Y.; Shi, J.; Gao, X.; Geng, L. Photodynamic Therapy in the Treatment of Basal Cell Carcinoma: A Systematic Review and Meta-Analysis. *Photodermatol. Photoimmunol. Photomed.* **2015,** *31* (1), 44–53. DOI: 10.1111/phpp.12148.
4. Kwiatkowski, S.; Knap, B.; Przystupski, D.; Saczko, J.; Kędzierska, E.; Knap-Czop, K.; Kotlińska, J.; Michel, O.; Kotowski, K.; Kulbacka, J. Photodynamic Therapy— Mechanisms, Photosensitizers and Combinations. *Biomed. Pharmacother.* **2018,** *106*, 1098–1107. DOI: 10.1016/j.biopha.2018.07.049.
5. Feldman, D. Polymer. *J. Macromol. Sci. A* **2016,** *53* (1), 55–62. DOI: 10.1080/10601325. 2016.1110459.

6. Asiri, A. M. In *Applications of Nanocomposite Materials in Drug Delivery. A Volume in Woodhead Publishing Series in Biomaterials*; Asiri, I. A. M., Mohammad, A., Eds.; 2018; 990 p. DOI: 10.1016/C2016-0-05075-1.

7. Wei, Menglian; Gao, Yungfeng; Li, X.; Serpe, M. J. Stimuli-Responsive Polymers and Their Applications. *Polym. Chem.* **2017**, *8* (1), 127–143. DOI: 10.1039/C6PY01585A.

8. Kutsevol, N.; Harahuts, I.; Nadtoka, O.; Naumenko, A.; Yeshchenko, O. Chapter 12: Hybrid Nanocomposites Synthesized into Stimuli Responsible Polymer Matrices: Synthesis and Application Prospects. In *Springer Proceedings in Physics*; Fesenko, O., Yatsenko, L., Eds.; 2019; Vol. 222, pp 167–185. DOI: 10.1007/978-3-030-17759-1_12.

9. Jain, K.; Vedarajan, R.; Watanabe, M.; Ishikiriyama, M.; Matsumi, N. Tunable LCST Behavior of Poly(*N*-Isopropylacrylamide/Ionic Liquid) Copolymers. *Polym. Chem.* **2015**, *6* (38), 6819–6825. DOI: 10.1039/C5PY00998G.

10. Guo, Z.; Li, S.; Wang, C.; Xu, J.; Kirk, B.; Wu, J.; Liu, Z.; Xue, W. Biocompatibility and Cellular Uptake Mechanisms of Poly(*N*-isopropylacrylamide) in Different Cells. *J. Bioact. Compat. Polym.* **2017**, *32* (1), 17–31. DOI: 10.1177/0883911516648969.

11. Narang, P.; Venkatesu, P. Unravelling the Role of Polyols with Increasing Carbon Chain Length and OH Groups on the Phase Transition Behavior of PNIPAM. *New J. Chem.* **2018**, *42* (16), 13708–13717. DOI: 10.1039/C8NJ02510J.

12. Chumachenko, V.; Kutsevol, N.; Harahuts, Yu.; Rawiso, M.; Marinin, A.; Bulavin, L. Star-Like Dextran-Graft-PNIPAM Copolymers. Effect of Internal Molecular Structure on the Phase Transition. *J. Mol. Liq.* **2017**, *235*, 77–82. DOI: 10.1016/j.molliq.2017.02.098.

13. Matvienko, T.; Sokolova, V.; Prylutska, S.; Harahuts, Y.; Kutsevol, N.; Kostjukov, V.; Evstigneev, M.; Prylutskyy, Y.; Epple, M.; Ritter, U. In Vitro Study of the Anticancer Activity of Various Doxorubicin-Containing Dispersions. *Bioimpacts* **2019**, *9* (1), 57–63. DOI: 10.15171/bi.2019.07.

14. Yeshchenko, O.; Naumenko, A.; Kutsevol, N.; Harahuts, Y. Laser Driven Structural Transformation in Dextran-Graft-PNIPAM Copolymer/Au Nanoparticles Hybrid Nanosystem: The Role of Plasmon Heating and Attractive Optical Forces. *RSC Adv.* **2018**, *8*, 38400–38409. DOI: 10.1039/C8RA07768A.

15. Yeshchenko, O. A.; Naumenko, A. P.; Kutsevol, N. V.; Maskova, D. O.; Harahuts, I. I.; Chumachenko, V. A.; Marinin, A. I. Anomalous Inverse Hysteresis of Phase Transition in Thermosensitive Dextran-Graft-PNIPAM Copolymer/Au Nanoparticles Hybrid Nanosystem. *J. Phys. Chem. C* **2018**, *122* (14), 8003–8010. DOI: 10.1021/acs.jpcc.8b01111.

16. Kutsevol, N.; Glamazda, A.; Chumachenko, V.; Harahuts, Y.; Stepanian, S. G.; Plokhotnichenko, A. M.; Karachevtsev, V. A. Behavior of Hybrid Thermosensitive Nanosystem Dextran-Graft-PNIPAM/Gold Nanoparticles: Characterization within LCTS. *J. Nanopart. Res.* **2018**, *20* (9), 236. DOI: 10.1007/s11051-018-4331-2.

17. Kim, H. S.; Lee, D. Y. Near-Infrared-Responsive Cancer Photothermal and Photodynamic Therapy Using Gold Nanoparticles. *Polymers* **2018**, *10* (9), 961. DOI: 10.3390/polym10090961.

18. Jain, S.; Hirst, D. G.; O'Sullivan, J. M. Gold Nanoparticles as Novel Agents for Cancer Therapy. *Br. J. Radiol.* **2012**, *85* (1010), 101–113. DOI: 10.1259/bjr/59448833.

19. Corbierre, M. K.; Cameron, N. S.; Sutton, M.; Laaziri, K.; Lennox, R. B. Nanocomposites: Dispersion of Nanoparticles as a Function of Capping Agent Molecular Weight and Grafting Density. *Langmuir* **2005**, *21* (13), 6063–6072. DOI: 10.1021/la047193e.

20. Chumachenko, V. A.; Shton, I. O.; Shishko, E. D.; Kutsevol, N. V.; Marinin, A. I.; Gamaleia, N. F. Branched Copolymers Dextran-Graft-Polyacrylamide as Nanocarriers

for Delivery of Gold Nanoparticles and Photosensitizers to Tumor Cells. In *Nanophysics, Nanophotonics, Surface Studies, and Applications*; Fesenko, O., Leonid Yatsenko, D. O. I., Eds.; *Springer Proceedings in Physics*; 2016; Vol. 183, pp 379–390. DOI: 10.1007/ 978-3-319-30737-4_32.

21. Kutsevol, N.; Naumenko, A.; Harahuts, Yu.; Chumachenko, V.; Shton, I.; Shishko, E.; Lukianova, N.; Chekhun, V. New Hybrid Composites for Photodynamic Therapy: Synthesis, Characterization and Biological Study. *Appl. Nanosci.* **2019,** *9* (5), 881–888. DOI: 10.1007/s13204-018-0768-y.

22. Makharza, Sami; Auisa, J.; Sharkh, S. A.; Ghabboun, J.; Faroun, M.; Dweik, H.; Sultan, W.; Sowwan, M. Structural and Thermal Analysis of Copper-Doped Poly(N-isopropylacrylamide) Films. *Int. J. Polym. Anal. Charact.* **2010,** *15* (4), 254–265. DOI: 10.1080/10236661003747031.

23. Gattuso, H.; Monariab, A.; Marazzi, M. Photophysics of Chlorin e6: From One- and Two-Photon Absorption to Fluorescence and Phosphorescence. *RSC Adv.* **2017,** *7*, 10992–10999. DOI: 10.1039/c6ra28616j.

24. Kutsevol, N.; Harahuts, Yu.; Shton, I.; Borikun, T.; Storchai, D.; Lukianova, N.; Chekhun, V. In Vitro Study of Toxicity of Hybrid Gold–Polymer Composites. *Mol. Cryst. Liq. Cryst.* **2018,** *671* (1), 1–8. DOI: 10.1080/15421406.2018.1542078.

25. Harahuts, Y.; Pavlov, V.; Mokrinskaya, E.; Davidenko, I.; Davidenko, N.; Kutsevol, N.; Pampukha, I.; Martynyuk, V. Chapter 3: Anomalous Change of Refractive Index for Au Sols Under Laser Illumination. In *Springer Proceedings in Physics*; Fesenko, O., Yatsenko, L., Eds.; 2019, pp 53–71. DOI: 10.1007/978-3-030-17755-3_3.

CHAPTER 15

Magnetic Nanoparticles in Healthcare

MERIN JOBY[1], MINU JOSE[2], SUNSU KURIAN[1], and ANJU K. NAIR[1]

[1]Department of Physics and Centre for Research, St. Teresa's College (Autonomous), Ernakulam, Kerala, India

[2]Department of Botany, St, Teresa's College (Autonomous), Ernakulam, Kerala, India

ABSTRACT

Magnetic nanomaterials have gained tremendous attention for their various biomedical applications including their function as biosensors, medical diagnostic agents, therapeutic agents, and for clinically relevant applications which include imaging-guided therapy, magnetic hyperthermia, targeted drug and gene delivery systems, manipulation of cellular functions, and so on. They exhibit size-dependent magnetic properties which are tunable that may be manipulated for plenty distinct magnetically controlled applications. Various preparation techniques have been explored to synthesize magnetic nanomaterials with desired properties.

15.1 INTRODUCTION

Magnetic nanoparticles (MNPs) are certainly promising substances to assemble flexible structures for biomedical applications because of their extremely small size.

For the use of magnetic nanomaterials in biomedical applications, they have to be biocompatible and synthesized with the aid of using easy

Mechanics and Physics of Porous Materials: Novel Processing Technologies and Emerging Applications.
Chin Hua Chia, Tamara Tatrishvili, Ann Rose Abraham, & A. K. Haghi (Eds.)

methods. Nanoscale gadgets smaller than 50 nm can enter most cells easily. Nanoparticles have the potential to detect diseases, control cells within the living system, and enable more effective treatment that was more or less impossible before. They can act as carriers for various drugs. Gao et al. reported that nanoparticles with dual functionality can be used for diagnostic and therapeutic purposes (e.g., Fe_2O_3Pt NPs). They can be used to minimize the toxicity of a therapeutic drug or for pH-triggered drug release. The nano-product AuroShell can kill cells at the tumor target site by emitting heat upon absorption of near-infrared (NIR) light.[1-3]

15.2 MAGNETIC NANOMATERIALS

The properties of magnetic nanomaterials mainly depend on the intrinsic properties of each particle, such as chemical composition, stoichiometry, crystallinity, shape, magnetic anisotropy, and their interaction. Since the interactions between particles have an important effect, controlling the distances between particles has to be considered when designing magnetic nanocomposites with specific properties. Because of their superparamagnetic behavior, these nanomaterials are known as superparamagnetic iron oxides (SPIOs). They consist of an iron oxide core with a diameter of a few nanometers and a subnanometric surface layer of iron oxyhydroxide groups. Different strategies have been applied to functionalize the SPIO surface and obtain platforms to couple a variety of biomedical devices to enable their localized and controlled delivery. Functionalization can be achieved with a simple acid:base reaction that incorporates surface charges on the particles or coating with molecular species of small molecules as well as surfactants or polymers. SPIO platforms can be coupled to bioagents, resulting in multifunctional complexes tailored to perform diagnostics and therapies simultaneously (i.e., "theranostics"). Theranostics are designed to diagnose and deliver therapy more efficiently than traditional methods, thereby minimizing costs and side effects. MNPs can be used to make biosensors more sensitive, faster, and cheaper. MNP-based biosensors are used to detect nucleic acids, enzymes, other proteins, drugs, pathogens, and tumor cells with excellent sensitivity. MNPs have been successfully used in diagnostic magnetic resonance imaging (DMR). The basic principle behind DMR is the use of MNPs as proximity sensors that modulate the transverse relaxation time of neighboring water molecules.[4,5]

15.3 MAGNETIC NANOMATERIALS FOR BIOMEDICAL APPLICATIONS

This can be categorized as *in vivo* or *in vitro* depending on whether they are used within or outside the body. *In-vitro* applications primarily include diagnostic separation, selection, and magnetorelaxometry, whereas *in-vivo* applications are divided into therapeutic (hyperthermia, gene therapy, drug-targeting) and diagnostic (magnetic resonance imaging [MRI], magneto-acoustic tomography [MAT], magnetic particle imaging [MPI], and nuclear magnetic resonance [NMR] imaging).[6–8]

Shifted loops are seen in MNPs following field cooling. These phenomena are caused by the magnetic conduct of independent nanoparticles which are influenced by finite size and surface effects. Frenkel and Dorfman were the first to anticipate that a ferromagnetic particle below a certain particle size (about 15 nm) would comprise a single magnetic field. Over a precise T (named as the blocking temperature), the magnetization conduct of these particles is similar to that of atomic paramagnets (superparamagnetism), with the exception that an exceedingly large moment and consequently large susceptibilities are involved. Particles having superparamagnetic behavior at room temperature are chosen for biomedical applications. For uses in biology, medical diagnosis and therapy it is required for magnetic particles to be stable in water at neutral pH and physiological salinity. The colloidal stability of this fluid will be determined by two factors: first, the particle size, which must be small enough to avoid precipitation owing to gravitational forces, and second, the charge and surface chemistry, which cause both steric and coulombic repulsions. Additional constraints on the types of particles that could be employed in biomedical applications are heavily influenced by whether the particles will be used *in vivo* or *in vitro*. MNPs that are used for *in-vivo* biomedical applications must be made of a nontoxic and nonimmunogenic material.[9,10]

Drugs will be able to bind to the polymer via covalent attachment, adsorption, or trapping on the particles. The nature of the magnetically responsive component, such as magnetite, iron, nickel, cobalt, neodymium–iron–boron, or samarium–cobalt, as well as the ultimate size of the particles, their core, and the coatings, are critical elements that affect the biocompatibility and toxicity of these materials. Magnetite (Fe_3O_4) and its oxidized maghemite (Fe_2O_3) are the most widely used iron oxide particles in biological applications. Cobalt and nickel, which are highly magnetic, are poisonous and prone to oxidation, and hence of little interest. Furthermore, utilizing particles smaller than 100 nm (so-called nanoparticles) has several advantages, including increased effective surface areas (easier ligand binding), and decreased sedimentation. They must also be highly magnetized so that their movement in the circulation can be regulated by a magnetic field and they can be immobilized near to the diseased tissue being targeted.[11,12] Figure 15.1 illustrate the action of magnetite nanoparticles over bacterial magnetosomes.

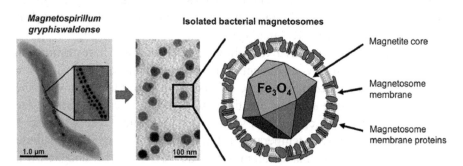

FIGURE 15.1   Action of magnetite nanoparticle over bacterial magnetosomes.
Source: Reprinted with permission from Ref. [13]. Copyright 2021. Elsevier.

15.3.1   IN-VITRO APPLICATIONS

15.3.1.1   DIAGNOSTIC SEPARATION AND SELECTION

Biomedical research usually involves the separation of specific biological entities from their natural environments. Superparamagnetic colloids are ideal for bioseparation, allowing the transit of biomaterials with a magnetic field. The nanometer-sized SPIO magnetic particles have been widely used to separate and purify cells and biomolecules. Because of their small size and high surface area, MNPs have considerable benefits over typical micrometer-sized resins or beads for various bioseparation applications, including superior dispersibility, rapid and effective biomolecule binding, and reversible and controllable flocculation. Antibodies allow for extremely efficient magnetic separation.[14–16]

15.3.1.2   MAGNETORELAXOMETRY

Magnetorelaxometry can be used to evaluate immunoassays as an analytical tool. The magnetic viscosity of a system of MNPs is measured using this technique, which is the relaxation of the net magnetic moment after removing a magnetic field. The advantages of lowering particle size to the nanoscale level for this application are similar to those stated for separation and selection applications.[17]

15.3.2 *IN-VIVO APPLICATIONS*

MNPs' size and surface functioning are crucial factors in their *in-vivo* uses. The diameters of SPIOs have a significant impact on *in-vivo* biodistribution. Particles having diameter of 10–40 nm, such as ultrasmall SPIOs, are critical for extended blood circulation because they can breach capillary walls and are frequently phagocytosed by macrophages that travel to the lymph nodes and bone marrow. Colloidal NPs are injected into the bloodstream and guided to the region of delivery with external manipulation by delivering a magnetic field gradient, notably in *in-vivo* applications, by employing molecular recognition to precisely attach to the target cells or both. Colloidal particles can also be injected directly into the desired location. The complexes generated by the SPIO particles act as theranostic agents once they are present in the target tissue. A theranostic agent is one that can diagnose and treat patients using a single platform. In addition to their biocompatibility, iron oxide nanoparticulate systems have traits like superparamagnetism and the capacity to reach the target cell, obtain a picture, and provide therapy all at the same time. SPIO devices were first utilized to improve the sensitivity of magnetic resonance signals as contrast agents. They are currently recognized as effective platforms for other imaging modes and therapeutic drugs to reach the target site. In 1996, the United States Food and Drug Administration (FDA) authorized dispersed SPIO particles as stable biocolloids. These particles have since been employed in clinical settings, such as to improve the contrast of MRI in diagnosis, SPIO-based colloids.[18-20]

15.3.3 *THERAPEUTIC APPLICATIONS*

15.3.3.1 *HYPERTHERMIA*

Tumor cells are more sensitive to a temperature increase than healthy ones. Hyperthermia in cancer treatment refers to the generation of heat at the site of the tumor. The heating induced by the MNPs present in the tumor is known as local magnetohyperthermia. The advantage of magnetic hyperthermia (MHT) is that it limits heating to the tumor area. The principle of MHT consists of the infusion of MNPs in the target tissue, followed by the generation of heat by applying an alternating magnetic field. Magnetic moments of the mono domains get aligned by thermal energy dissipated from the magnetic NPs through Néel and Brownian relaxation processes in response to the application of this alternating magnetic field. The local temperature can reach 40–42°C, which is sufficient to destroy target cells. Research has also verified that MHT therapy induces an antitumor-immune effect. SPIO complex-assisted chemotherapy attempts to inhibit the growth of cancer cells by providing agents that inhibit their cellular function.[21,22]

15.3.3.2   TARGETED DRUG DELIVERY

In biomedical applications, MNPs can be used as effective drug-delivery agents. The multifunctionality of magnetic NPs improves drug delivery, enables local administration of higher doses, and minimizes the side effects induced when the drug is administered conventionally. Iron oxide MNPs' potential for drug targeting has become one of the most promising therapy alternatives for cancer patients in recent years. Once the drug has reached the patient's bloodstream, a magnetic field is applied to retain the particles in the targeted site of the tumor. When used with an external magnetic field, these nanoparticles can be controlled to carry drug to its site, fix these particles in place while medicine is released, and perform magnetic drug targeting. Unlike in conventional cancer therapies such as radiation therapy or chemotherapy, MNPs in localized or targeted drug delivery can reduce the risk of certain side effects of treatment. In addition to this, drug targeting can reduce the dosage used in the treatment. Targeted drug delivery using magnetic systems is mainly based on the competition between the forces exerted by the blood compartment on the particles and the magnetic forces generated from the magnet.[8,23] Figure 15.2 depicts the notion of targeted drug delivery employing paramagnetic nanoparticles as therapeutic nano-agents. The experimental results incorporate the improved tumbling motion caused by the gradient and rotating magnetic fields, which is quantified by the analysis.[24]

15.3.3.3   PHOTODYNAMIC THERAPY

The photosensitizer produces an oxygen radical. The utilization of photosensitizers adsorbed in SPIO particles resulted in the development of a new class of complex materials that allow photodynamic therapy (PDT) and magneto hyperthermia to be combined. Combining MRI contrast improvement with localized PDT, this multifunctional magnetic complex can simultaneously increase MRI contrast and provide real-time observation of tumor removal.[25,26]

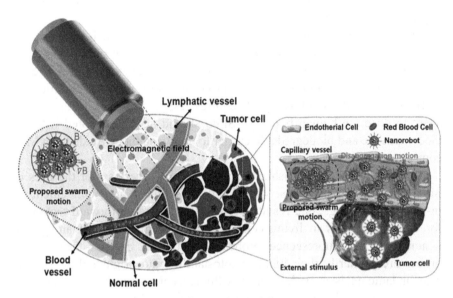

FIGURE 15.2 Conceptual scheme of targeted drug delivery.

Source: Reprinted from Ref. [24]. http://creativecommons.org/licenses/by/4.0/

15.3.4 DIAGNOSTIC APPLICATIONS

15.3.4.1 MAGNETIC RESONANCE IMAGING

MRI is a noninvasive medical imaging technique which makes use of the magnetic field and computer-generated radio waves to create three-dimensional detailed anatomical images. Strong magnets are used in MRI that produces a strong magnetic field which forces the body's protons into alignment with this field. The integration of MNPs and tumor-targeting molecules can generate contrast-enhanced MRI signal in the tumor region, which in turn output anatomical information for accurate tumor diagnosis. A tumor-targeting MNP typically consists of two primary components: a core of MNP (e.g., iron oxide) that serves as both a carrier for targeting agent and contrast agent for MRI and a coating on that decorates with targeting agents enabling to interact with biological targets on or within tumor cells. MRI can distinguish between white matter and gray matter in the brain and can also be used to diagnose aneurysms and tumors. MRI is preferred when frequent imaging is needed for diagnosis or therapy, particularly in the brain, although it is expensive than X-ray imaging and CT scanning. Noninvasive MRI aids the transfection of cells using magnetic markers, providing a technique to improve the experimental control of cell treatments.

SPIO NPs are proven to be a class of new probes suitable for *in-vitro* and *in-vivo* cellular and molecular imaging at the intersection of nanomaterials and medical diagnostics. In comparison to paramagnetic contrast agents, superparamagnetic contrast agents produce more proton relaxation in MRI. As a result, an SPIO agent requires less dose than a paramagnetic agent to dose the human body. An SPIO should be disseminated within a biocompatible and biodegradable carrier to apply magnetic fluids to an MRI contrast agent.[27]

Muller et al. have published a comprehensive overview of the uses of SPIO NPs as contrast agents. MRIs, on the other hand, are inconvenient for in-place monitoring. As a result, a sensitive and straightforward technique for monitoring NPs in living cells *in situ* is desired. The advantages of nanometer-sized fluorescence probes over typical organic fluorescence probes are primarily their higher photostability and brighter fluorescence. The fundamental issue with employing fluorescent nanoprobes to image cells is that background noise from the cells, matrix, and nonspecific scattering lights can readily degrade the fluorescence signal. It is difficult to achieve a high signal-to-noise ratio.[28,29]

15.3.4.2   MAGNETIC-ACOUSTIC TOMOGRAPHY

Nanoparticles are getting a lot of attention as contrast agents in a variety of imaging modalities. They can be employed onto the desired target sites and it is also feasible to image events at the cellular and molecular level. Due to its low spatial resolution and contrast, ultrasonic imaging is unable to directly view nanostructures. Pulsed magneto-acoustic (PMA) imaging, a new approach for imaging MNPs makes use of a high-intensity pulsed magnetic field to produce motion within magnetically labeled tissue, and ultrasonography is employed to detect internal tissue motion. PMA imaging could be employed as an imaging tool capable of visualizing the cellular and molecular composition of deep-lying structures. Magneto-acoustic tomography with magnetic induction is a noninvasive imaging method developed to map electrical conductivity of biological tissue with millimeter-level spatial resolution. These ultrasonic signals can then be used to create a map of the electrical conductivity contrast of the sample.[30]

15.3.4.3 MAGNETIC PARTICLE IMAGING

Because of its signal linearly proportional to the tracer mass, capacity to provide positive contrast, minimal tissue background, infinite tissue penetration depth, and lack of ionizing radiation, MPI has lately emerged as a promising noninvasive imaging approach. MPI's sensitivity and resolution are heavily dependent on the properties of magnetic nanoparticles (MNPs); hence, tracer design and manufacture have received a lot of attention.[31]

15.3.4.4 NUCLEAR MAGNETIC RESONANCE IMAGING

NMR imaging is a type of imaging that uses NMR. The need for a new class of pharmaceuticals, known as magneto-pharmaceuticals, has arisen as a result of the development of NMR imaging for clinical diagnosis. To improve the visual contrast between normal and sick tissue, these medications must be given to the patient. It can also tell you how well your organs are working or how well your blood is flowing. A variety of compounds have been proposed as possible NMR contrast agents. To date, the majority of contrast chemicals utilized in NMR imaging studies have been paramagnetic. Commercial maghemite iron oxide nanoparticles (Endorem® and Resovit®) have been used as contrast agents in NMR imaging for the location and diagnosis of brain and cardiac infarctions, liver lesions, and tumors, where the MNPs tend to accumulate at higher levels due to differences in tissue composition and/or endocytotic uptake processes. Improvements in the sensitivity of detection and delineation of diseased structures, such as primary and metastatic brain tumors, inflammation, and ischemia, have shown to be particularly promising. Proteins like transferrin, peptides like the membrane translocating tat peptide of the HIV tat protein and oligonucleotides of various sequences have been added to aminated cross-linked iron oxide nanoparticles for this purpose.[32,33]

15.4   TOXICITY OF MAGNETIC NANOPARTICLES

Nanoparticles' nanoscale size can result in a higher surface-to-volume ratio and higher activity when compared to bulk materials. One of the most often recognized factors for the cytotoxic effect of nanomaterials is the administered dose. A higher dose of nanoparticle exposure to cells could cause considerable cell growth suppression. However, due to nanomaterials' strong colloidal stability, most cells can survive even in the presence of potentially hazardous components like quantum dots at low concentrations of nanoparticles. As a result, numerous other factors, including as colloidal stability, component inertness, size, shape, charge, surface ornamentation, and magnetism, could affect MNP toxicity. Cell autophagy was discovered to be induced by the aggregation state of MNPs in cell culture medium, for example. Based on the composition and cell lines, the induced cytotoxic effect of nanoparticles can vary depending on their form and size. Furthermore, MNPs' biological inertness is determined by surface ornamentation, which influences their ability to absorb biological components from the environment (e.g., proteins).[34,35]

15.5   DESIGN CONSIDERATIONS FOR BIOCOMPATIBLE MAGNETIC NANOPARTICLES

The biocompatibility of MNPs is of great importance, especially while using them as diagnostic probes in biomedical applications. As a matter of safety, it is highly desirable to use biodegradable, noninvasive, and nontoxic probes so that less harm is caused to the patients. It must be chemically stable under physiological conditions and also render minimal or no transfer of by-products to surrounding tissue/cells. For this, the physiochemical properties of MNPs, such as the shape, size, charge, hydrophobicity, and surface property must be taken into consideration. Little or no toxicity can be expected from MNPs with favorable metabolic route to clear out of body after diagnostic imaging. As an example, engineering on the size, charge, and surface property of iron-oxide-based nanoparticles can result in fast renal clearance after tumor imaging. It has been reported that a magnetolytic therapeutic method can be used to regulate the toxicity of MNPs as a novel cancer-killing strategy using magnetic field modulation. MNPs with a favorable metabolic pathway to clear out-of-body after diagnostic imaging should cause little or no harm. Engineering on the size, charge, and surface properties of iron-oxide-based nanoparticles, for example, can result in rapid renal clearance following tumor imaging. A magnetolytic treatment technique has been reported as an unique cancer-killing strategy using magnetic field control to manage the toxicity of MNPs.[36–38]

15.6 ENGINEERED MAGNETIC NANOPARTICLES WITH MULTIPLE FUNCTIONS

Beyond magnetism-based imaging modalities, the integration of MNPs with other components promises diverse imaging capacities, which are potential candidates for bio-imaging and diagnosis in nanomedicine contexts. One of the most significant advantages of multifunctional MNPs is the ability to do multiparameter analysis utilizing a variety of imaging capabilities. MRI, for example, has high anatomical resolution but low sensitivity, whereas fluorescence imaging (FL) has high signal sensitivity but low spatial resolution. The use of both MRI and FL imaging modalities together would provide advantages over using either mode alone.[39,40] By regulating the interplay between multiple modalities, these bimodal imaging platforms are also attractive candidates for wisely constructed systems. Through photo uncaging ligands, Biju and colleagues revealed that QDs adorned magnetic iron oxide nanoparticles may enable UV light induced MRI and FL bimodal imaging of animal organs. PET/CT and MRI multimodal imaging of sentinel lymph nodes with great sensitivity was achieved using polymeric nano-beacons that contained positron emitter 89-zirconium and MRI-contrast agent gadolinium. Furthermore, because of their inherent Cerenkov luminescence (CL), the ^{89}Zr can be used to guide the surgical resection of sentinel lymph nodes, making them a new multifunctional theranostic nanotechnology platform for the clinic.[41]

MNPs are intriguing options for material chemists, biologists, and pharmacologists to develop flexible nanoplatforms as therapeutically translatable medical probes because of their many imaging applications. The one-by-one combination with additional imaging agents has made a significant step toward bridging the gap between preclinical research and clinical use. Lovell and coworkers reported a hexamodal imaging system based on porphyrin phospholipid-coated upconversion nanoparticles that combined NIR fluorescence, NIR-to-NIR upconversion luminescence (UC), photoacoustic (PA) imaging, CL, computed tomography, and positron emission tomography techniques for multimodal imaging of a mouse lymphatic model. However, because various imaging techniques have very different spatial and temporal resolutions, it would be difficult to compare the acquisitions and images.[42,43]

KEYWORDS

- **magnetic**
- **biomedical**
- **biosensor**
- **magnetic hyperthermia**
- **drug delivery**
- **tunable**

REFERENCES

1. Akbarzadeh, A.; Samiei, M.; Davaran, S. Magnetic Nanoparticles: Preparation, Physical Properties, and Applications in Biomedicine. *Nanoscale Res. Lett.* **2012,** *7* (1), 144. DOI: 10.1186/1556-276X-7-144.

2. Zhu, K.; Ju, Y.; Xu, J.; Yang, Z.; Gao, S.; Hou, Y. Magnetic Nanomaterials: Chemical Design, Synthesis, and Potential Applications. *Acc. Chem. Res.* **2018,** *51* (2), 404–413. DOI: 10.1021/acs.accounts.7b00407.

3. Liu, G.; Li, R.-W.; Chen, Y. Magnetic Nanoparticle for Biomedicine Applications. *Nanotechnol. Nanomed. Nanobiotechnol.* **2015,** *2* (1), 1–7. DOI: 10.24966/NTMB-2044/100003.

4. Acidereli, H.; Karataş, Y.; Burhan, H.; Gülcan, M.; Şen, F. Magnetic Nanoparticles. *Nanoscale Process.* **2020.** DOI: 10.1016/B978-0-12-820569-3.00008-6.

5. Stueber, D. D.; Villanova, J.; Aponte, I.; Xiao, Z.; Colvin, V. L. Magnetic Nanoparticles in Biology and Medicine: Past, Present, and Future Trends. *Pharmaceutics* **2021,** *13* (7). DOI: 10.3390/pharmaceutics13070943.

6. Wu, M.; Huang, S. Magnetic Nanoparticles in Cancer Diagnosis, Drug Delivery and Treatment. Review. *Mol. Clin. Oncol.* **2017,** *7* (5), 738–746. DOI: 10.3892/mco.2017.1399.

7. Indira, T. Magnetic Nanoparticles: A Review. *Int. J. Pharm.* **2010,** *3* (3), 1035–1042.

8. Kianfar, E. Magnetic Nanoparticles in Targeted Drug Delivery: A Review. *J. Supercond. Novel Magn.* **2021,** *34* (7), 1709–1735. DOI: 10.1007/s10948-021-05932-9.

9. Matschegewski, C.; Kowalski, A.; Müller, K.; Teller, H.; Grabow, N.; Großmann, S.; Schmitz, K.; Siewert, S. Biocompatibility of Magnetic Iron Oxide Nanoparticles for Biomedical Applications. *Curr. Dir. Biomed. Eng.* **2019,** *5* (1), 573–576. DOI: 10.1515/cdbme-2019-0144.

10. Ganapathe, L. S.; Mohamed, M. A.; Mohamad Yunus, R. M.; Berhanuddin, D. D. Magnetite (Fe_3O_4) Nanoparticles in Biomedical Application: From Synthesis to Surface Functionalisation. *Magnetochemistry* **2020,** *6* (4). DOI: 10.3390/magnetochemistry 6040068.

11. Mohammed, L.; Gomaa, H. G.; Ragab, D.; Zhu, J. Magnetic Nanoparticles for Environmental and Biomedical Applications: A Review. *Particuology* **2017,** *30*, 1–14. DOI: 10.1016/j.partic.2016.06.001.

12. Shabatina, T. I.; Vernaya, O. I.; Shabatin, V. P.; Melnikov, M. Y. Magnetic Nanoparticles for Biomedical Purposes: Modern Trends and Prospects. *Magnetochemistry* **2020,** *6* (3). DOI: 10.3390/magnetochemistry6030030.

13. Rosenfeldt, S.; Mickoleit, F.; Jörke, C.; Clement, J. H.; Markert, S.; Jérôme, V.; Schwarzinger, S.; Freitag, R.; Schüler, D.; Uebe, R.; Schenk, A. S. Towards Standardized Purification of Bacterial Magnetic Nanoparticles for Future In Vivo Applications. *Acta Biomater.* **2021,** *120,* 293–303. DOI: 10.1016/j.actbio.2020.07.042.

14. Belanova, A. A.; Gavalas, N.; Makarenko, Y. M.; Belousova, M. M.; Soldatov, A. V.; Zolotukhin, P. V. Physicochemical Properties of Magnetic Nanoparticles: Implications for Biomedical Applications In Vitro and In Vivo. *Oncol. Res. Treat.* **2018,** *41* (3), 139–143. DOI: 10.1159/000485020.

15. Rego, G. N. A.; Nucci, M. P.; Mamani, J. B.; Oliveira, F. A.; Marti, L. C.; Filgueiras, I. S.; Ferreira, J. M.; Real, C. C.; Faria, D. P.; Espinha, P. L.; Fantacini, D. M. C.; Souza, L. E. B.; Covas, D. T.; Buchpiguel, C. A.; Gamarra, L. F. Therapeutic Efficiency of Multiple Applications of Magnetic Hyperthermia Technique in Glioblastoma Using Aminosilane Coated Iron Oxide Nanoparticles: In Vitro and In Vivo Study. *Int. J. Mol. Sci.* **2020,** *21* (3). DOI: 10.3390/ijms21030958.

16. Wu, W.; Wu, Z.; Yu, T.; Jiang, C.; Kim, W. S. Recent Progress on Magnetic Iron Oxide Nanoparticles: Synthesis, Surface Functional Strategies and Biomedical Applications. *Sci. Technol. Adv. Mater.* **2015,** *16* (2), 023501. DOI: 10.1088/1468-6996/16/2/023501.

17. Coene, A.; Leliaert, J.; Liebl, M.; Löwa, N.; Steinhoff, U.; Crevecoeur, G.; Dupré, L.; Wiekhorst, F. Multi-color Magnetic Nanoparticle Imaging Using Magnetorelaxometry. *Phys. Med. Biol.* **2017,** *62* (8), 3139–3157. DOI: 10.1088/1361-6560/aa5e90.

18. Senthilkumar, N.; Sharma, P. K.; Sood, N.; Bhalla, N. Designing Magnetic Nanoparticles for In Vivo Applications and Understanding Their Fate Inside Human Body. *Coord. Chem. Rev.* **2021,** *445.* DOI: 10.1016/j.ccr.2021.214082.

19. Duguet, E.; Vasseur, S.; Mornet, S.; Goglio, G.; Demourgues, A.; Portier, J.; Grasset, F.; Veverka, P.; Pollert, E. Towards a Versatile Platform Based on Magnetic Nanoparticles for In Vivo Applications. *Bull. Mater. Sci.* **2006,** *29* (6), 581–586. DOI: 10.1007/s12034-006-0007-0.

20. Nedyalkova, M.; Donkova, B.; Romanova, J.; Tzvetkov, G.; Madurga, S.; Simeonov, V. Iron Oxide Nanoparticles—In Vivo/In Vitro Biomedical Applications and In Silico Studies. *Adv. Colloid Interface Sci.* **2017,** *249,* 192–212. DOI: 10.1016/j.cis.2017.05.003.

21. Ramazanov, M.; Karimova, A.; Shirinova, H. Magnetism for Drug Delivery, MRI and Hyperthermia Applications: A Review. *Biointerface Res. Appl. Chem.* **2021,** *11* (2), 8654–8668. DOI: 10.33263/BRIAC112.86548668.

22. Hepel, M. Magnetic Nanoparticles for Nanomedicine. *Magnetochemistry* **2020,** *6* (1). DOI: 10.3390/magnetochemistry6010003.

23. McBain, S. C.; Yiu, H. H. P.; Dobson, J. Magnetic Nanoparticles for Gene and Drug Delivery. *Int. J. Nanomed.* **2008,** *3* (2), 169–180. DOI: 10.2147/ijn.s1608.

24. Nguyen, K. T.; Go, G.; Zhen, J.; Hoang, M. C.; Kang, B.; Choi, E.; Park, J. O.; Kim, C. S. Locomotion and Disaggregation Control of Paramagnetic Nanoclusters Using Wireless Electromagnetic Fields for Enhanced Targeted Drug Delivery. *Sci. Rep.* **2021,** *11* (1), 15122. DOI: 10.1038/s41598-021-94446-4.

25. Wang, D.; Fei, B.; Halig, L. V.; Qin, X.; Hu, Z.; Xu, H.; Wang, Y. A.; Chen, Z.; Kim, S.; Shin, D. M.; Chen, Z. G. Targeted Iron-Oxide Nanoparticle for Photodynamic Therapy

and Imaging of Head and Neck Cancer. *ACS Nano* **2014,** *8* (7), 6620–6632. DOI: 10.1021/nn501652j.

26. Halig, L. V.; Wang, D.; Wang, A. Y.; Chen, Z. G.; Fei, B. Biodistribution Study of Nanoparticle Encapsulated Photodynamic Therapy Drugs Using Multispectral Imaging. In *Medical Imaging 2013: Biomedical Applications in Molecular, Structural, and Functional Imaging,* 2013, 8672. DOI: 10.1117/12.2006492.

27. Hola, K.; Markova, Z.; Zoppellaro, G.; Tucek, J.; Zboril, R. Tailored Functionalization of Iron Oxide Nanoparticles for MRI, Drug Delivery, Magnetic Separation and Immobilization of Biosubstances. *Biotechnol. Adv.* **2015,** *33* (6 Pt. 2), 1162–1176. DOI: 10.1016/j.biotechadv.2015.02.003.

28. Benz, M. Superparamagnetism: T. *Hybrid Nanostructures for Cancer Theranostics* 2018.

29. Parveen, S.; Misra, R.; Sahoo, S. K. Nanoparticles: A Boon to Drug Delivery, Therapeutics, Diagnostics and Imaging. *Nanomedicine* **2012,** *8* (2), 147–166. DOI: 10.1016/j.nano.2011.05.016.

30. Mariappan, L.; Shao, Q.; Jiang, C.; Yu, K.; Ashkenazi, S.; Bischof, J. C.; He, B. Magneto Acoustic Tomography with Short Pulsed Magnetic Field for In-Vivo Imaging of Magnetic Iron Oxide Nanoparticles. *Nanomedicine* **2016,** *12* (3), 689–699. DOI: 10.1016/j.nano.2015.10.014.

31. Lu, C.; Han, L.; Wang, J.; Wan, J.; Song, G.; Rao, J. Engineering of Magnetic Nanoparticles as Magnetic Particle Imaging Tracers. *Chem. Soc. Rev.* **2021,** *50* (14), 8102–8146. DOI: 10.1039/d0cs00260g.

32. Du, Y.; Mao, J. Effect of Magnetite Nanoparticles on Nuclear Magnetic Resonance Imaging. *J. Nanoelectron. Optoelectron.* **2017,** *12* (9), 961–965. DOI: 10.1166/jno. 2017.2224.

33. Burtea, C.; Laurent, S.; Vander Elst, L.; Muller, R. N. Contrast Agents: Magnetic Resonance. *Handb. Exp. Pharmacol.* **2008,** *185* (Pt 1), 135–165. DOI: 10.1007/978-3-540-72718-7_7.

34. Liu, G.; Gao, J.; Ai, H.; Chen, X. Applications and Potential Toxicity of Magnetic Iron Oxide Nanoparticles. *Small* **2013,** *9* (9–10), 1533–1545. DOI: 10.1002/smll.201201531.

35. Malhotra, N.; Lee, J. S.; Liman, R. A. D.; Ruallo, J. M. S.; Villaflores, O. B.; Ger, T. R.; Hsiao, C. D. Potential Toxicity of Iron Oxide Magnetic Nanoparticles: A Review. *Molecules* **2020,** *25* (14). DOI: 10.3390/molecules25143159.

36. Mulens, V.; Morales, M. del P.; Barber, D. F. Development of Magnetic Nanoparticles for Cancer Gene Therapy: A Comprehensive Review. *ISRN Nanomater.* **2013,** *2013*, 1–14. DOI: 10.1155/2013/646284.

37. Das, P.; Colombo, M.; Prosperi, D. Recent Advances in Magnetic Fluid Hyperthermia for Cancer Therapy. *Colloids Surf. B: Biointerfaces* **2019,** *174*, 42–55. DOI: 10.1016/j. colsurfb.2018.10.051.

38. Dobson, J. Magnetic Nanoparticles for Drug Delivery. *Drug Dev. Res.* **2006,** *67* (1), 55–60. DOI: 10.1002/ddr.20067.

39. Avasthi, A.; Caro, C.; Pozo-Torres, E.; Leal, M. P.; García-Martín, M. L. Magnetic Nanoparticles as MRI Contrast Agents. *Top. Curr. Chem. (Cham)* **2020,** *378* (3), 40. DOI: 10.1007/s41061-020-00302-w.

40. Avasthi, A.; Caro, C.; Pozo-Torres, E.; Leal, M. P.; García-Martín, M. L. Correction to: Magnetic Nanoparticles as MRI Contrast Agents. *Top. Curr. Chem. (Cham)* **2021,** *379* (4), 30. DOI: 10.1007/s41061-021-00340-y.

41. Martins, P. M.; Lima, A. C.; Ribeiro, S.; Lanceros-Mendez, S.; Martins, P. Magnetic Nanoparticles for Biomedical Applications: From the Soul of the Earth to the Deep History of Ourselves. *ACS Appl. Bio Mater.* **2021,** *4* (8), 5839–5870. DOI: 10.1021/acsabm.1c00440.

42. Thuy, T.; Maenosono, S.; Thanh, N. Next Generation Magnetic Nanoparticles for Biomedical Applications. *Magn. Nanopart.* **2012**. DOI: 10.1201/b11760-7.

43. Jun, Y. W.; Choi, J. S.; Cheon, J. Heterostructured Magnetic Nanoparticles: Their Versatility and High Performance Capabilities. *Chem. Commun. (Camb.)* **2007,** *12,* 1203–1214. DOI: 10.1039/b614735f.

Index

Printed in the United States
by Baker & Taylor Publisher Services